eBook Bonus.

With the purchase of this book, you also receive access to its electronic version. Use the following personal code to download the electronic pages.

airug-l26r5-
01800-1o1tb

Take advantage of the eBook Bonus on your computer, smartphone, or tablet. Go to **www.hanserpublications.com/ebookbonus** and register with your personal access code.

Tosello

Micro Injection Molding

Guido Tosello

Micro Injection Molding

HANSER

Hanser Publishers, Munich

Hanser Publications, Cincinnati

The Author:

Guido Tosello, PhD
Associate Professor
Technical University of Denmark
Department of Mechanical Engineering
Produktionstorvet, DTU building 427A
DK-2800 Kgs. Lyngby
Denmark

FSC
www.fsc.org
MIX
Papier aus verantwortungsvollen Quellen
FSC® C016439

Distributed in the Americas by:
Hanser Publications
6915 Valley Avenue, Cincinnati, Ohio 45244-3029, USA
Fax: (513) 527-8801
Phone: (513) 527-8977
www.hanserpublications.com

Distributed in all other countries by:
Carl Hanser Verlag
Postfach 86 04 20, 81631 München, Germany
Fax: +49 (89) 98 48 09
www.hanser-fachbuch.de

Cataloging-in-Publication Data is on file with the Library of Congress

Editor: Dr. Julia Diaz Luque
Production Management: Jörg Strohbach
Coverconcept: Marc Müller-Bremer, www.rebranding.de, München
Coverdesign: Stephan Rönigk
Typesetting: Kösel Media GmbH, Krugzell
Printed and bound by Druckerei Hubert & Co GmbH und Co KG BuchPartner, Göttingen
Printed in Germany

ISBN: 978-1-56990-653-8
E-Book ISBN: 978-1-56990-654-5

Acknowledgments

The contributors to the "Micro Injection Molding" book are greatly acknowledged for their timely and professional inputs in connection with their respective chapters. They are Dr. Maksims Babenko, Dr. Giuliano Bissacco, Dr. Elaine Brown, Dr. Matteo Calaon, Dr. Ali Davoudinejad, Dr. Gualtiero Fantoni, Ms. Carla Flosky, Dr. Donata Gabelloni, Dr. Christian Andrew Griffiths, Prof. Hans Nørgaard Hansen, Dr. Aminul Islam, Prof. Hiroshi Ito, Dr. Ryo Kaneda, Dr. Giovanni Lucchetta, Dr. David Maximilian Marhöfer, Dr. Davide Masato, Dr. Paritat Muanchan, Dr. David Bue Pedersen, Dr. Volker Piotter, Dr. Danilo Quagliotti, Dr. Oltmann Riemer, Dr. Marco Sorgato, and Dr. Ben Whiteside. Gratitude and appreciation is extended to Dr. Mark Smith at Carl Hanser Verlag for his support throughout the book's development, and to his coworkers for copyediting and typesetting the final version of the book.

Much of the early work presented in this book generates from research carried out during the period 2004–2009 by the Polymer Micro Technology Division of the European Network of Excellence 4M (Multi-Material Micro Manufacture), of which I have been fortunate to be an active member. The contributions during those early days from the core members and colleagues of the Polymer Micro Technology Division (Mr. Sabino Azcarate, Dr. Pieter J. Bolt, Prof. Stefan Dimov, Dr. Bertrand Fillon, Dr. Chris Griffith, Dr. Mathias Heckele, Prof. Lars Mattsson, Dr. Stephan Roth, Dr. Andreas Schneider, and Dr. Andreas Schoth) are greatly acknowledged. Furthermore, the contributions to the advancement of micro injection molding science and technology achieved during the European project COTECH (converging technologies for micro systems manufacturing) in the period 2008–2012, in which I was responsible for micro molding process control, metrology, and simulation, have also set the basis of much of the content present in this book. Key professionals and great colleagues in the COTECH Consortium that are greatly acknowledged for inspiration and scientific insights are: Mr. Sabino Azcarate, Prof. Phil Coates, Dr. Franco Costa, Prof. Stefan Dimov, Dr. Chris Griffith, Mr. Helmut Loibl, Dr. Humbert Noll, Mr. Martin Philipp-Pichler, Dr. Manfred Prantl, Dr. Andreas Schoth, and Dr. Ben Whiteside. I would like to thank them for sharing their expertise, for valuable discussions, and for a great friendship that has lasted throughout the years.

On the personal side, I would like to thank my family for their constant love, support, encouragement, and understanding. This book is dedicated to the loving memory of my mother.

Guido Tosello

Technical University of Denmark

Department of Mechanical Engineering

Kgs. Lyngby, Copenhagen, Denmark

Preface

Micro injection molding is the preferred process technology for the mass production of polymer micro components. Micro molded parts are characterized by dimensional tolerances in the micrometer range, high surface finish in the submicrometer down to optical range, and high geometrical complexity. Micro injection molding is used to manufacture highly valued micro medical components (sensors, implants, tubes, catheter tips), micro optics, microfluidic systems, and micro mechanical systems, among others. The global market of micro injection molding is expected to reach USD 1 billion by 2020 with a compound annual growth rate between 10–15% in the period 2013–2020. High demand for micro molded parts is seen from the medical, healthcare, automotive, and electronics sectors.

To exploit the micro molding technology process capabilities and take advantage of research and market opportunities, there was a need for a dedicated book that deals exclusively with micro injection molding and the knowledge to overcome the challenges of successfully managing and processing polymer materials at these ultra-small scales. This book meets that need. Micro injection molding is indeed not just a simple downscaling of conventional injection molding, and specific material-process-product interactions must be understood in order to achieve near zero-defect net-shape micro molded products.

"Micro Injection Molding" is written for engineers, researchers, project managers, consultants, and other professionals involved in precision polymer processing and micro manufacturing. The book provides a comprehensive, up-to-date, and detailed treatment of the main topics related to micro molding. It includes the underlying physics of micro polymer processing and replication at the micro/nano scales, as well as fundamentals of micro molding machine construction and tool making technologies (micro machining, surface treatments). The book also discusses in detail supporting technologies for high-precision micro molding such as quality control of micro parts, micro additive manufacturing for micro product prototyping, and process simulation. Process variations such as vacuum-assisted micro molding and multimaterial processing (micro two-component injection molding, metal and ceramic micro molding) bringing further opportunities for advanced micro manufacturing are presented in detail as well.

The book is composed of 13 chapters organized in four parts (see table below):

- Part 1 – Polymer Materials and Process Micro Technology
- Part 2 – Tooling Technologies for Micro Mold Making
- Part 3 – Micro Molding Key-Enabling Technologies
- Part 4 – Multimaterial Micro Processing

"Micro Injection Molding" Book Chapters Organized in Four Parts

Part	Chapters	Illustrations
1. Polymer Materials and Process Micro Technology	1. Micro Injection Molding Machines Technology 2. Micro Molding Process Monitoring and Control 3. Polymer Materials Structure and Properties in Micro Injection Molding Parts 4. Surface Replication in Micro Injection Molding	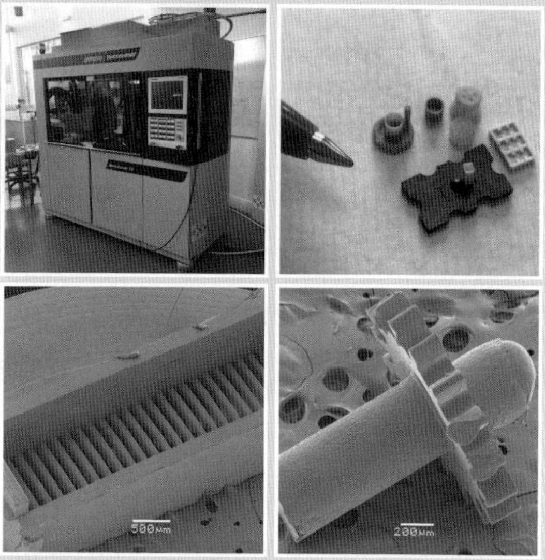
2. Tooling Technologies for Micro Mold Making	5. Micro Machining Technologies for Micro Injection Mold Making 6. Ultra-Precision Machining Technologies for Micro Injection Mold Making 7. Surface Treatment of Mold Tools in Micro Injection Molding	

3. Micro Molding Key-Enabling Technologies	8. Vacuum-Assisted Micro Injection Molding 9. Modeling and Simulation of Micro Injection Molding 10. Metrological Quality Assurance in Micro Injection Molding 11. Additive Manufacturing for Micro Tooling and Micro Part Rapid Prototyping	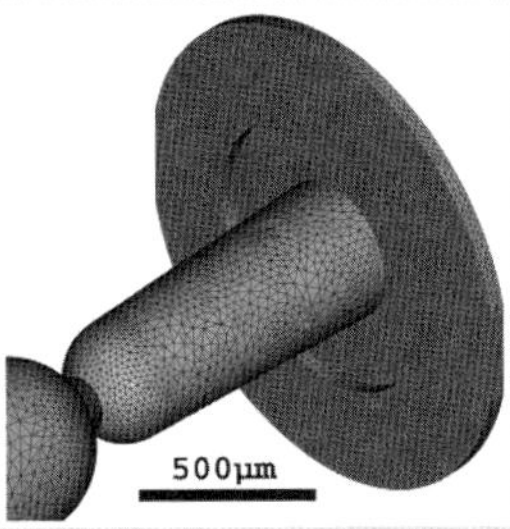 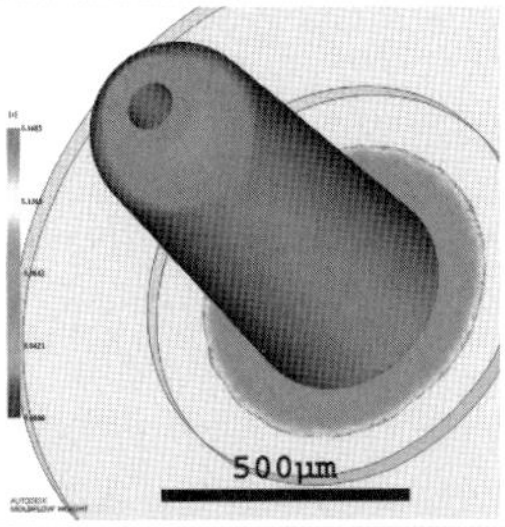 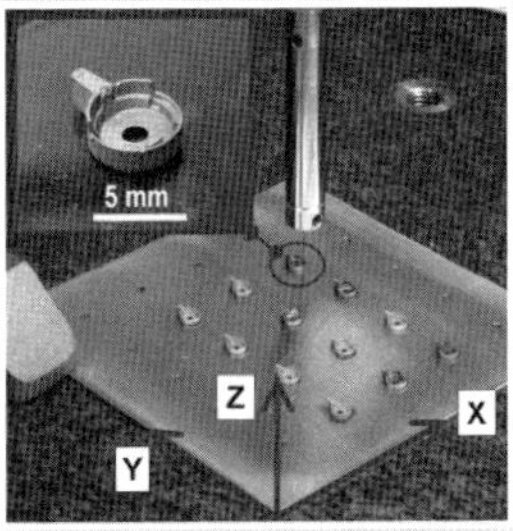 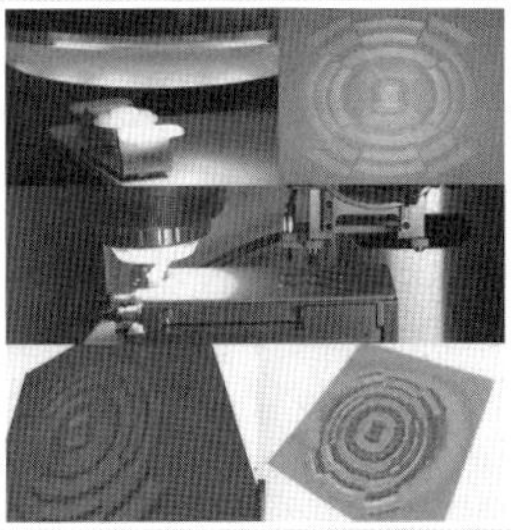
4. Multimaterial Micro Processing	12. Micro Powder Injection Molding 13. Multimaterial Micro Injection Molding	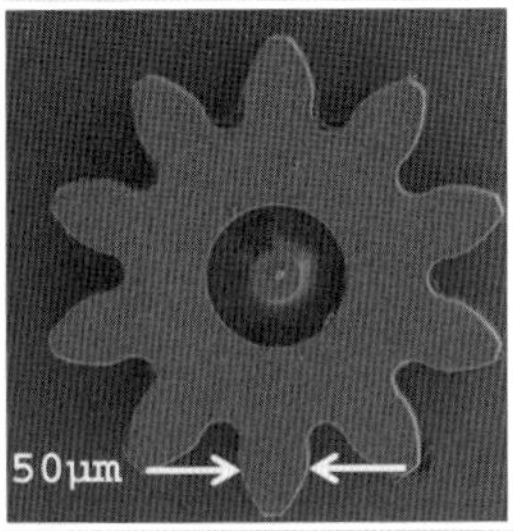  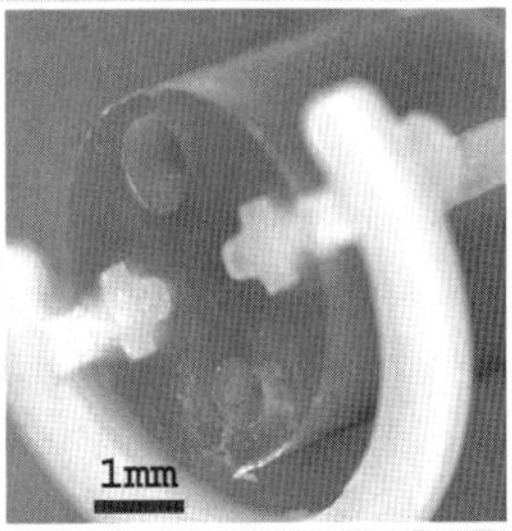 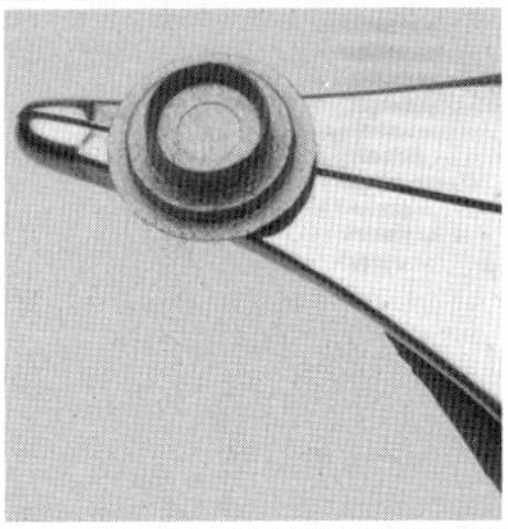

About the Author

Guido Tosello, PhD, is Associate Professor at the Technical University of Denmark, Department of Mechanical Engineering, Section of Manufacturing Engineering. He is senior lecturer, research manager, supervisor of PhD, MSc, and BSc projects, and industrial and management consultant. Guido's principal research interests are the analysis, characterization, monitoring, control, optimization, and simulation of precision molding processes at micro/nano scales of thermoplastic materials. Technologies supporting precision/micro/nano molding processes are of research interest, for example: advanced process chain for micro/nano tools manufacturing, quantitative validation of injection molding simulation, additive manufacturing, dimensional and surface micro/nano metrology, measurement calibration and uncertainty, statistical process control, design of experiment, polymer materials characterization, design and manufacture of 3D precision/micro components and micro/nano structured surfaces, and multimaterial and micro insert molding.

Guido Tosello is the recipient of the "Technical University of Denmark Best PhD Research Work 2008 Prize" for his PhD thesis "Precision Moulding of Polymer Micro Components"; of the 2012 Alan Glanvill Award by The Institute of Materials, Minerals and Mining (IOM3) (UK), given as recognition for research of particular merit in the field of polymeric materials; of the Young Research Award 2014 from the Polymer Processing Society (USA), in recognition of scientific achievements and research excellence in polymer processing within six years from PhD graduation; and of the Outstanding Reviewer Award 2016 of the Institute of Physics (UK) for his contribution to the Journal of Microengineering and Micromechanics.

List of Contributors

Dr. Maksims Babenko	University of Bradford, RKT Centre for Polymer Micro and Nano Technology, UK
Dr. Giuliano Bissacco	Technical University of Denmark (DTU), Department of Mechanical Engineering, Lyngby, Denmark
Dr. Elaine Brown	University of Bradford, RKT Centre for Polymer Micro and Nano Technology, UK
Dr. Matteo Calaon	Technical University of Denmark (DTU), Department of Mechanical Engineering, Lyngby, Denmark
Dr. Ali Davoudinejad	Technical University of Denmark (DTU), Department of Mechanical Engineering, Lyngby, Denmark
Dr. Gualtiero Fantoni	University of Pisa, Department of Civil and Industrial Engineering, Italy
Ms. Carla Flosky	University of Bremen, LFM Laboratory for Precision Machining, Institute of Manufacturing Technologies, Bremen, Germany
Dr. Donata Gabelloni	Erre Quadro s.r.l., Pisa, Italy
Dr. Christian Andrew Griffiths	Swansea University, College of Engineering, Swansea, UK
Prof. Hans Nørgaard Hansen	Technical University of Denmark (DTU), Department of Mechanical Engineering, Lyngby, Denmark
Dr. Aminul Islam	Technical University of Denmark (DTU), Department of Mechanical Engineering, Lyngby, Denmark
Prof. Hiroshi Ito	Yamagata University, Graduate School of Organic Materials Science, Department of Organic Materials Science, Yamagata, Japan
Dr. Ryo Kaneda	Yamagata University, Graduate School of Organic Materials Science, Department of Organic Materials Science, Yamagata, Japan
Dr. Giovanni Lucchetta	University of Padova, Department of Industrial Engineering, Padova, Italy
Dr. David Maximilian Marhöfer	Technical University of Denmark (DTU), Department of Mechanical Engineering, Lyngby, Denmark
Dr. Davide Masato	University of Padova, Department of Industrial Engineering, Padova, Italy
Dr. Paritat Muanchan	Yamagata University, Graduate School of Organic Materials Science, Department of Organic Materials Science, Yamagata, Japan
Dr. David Bue Pedersen	Technical University of Denmark (DTU), Department of Mechanical Engineering, Lyngby, Denmark
Dr. Volker Piotter	Karlsruhe Institute of Technology, Material Process Technology, Eggenstein-Leopoldshafen, Germany
Dr. Danilo Quagliotti	Technical University of Denmark (DTU), Department of Mechanical Engineering, Lyngby, Denmark
Dr. Oltmann Riemer	University of Bremen, LFM Laboratory for Precision Machining, Institute of Manufacturing Technologies, Bremen, Germany
Dr. Marco Sorgato	University of Padova, Department of Industrial Engineering, Padova, Italy
Dr. Guido Tosello	Technical University of Denmark (DTU), Department of Mechanical Engineering, Lyngby, Denmark
Dr. Ben Whiteside	University of Bradford, RKT Centre for Polymer Micro and Nano Technology, UK

Contents

PART 1

Polymer Materials and Process Micro Technology

1 Micro Injection Molding Machines Technology

Gualtiero Fantoni, Donata Gabelloni, Guido Tosello,
Hans N. Hansen

1.1 Introduction

The present chapter introduces the design of micro molding machines from two different perspectives: analysis of patents on the technology, and design architectures and solutions for the micro injection molding machines.

The first analysis is oriented to provide a quantitative scenario on the attempts to exploit research for commercial uses, and it is performed taking into account patents applications worldwide. It helps in situating the micro injection molding domain in the industrial market (particularly on the machine manufacturing side). Conversely, the second part of the present chapter focuses on some relevant architecture aspects of micro injection molding machines that have been analyzed from a functional perspective. The choice of the machines is based on interesting architectures that have been the target of various researches in the recent past and that are often cited within the present book.

1.2 Patent Analysis

Patents data can provide an overview of the main technologies related to an artifact. A landscape study of the patents concerning micro injection molding machines and their components has been performed. Such study identified a patent set selected from a patent database provided by Erre Quadro s.r.l. *(http://www.errequadrosrl.com/).*

The patent set contains 2809 individual patents. They cover a period of 135 years and have been filed by more than 500 assignees coming from 17 countries. Many of them (>75%) belong to the last two decades, when patents and their use increased thanks to the internet adoption and patent spread.

Different analyses have been carried out on the selected patent set.

First, a linguistic analysis has been performed to identify expressions (i.e., multi-word terms) that are specific to the particular patent set. Figure 1.1 illustrates the most relevant expressions used in the patent set on micro injection molding machines. The size of the expressions within the tag cloud is proportional to their frequency of occurrence in the selected patent set.

The tag cloud reports many of the components of the machines and a particular interest in the mold cavity as a key element in the specific case of micro injected components.

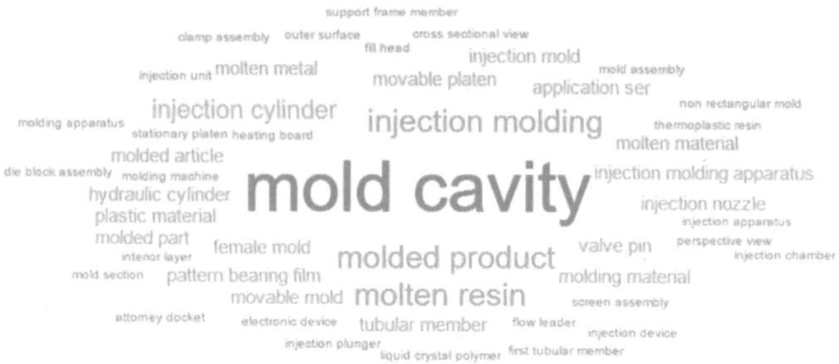

Figure 1.1 Tag cloud in the patent set on micro injection molding machines

Patents filed in states that are members of the World Intellectual Property Organization (WIPO) are assigned, by the patent office, to one or more IPC[1] classes. Thus, each IPC class can be interpreted as a corresponding technological area. Table 1.1 provides a list of the 10 largest IPC classes within the investigated set of patents. As expected, the main topics are those related to techniques for "working plastic" and "organic macromolecular compounds" of various origins.

Table 1.1 Main 10 IPC Classes Sorted According to Occurrence within Patent Set

Class	Class description
B29C	working of plastics; working of substances in a plastic state in general, shaping or joining of plastics; shaping of substances in a plastic state in general; after-treatment of the shaped products
B22D	casting; powder metallurgy; casting of metals; casting of other substances by the same processes or devices
B29B	working of plastics; working of substances in a plastic state in general; preparation or pretreatment of the material to be shaped; making granules or preforms; recovery of plastics or other constituents of waste material containing plastics

[1] International Patent Classification

Class	Class description
B29L	working of plastics; working of substances in a plastic state in general indexing scheme associated with subclass B29C, relating to particular articles
C08J	organic macromolecular compounds; their preparation or chemical working-up; compositions based thereon working-up; general processes of compounding
B29K	working of plastics; working of substances in a plastic state in general indexing scheme associated with subclasses B29B, B29C, or B29D, relating to molding materials or to materials for reinforcements, fillers, or preformed parts
C08L	organic macromolecular compounds; their preparation or chemical working-up; compositions based thereon; compositions of macromolecular compounds
C08K	organic macromolecular compounds; their preparation or chemical working-up; compositions based thereon; use of inorganic or non-macromolecular organic substances as compounding ingredients
H01L	basic electric elements; semiconductor devices; electric solid state devices
B29D	working of plastics; performing operations; transporting/producing particular articles from plastics or from substances in a plastic state (making granules B29B 9/00; making preforms B29B 11/00)

Figure 1.2 shows how patents are distributed in time according to filing date. Referring to such figure, the distribution shows a growing trend (more regular with respect to other machines and disciplines) but with several fluctuations and some anomalies (e.g., the year 2000).

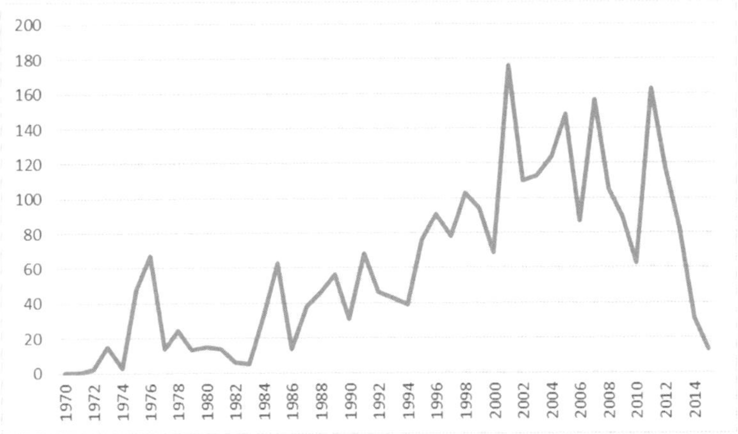

Figure 1.2 Temporal distribution according to filing date

An important remark concerning Figure 1.2: patents become accessible to the public only after publication, which on average takes place about 18 months after a patent has been filed. During this period patents undergo examination by a patent office. This may explain, at least in part, the noticeable drop that can be seen in the last two years.

Some of the most important information that can be extracted from patents metadata is related to the companies that own the property of the patents, i.e., the assignees. The top 10 assignees found in the patent set are shown in Table 1.2.

Table 1.2 Assignees: Top 10 List

Top assignee	% Patents
Nissei Plastic Ind. Company	4.2%
Toshiba Mach. Company	2.7%
Sumitomo Heavy Industries	2.6%
Moldmasters	2.4%
Husky	1.5%
Fanuc	1.4%
Ube Ind.	1.2%
Dai Nippon Printing Company	1.1%
Yoshida Kogyo	1.0%
Toyota Motor Corporation	1.0%

The values of the percentages of patents owned by each assignee show that the patent ownership is rather spread and there is not a big player that has a remarkable fraction of patents. However, Nissei Plastic Ind. Company is the most relevant, followed by Toshiba Mach. Company and Sumitomo Heavy Industries. Such distribution is typical of innovative products built on the basis of pre-existing technologies and where innovation and disruption happen mainly within the companies' R&D division.

Additional interesting information concerns where the assignees of the patents are located (such data are different from the headquarters of multinational companies but often refer to the plant or division where the invention has been conceived and tested). However, before presenting the geographical distribution, it is important to notice that assignees can be further distinguished in two categories: public and private. Public assignees are institutions or organizations like universities and research centers. Private assignees are companies and individuals.

Figure 1.3 and Figure 1.4 show the concentration of private and public assignees, respectively. The different colors in the images represent the different concentrations of assignees. Therefore, in Figure 1.3 the light blue areas represent a smaller number of private assignees than the dark blue regions. Similarly, in Figure 1.4, the red areas reveal the countries where a high number of public assignees filed patents, with respect to the green ones.

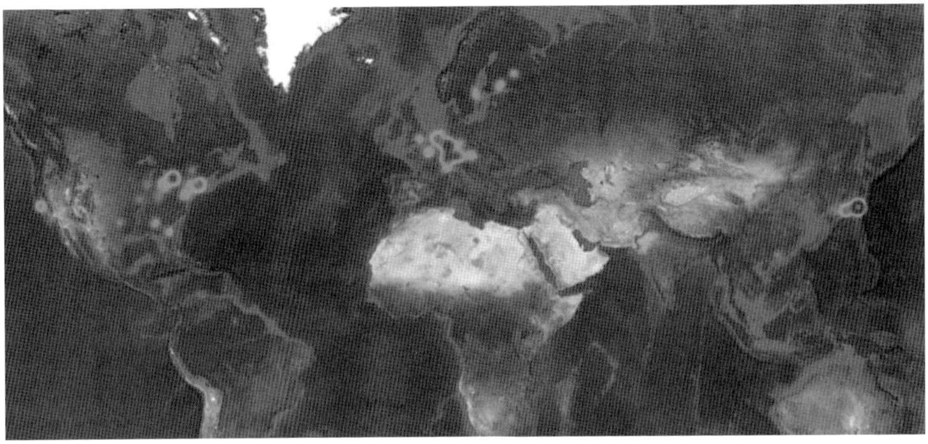

Figure 1.3 Geographical distribution of private assignees in the reference patent set. The figure is presented in the color supplement

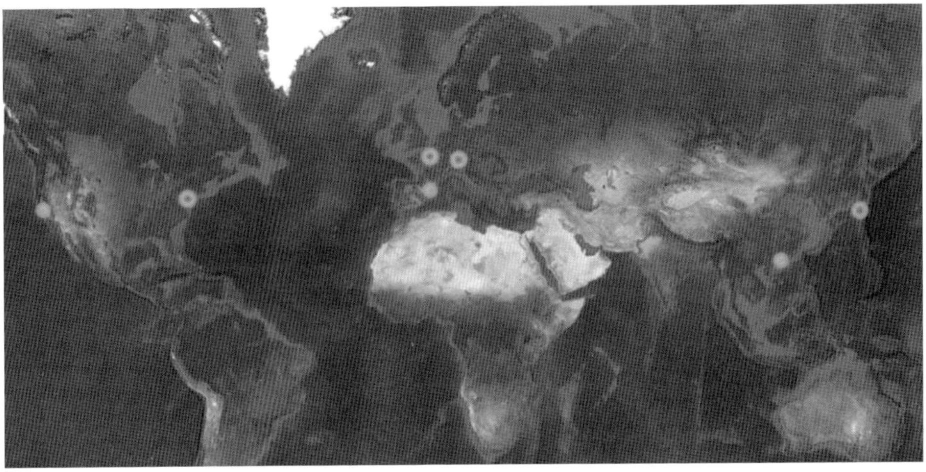

Figure 1.4 Geographical distribution of public assignees in the reference patent set. The figure is presented in the color supplement

The analysis reveals that the highest concentration of private assignees–in terms of countries–is in Japan. However, both north-central Europe and the north-west of the U.S. show several private companies playing an important role. On the other hand, the majority of public assignees is concentrated in small areas. It is interesting to notice that in China there is a higher concentration of public assignees than that of private ones. However, it is worth underlining that only Chinese patents with a translation in English have been considered in the present analysis and therefore the image in Figure 1.4 underestimates the Chinese volume.

Independently from the geographical origin of the assignees and their public or private nature, they can be active in different jurisdictions. Such information is highly relevant, since it can be related to the market where assignees produce and distribute their products.

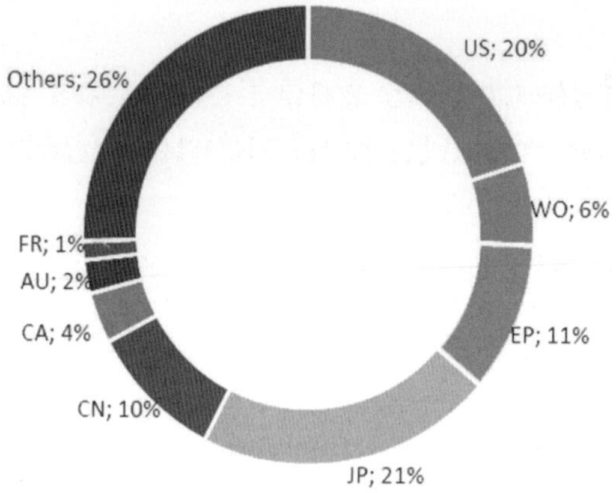

Figure 1.5 Distribution of jurisdictions within the patents in the reference patent set

Figure 1.5 shows that most of the patents are filed in the United States and in Japan, followed by the European and Chinese jurisdictions. Such a distribution is quite predictable, since they are the countries where both the market and the production have a higher probability to succeed.

Other metadata that can be found in patents are the inventors of the innovations patented. Table 1.3 shows a list of 10 top inventors, whose patents have been found in the patent set, including information about their presumed employment company (at the time of filing).

Table 1.3 Inventors: Top 10 List and Their Presumed Employment Company

Top Inventor	Company
Ganz, Martin	Ettlinger Kunststoffmaschinen GmbH, Wittmann Battenfeld
Taniguchi, Yoshiya	Toyo Machinery & Metal
Olaru, George	Moldmasters
Anger, Anton	Anton Anger Maschinenbau
Laveau, Richard	Techmire
Babin, Denis	Moldmasters
Thibault, Carl	Techmire
Bourbonnais, Alain	Techmire
Mundwiler, Ulrich	Straumann Holding
Gruber, Peter A.	International Business Machines Corporation

The first two inventors in the table do not belong to any of the top 10 assignees, while there are three inventors from Moldmasters, that is, the assignee at the third place in Table 1.1. Only Martin Ganz worked for more than one company that operates in the development of injection molding machines.

1.3 Architectures and Solutions for Micro Injection Molding Machines

While the previous section focused on aggregate analysis of patents on the field, the present section focuses on the study of the evolution of injection molding machines from a design perspective.

Since the present book focuses mainly on the micro injection molding process, and due to the fact that, on the other hand, a number of key-enabling technologies are necessary in order to develop the process to its full potential, an introduction to functional modelling has been considered a fundamental topic and it is presented here. The analysis is based on both a theoretical functional analysis and the performance evaluation of two different injection molding machines: a conventional machine and a micro injection molding machine. Firstly, the study starts from the working phases of a machine, followed by the functional analysis to decompose the phases, and then attributes functions to part features. Later, part features are critically analyzed to understand the causal chains leading to the wanted behavior or to the failure to be avoided. Finally, the assessment of the design solution is performed by gathering quantitative data from experiments and correlating failures and advantages with the design of the machine parts.

1.3.1 Introduction to Functional-Based Modelling

Functional analysis is a powerful tool for the analysis of artifacts. It allows the identification of the functions performed by the product and by the components of the structure that carry on these functions. Functional analysis is an effective tool to both design new products and redesign existing products to improve their functionality.

Functional analysis started from the value analysis developed by Miles [1], who identified the functions performed by the products and analyzed the value of each function, given by the ratio of the benefits provided by the functionality of the components to their cost. Hence, the designer could decide to eliminate the functions (and so the components linked to the functions) with lower value in order to reduce the final cost of the product.

The adopted modelling procedure uses a few well-established models to analyze in a systematic approach the technical system and to maximize the repeatability of the overall method. Among those models we can cite the EMS (Energy Material Signal) functional modelling methodology used to analyze the expected functions in a technical system [2]. Pahl and Beitz described the functioning of a product by modelling functions acting on the flows of EMS: energy, material, and signals. The identified functions are black boxes that transform or transmit EMS input into EMS output. More in detail, each function can be expressed by a using verb, in active form, that describes the transfer function of the input into output.

To increase the repeatability of the analysis, a standard language to describe functions has been developed by Stone and Wood [3] and then enhanced by Hirtz et al. [4]. In this function and flow taxonomy, called "functional basis", the basic functions are split into eight classes while flows are classified into three groups. Moreover, since it reconciles and integrates two independent research efforts–the functional basis of Stone and Wood [3], and the technological dictionary developed by the National Institute of Standards and Technology [5]–it has been considered as a reference for the present work.

As far as the decomposition is concerned, the employed principle is similar to the one presented in [6], while for the representation we adopted an approach linked to both EMS and Subject-Action-Object, derived from TRIZ, in analogy with the one presented in [7].

Furthermore, the model applied in the present work takes advantage of several past contributions:

FBS framework: Umeda et al.'s model [8] describes the relationships between structure, behaviors, and functions. Here, the analysis of the product features is firstly performed to determine the behaviors related to each feature, and then each derived behavior is transformed into the interpreted functions.

Axiomatic Design: Axiomatic Design [9] aims at correlating functions with requirements. Functional Requirements (FRs) are defined as the "minimum set of independent requirements that completely characterize the functional needs of the product in the functional domain". The entire theory is based on axioms and theorems. Among the different formulations, we chose the following: for the axiom 1 (Independence Axiom), "An optimal design maintains the independence of functional requirements", while for the axiom 2 (Information Axiom), "The best design is a functionally un-coupled design that has the minimum information content" was selected. The two axioms can be read together as "Among all the designs that satisfy the independence axiom, the one with the minimum information content is the best design". This means that Design Parameters (DPs), which are the physical variables characterizing the physical entities, are related to Functional Requirements (FRs) in such a way that specific DPs can be adjusted to satisfy their corresponding FR without affecting other FRs.

The axiomatic design has been previously employed to study the injection molding process. Suh et al. [9] discussed the application of axiomatic design principles to gain controllability of the injection molding process by considering a dynamic temperature control that enables temporal decoupling of the injection and solidification stages to increase the process performance. Also, studies have been conducted on injection molding applying axiomatic design and its principles to improve the quality and the performance of the process [10, 11]. However, the functional analysis of injection molding machines and its implications on micro molding applications have not been addressed so far and are therefore discussed in the present chapter.

The functional analysis can be successfully applied to perform a comparison of two different injection molding machines for the production of polymer-based micro products. The two machines are based on two different molding technologies. The application of functional analysis is instrumental for the understanding and assessment of the machine design solutions. Functional analysis allows the investigation of the design motivations that contributed to the improvements of the machine design that affect the characteristics of the injected micro products. The results of the functional analysis are supported by experimental data obtained from the two different injection molding machines.

Before starting with the analysis, it is worth noticing some of the reasons that motivate it:

- Micro injection molding has become a key manufacturing process for a large variety of micro components made of one polymer [12–14], two polymers (i.e., two component micro injection molding) [15], both polymer and metal (i.e., micro insert molding, micro assembly injection molding) [16, 17], metal and ceramic (powder injection molding, to produce, e.g., piezoelectric ceramic 3D micro parts) [18–22], or a combination of these micro injection molding-based processes [23, 24]. However, a dominant design does not exist yet.
- The design of micro injection molding machines involves a radical dimensional scale shift, and, in this design process, a barrier by simply downscaling the injection molding design from the macro dimensional scale is generated [25].
- Without a reference/dominant design for the micro injection molding machines, the interest in machine design remains high and it is still growing, as demonstrated also by the growing trend of patents and by the fact that new machines are appearing in the market with a regular pace. Different micro injection molding machine design principles are applied in order to establish a micro injection molding process capable of meeting the high accuracy required by micro products manufacture (e.g., machines from DesmaTec, Battenfeld, Arburg, Sodick Plustech, Männer, Ultrasion, among others).

1.3.2 Method

The analysis of the micro injection molding machines starts from their functional decomposition. First, the machines have been split into their groups, each group into components/parts, and each part into its features. Features can be defined as the specific characteristics of a single part of the product, in terms of the geometrical entities that define it and in terms of the properties of its material (e.g., the Young's modulus, resistance to acids or to flames, transparency to certain light frequencies, thermal or electrical conductivity, porosity, etc.) [26]. Such concept of feature is very close to Suh's Design Parameter [9]. However, features are more general and closer to the designer's lexicon and common understanding.

Following axiom 2, each feature is related to at least one functional reason. Therefore, all the features carry on at least one function in a certain moment of the product lifecycle. More likely, each feature plays different roles in different moments, and very often it happens that some features concur in performing a single function. In a good design, each and every feature of a product performs at least one function or contributes to it [26]. Indeed, if no function is performed by the feature, such design contains at least one piece of information (e.g., the one necessary to define the feature) without a functional rationale. Thus, such useless information or feature can be removed without affecting the product performance and actually saving manufacturing/production/assembly time and/or cost. Therefore, features are the starting elements for the functional analysis.

Another important element is represented by the working phases of the machines. Phases are homogeneous sets of functions belonging to—or performed by—the same components and characterized by the same physics/chemistry/logics. Therefore, a phase can be defined as the set of statuses (e.g., of environment, machine, and product) [27] in which certain logical or physical structures persist [26]. During the same phase the involved physical behaviors remain the same and the flows involved do not change (except for their values).

Both of these elements help improve the understanding of the general functioning of the machines, and which parts/components carry out the functions in the different phases of usage of the machines.

Therefore, the first four steps of the method shown in [27] are here applied to both injection molding machines (i.e., conventional and micro molding machines). They consist of:

- Analysis of the process phases: When the input is the same, the physics/chemistry/logics of the process does not change, and the output has to be similar. The procedure imposes distinguishing the process phases or the use phases.
- Functional analysis of the machine: Different technical choices affect differently the details of the process and, in some way, also the result of the process itself. The functional breakdown is necessary to correlate design and performance.

- Identification of the key features performing each function: The performance (namely the level of the flows) of each function is due to a subset of features directly linked to it through causal chains.
- Analysis of the functional map: The map describing the machine can be used to highlight the negative function and those functions that are in contradiction. This is the main step of the procedure to investigate the failures and the useless functions performed by the machine.

After the functional analysis of the two machines is performed, a further step regarding the comparative analysis of the two functional maps is introduced. The comparison is also carried out by means of quality parameter measures obtained from actual experiments with both the injection and the micro injection molding machines. From the comparative analysis it will be possible to see how the maps describing the two different machines share some common patterns but also have different functional sequences. The analysis of the differences between the two functional maps allows for a direct correlation between functional design reasons and different performances. The analysis of differences can produce hypotheses for machine redesign.

1.3.3 Functional Analysis

The micro injection molding process operated with two different machines (a conventional injection molding machine and a micro injection molding machine) has been studied and compared in order to fully understand the reasons of their design and to correlate design choices and final results (positive features as well as drawbacks and failures).

The following two different machines have been studied and compared:

- A small injection molding machine based on the conventional injection molding technology using the reciprocating plastication screw [28]. This machine is interesting, since it is an example of simple downscaling of the components, not a machine properly designed to manufacture micro parts.
- Micro injection molding machine with two pistons and a plastication screw [29–31]. The machine has an innovative design and its architecture seems to better meet the requirements for manufacturing micro parts through the injection molding process.

Process Phases

In order to identify all the functions of the machine, it is necessary to consider the various phases that occur during the process. Therefore, the study of the evolution of the flows leads to the determination of the transitions from one state to another, while homogenous sets of states represent phases. Following this approach, the main phases of the micro injection process, described in the literature, are identified and summarized in Figure 1.6.

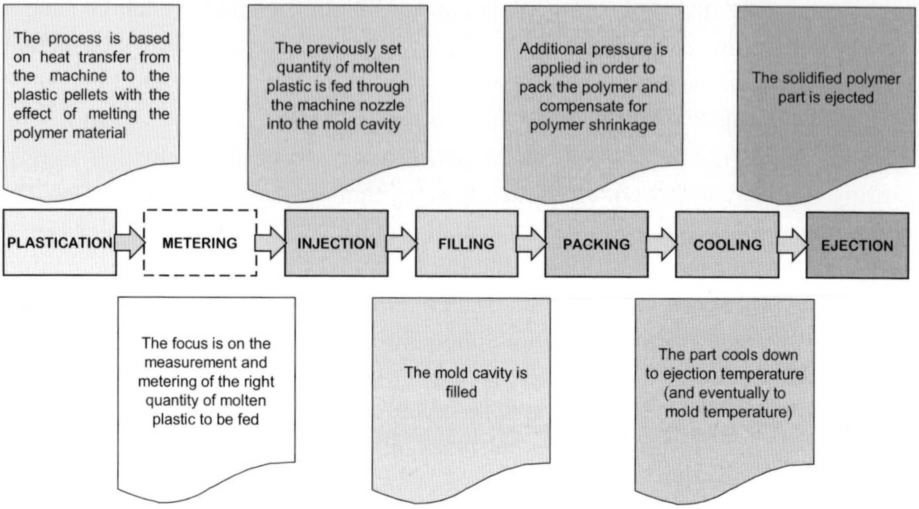

Figure 1.6 The main phases in the injection and micro injection molding processes

Figure 1.6 shows the main phases in the injection and micro injection molding processes (the dotted line block is absent in conventional injection machines, and present in micro molding machines).

It is worth noticing that the metering phase is explicitly described only in the machine with a screw and two pistons, and it is made possible by the control of a suitable valve. Conversely, in the conventional injection molding machine, as it will be shown more clearly in the next subsection, the metering phase is embedded in the contiguous phases.

Although the standard phases are those described above, according with the definition of phase given above, a more detailed process workflow is proposed (see Figure 1.7).

Actually, even from a quick analysis of the phases it is possible to notice that a new chain of functions related to the metering of the right amount of material has been introduced. On the other hand, the "screw's setting" phase, which contained the metering in the traditional machine, is eliminated.

The detailed functional analysis shows the consequences coming from the introduction/elimination of some phases in the micro injection molding machine with two pistons and one screw.

TRADITIONAL INJECTION MOLDING	MICRO INJECTION MOLDING 2 pistons and a plastication screw
1. Plastication	1. Plastication
2. (Ejection')	2. Metering (NEW PHASES)
3. Injection	3. Injection chamber's filling (NEW PHASES)
4. Filling	4. Pushing (NEW PHASES)
5. Packing	5. Injection
6. Cooling	6. Filling
7. Screw's setting	7. Packing
8. (Plastication)	8. Cooling
9. Ejection	9. Ejection

Figure 1.7 The different phases of conventional and micro injection molding processes

Functional Analysis of the Machines

After the identification of the phase of usage, the functional analysis of the two injection molding machines can be performed.

Each machine and its related detailed process are described in natural language. The functional description has been tagged as follows: bold for verbs and italic for flows, respectively.

Firstly, a conventional plastication-screw injection molding machine (see Figure 1.8) adapted for micro parts injection has been studied and analyzed. The machine and related detailed process have been described in natural language.

Structural description: A conventional injection molding machine (e.g., screw diameter = 15 to 18 mm, clamping force = 250–350 kN, maximum injection speed = 100–150 mm/s) can be adapted for manufacturing micro parts. An injection molding machine consists of a material hopper, an injection ram or screw-type plunger, and a heating unit. Injection molding machines hold the molds in which the components are shaped. Injection molding machines are rated by tonnage, which expresses the amount of clamping force that the machine can exert. This force keeps the mold closed during the injection process [28].

Functional description in natural language: First of all, *pellets* **moved** by *gravity* are **fed** to the *screw system*; during the plastication phase, the *screw* is **rotating** to build up the *melt polymer* necessary for the injection phase. The *volume* close to the tip of the screw is **heated** by *electrical units* and the *pellets* **melt**. During the rotation, the *molten polymer* is **fed** downward. The pressure **pushes** the *screw* backward. When sufficient *polymer* has built up (i.e., shot volume is plasticated), rotation **stops**; when the *mold* is **closed**, the *screw* is **pushed** (injection), the *melt polymer* **fills** the *sprue*, the *runners*, and the mold *cavity* (filling), and the *screw* begins **rotating** again to build up more polymer (packing). **Cooling** and ejection: after the *polymer* is **solidified** (cooling), the *mold* **opens** and *ejector pins* **remove** the *molded part* (ejection).

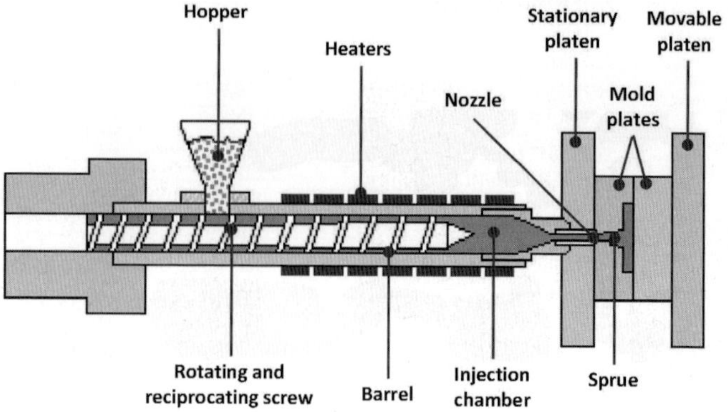

Figure 1.8 Elements of a conventional injection molding machine

The second injection molding machine (a micro injection molding machine, in this case a Battenfeld Microsystem 50 is selected) is equipped with a three-stage plastication-metering-injection unit specifically designed for micro molding (see Figure 1.9).

Structural description: The micro injection molding machine is equipped with a three-stage plastication-metering-injection unit particularly designed for micro molding applications. The first stage includes a 14 mm diameter screw (mounted at an angle of 45°) for the homogeneous preparation of the melt. The second stage is represented by a vertical metering piston for the accurate metering of the material to be injected. The third stage is the injection chamber, where a 5 mm diameter servoelectric-driven piston injects the dose coming from the metering unit directly into the cavity. The piston is capable of a maximum injection speed of 960 mm/s and it is capable of injecting up to the parting line of the micro cavity. In this way no sprue is attached to the mold. The maximum clamping force of the machine is 5 kN.

Functional description in natural language: Initially, the rotating *valve* is **closed** and, as in the traditional machine, *pellets* **moved** by *gravity* are **fed** to the *screw system*; during the plastication phase, the *screw* is **rotating** to build up the *melt polymer* necessary for the injection phase. After that, the *valve* **rotates** and it **opens**, **allowing** the passage of the melting *polymer*. Progressively the *liquid* **enters** in the *dosage barrel*, **blocked** by the *injection plunger* from below, and the *dosage plunger* **goes** up. After the desired amount of *liquid* is **imported/stored** in the *dosage barrel*, the rotating *valve* **rotates** and **closes**, and the *injection plunger*

moves back. This phase allows for the passage of the *liquid* in the *injection barrel.* Therefore, the latter **pushes** the *dose* coming from the metering unit directly into the *cavity*. The piston is capable of a high injection speed and can inject up to the parting line of the mold. The remaining phases are carried out as in the conventional machine previously analyzed.

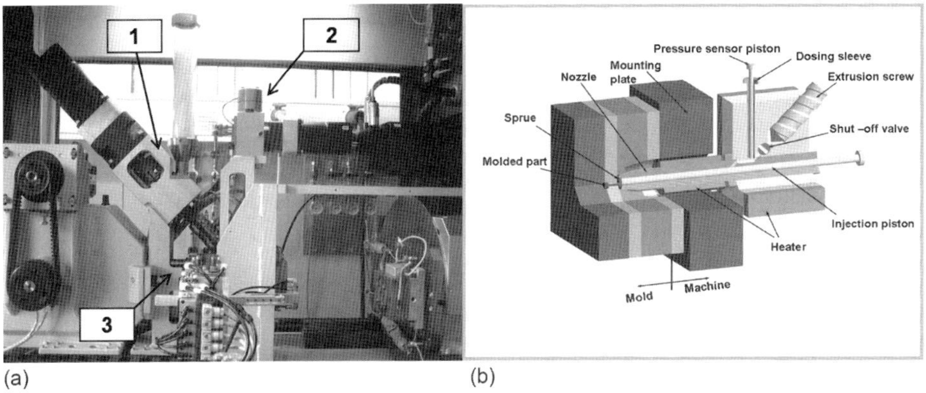

Figure 1.9 (a) Micro injection molding machine and the three-stages unit: (1) plastication, (2) metering, (3) injection [13]; (b) micro injection unit elements [32]

The machine descriptions in natural language (further detailed) are transferred into two functional maps. The full map corresponding to the first machine is shown in the Appendix (Section 1.4), while a comparison of the two maps (where only the most relevant phases and functions are shown) is presented in Figure 1.10a and b.

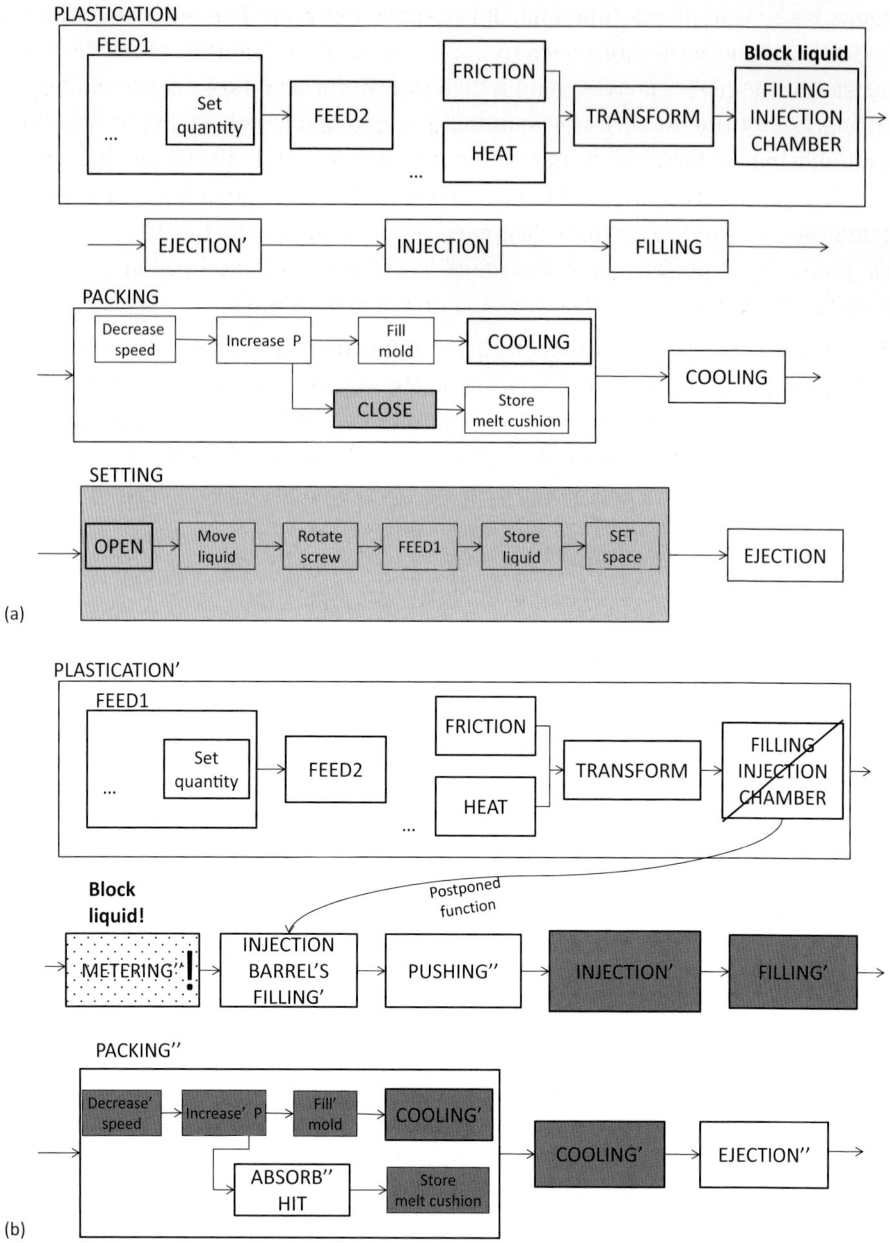

Figure 1.10 Synthesis of the functional analysis of (a) the conventional injection molding machine (equipped with a reciprocating screw) and (b) the micro injection molding machine (equipped with the two pistons injection unit)

Figure 1.10 presents the functions of the screw machine. The functions related to the setting of the screw are shown in light gray. Such functions are subsidiary with respect to the main function and are completely coupled (see axiom 1). They are responsible for a set of operations oriented to feed and meter the quantity of liquid to be injected. Unfortunately, due to the coupled design, both the feeding and the metering phase are not performed correctly, given the existence of backflow of molten plastic. These functions disappear in the second machine, where the injection phase is not performed by the screw but by the horizontal piston. Conversely, those functions remaining the same are shown in dark gray.

The most different blocks between the two machines are those whose box has a dashed pattern, corresponding to the metering phase. In the first machine there is only a function dedicated to dose the right quantity of liquid (set quantity), one of the 38 functions performed by the screw system during the feeding phase. Instead, in the second machine, the metering is an independent phase whose responsible part is the vertical piston (in cooperation with the valve and the horizontal piston). Such architecture and sequence follow the separation in space and time principle [33], and allow for a smart decoupling of the coupled functions.

Analyzing in detail the energy-material-signal (EMS) flows in the maps, another key difference regards the presence of the sprue in the first machine. Indeed, during the conventional injection molding process, the sprue solidifies on the nozzle of the injection chamber. The sprue is then extracted with the molded part during the phase of ejection. As a consequence, the process is associated with a loss of material. On the other hand, the sprue plays a necessary function: in cold runner molds, it blocks the fluid and prevents it from entering into the mold before the filling phase. In the second machine the sprue is absent, since (i) the piston travels up to the parting line of the mold and (ii) the exact amount of polymer melt is measured and injected (i.e., there is no necessity of a melt cushion). Therefore, nearly no loss of material is associated with the novel process.

As a consequence of these observations based on the functional maps, it can be seen that one of the main problems of the first type of machines is that the process is not controlled with respect to the amount of plastic to be injected in each cycle (actually, a specific control unit and a dedicated device are missing). Indeed, there is only one function dedicated to deliver the right quantity of liquid (set quantity), but it is coupled with other functions of the screw (e.g., feeding). On the contrary, in the second machine each phase is performed separately by a dedicated component (separation in space); furthermore, the sequence of functions in the workflow allows also for a temporal decoupling (separation in time). This difference depends on the architecture and on component features, and it will be shown in detail in the next paragraphs.

Identification of the Key Features Performing Critical Functions

The procedure pushes the investigation toward the analysis of the component features involved in performing each function. The key elements are the plastication screw in the conventional machine and the metering piston in the micro injection molding machine (see Figure 1.11). Each machine has a different and specifically designed valve associated with these two elements. The different functions of the valves are discussed in the following.

Conventional injection molding machine

The screw is involved in many phases of the process and performs several different functions through its features. However, its functional analysis shows that it is not able to fulfill all the foreseen (designed) functions, especially in micro injection molding applications. Indeed, one of the most critical sub-phases concerns the crossing of the liquid through the valve (Figure 1.11a, b, and c). Its closure, during the injection phase, blocks the backflow of the molten polymer. For this purpose, a valve is integrated in the screw and its opening/closing phases are regulated by the screw position.

During the plastication phase the screw rotates and the liquid begins to fill the injection chamber. In this case, the features that allow the liquid to cross are the diameter of the restriction of the screw (see part a in Figure 1.10) and the internal diameter of the ring (see Figure 1.11a). These features and their relative position regulate the functions of opening and closing of the valve.

The feeding of the material toward the volume in front of the screw pushes back the screw itself. Afterward, the injection phase starts: the screw stops rotating, then moves toward the mold, and the area where the liquid passes through narrows until the valve is completely closed (end of the packing phase). Finally, while the part is first cooled and then ejected, the screw moves back up to the starting position, the valve opens again, and the cycle can restart.

The functions "to couple" and "to block" are here performed by the diameter of the screw and by the internal diameter of the ring.

Micro injection molding machine

Alternatively, in the second machine, the function "to block" is separated in space and time–TRIZ separation principle [33]–if compared with the first machine, and it is executed by the horizontal injection piston when it is located under the dosage barrel (see Figure 1.11d and e).

In the micro injection molding machine with two pistons and one screw, the plastication phase is still performed by the screw, while metering and injecting are done by the valve+vertical piston and valve+horizontal piston, respectively (see Figure 1.11d and e). In particular, by separating melt plastication and melt injection the machine controls better both the metering accuracy and the homogeneity of the very small quantities of melt.

A small injection plunger, whose diameter measures 5 mm, is used for melt injection to control metering accuracy. Thus the gap between pistons and the respective barrels is much lower than the one between the screw and the barrel in the traditional machine. This means that there is no backflow of the liquid polymer and, therefore, a better control of the metering accuracy is performed through the vertical piston.

The separation brings also another consequence: the screw can have a sufficient channel depth to properly handle standard plastic pellets, assuring a better homogeneity for the polymer fluid.

Moreover, the valve in the first machine does not have a rigorous on-off behavior, while the rotating valve in the micro injection machine is automatically actuated, so that both the closing and the opening functions are executed with higher precision. Due to such elements and features (and also related parameters as the clearance, tolerances, etc.), the process is more accurate and the setting of the fluid to be injected is more precise.

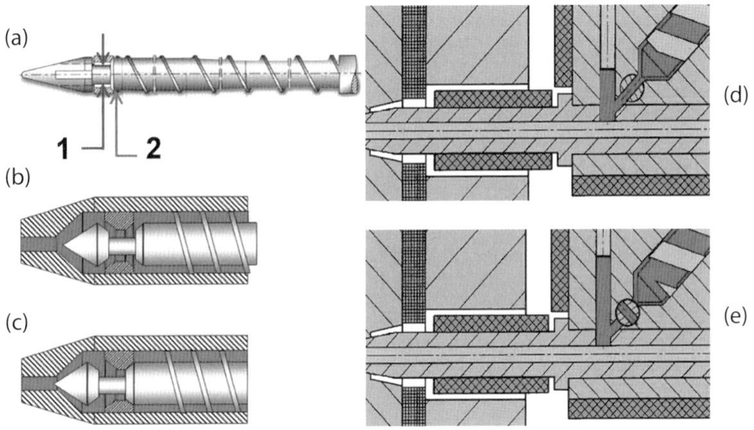

Figure 1.11 Left: (a) screw and valve in a conventional injection molding machine, (b) configuration during plastication, (c) configuration during injection; right: metering piston, injection piston, and valve configurations (d) during and (e) at the end of metering in the micro injection machine

Comparative Assessment of the Machines' Performance

For the comparative assessment of the machines' performance, molding experiments have been carried out and they allowed for an understanding of the different performances of the two machines, to underline their advantages, drawbacks, characteristics, and failures. The focus of this step is to correlate how the process is carried out by the machines with their design. A rigorous functional analysis of the two machines made possible to highlight the physical reasons for the different components and their features, while experimental results provided the measurement of the adopted solutions.

In detail, the first machine (conventional reciprocating screw employed on molding micro parts) presents the following problems:

- It is difficult to control the melt metering accuracy as a result of the screw structure and the limitation to reduce screw size (position accuracy of typically ±(0.010–0.025) mm leads to an uncertainty in metering size of ±(2–10) mg depending on material density and screw diameter, usually 15–25 mm for micro and miniature molding applications).
- A problem that occurs with the small shot weight typical of micro parts is related to the size of used pellets. Conventional injection molding machines utilize screws with diameters down to 14 mm. Thus the depth of the screw channels should have at least the dimensions of a single grain. Hence, when the screw moves just 1 mm, about 150–200 mg of plastic are already injected. Even one single pellet of plastic has a weight typically in the range of 20–30 mg, and for highly miniaturized 3D micro products this often exceeds the part weight.
- The machine can produce precise components but large sprues are necessary to achieve the minimum shot weight to perform properly the process. Very often, over 90% of the polymer is wasted and this waste can be an important cost factor (considering, e.g., plastic material for optical or medical applications, the cost can be in the range of 30 €/kg for a cyclo-olefin polymer to 70 €/kg for poly-ether ether ketone, respectively).
- The large sprue increases cooling time and therefore cycle time.
- The small shot volumes needed for micro molding and the relatively large screw diameter reduce drastically the injection shot lengths, and also, due to the large mass of the screw, it is difficult to accelerate and decelerate with high precision in short times.
- As a consequence, the actual injection speed is difficult to be controlled and the screw never reaches a constant speed value during cavity filling (equal to the nominal parameter setting of the machine).

In the second machine (i.e., the micro injection machine) the following could be observed:

- The very small amount of plastics needed is plasticized by a plasticizing small screw (diameter of 12–14 mm), measured by the metering piston and then fed into the injection cylinder by a plunger (diameter of 5–6 mm).
- With the reduced plunger diameter, higher process resolution in terms of minimum shot volume (below 1 cm^3) and repeatability (≈0.1–0.5 mm^3) can be achieved. In particular, there is no need for a large sprue to ensure process stability and realization [34].

- The second plunger with a diameter of just 5 mm, even down to 2 mm (depending on the machine configuration), injects the molten material into the cavity. It is driven by an electric motor and a precise linear drive. Typically, the shot weight can vary between 5 and 500 mg with a repeatability of piston position accuracy in the range of 2–10 μm (equivalent to a mass repeatability of 0.1–0.4 mg, depending on polymer density and plunger diameter [34]).
- As a consequence, cooling time is now shorter and therefore cycle time is to a minimum.
- By using the small injection piston, acceleration time is very much reduced as compared with the conventional injection screw. The set injection speed is quickly reached and maintained during filling, improving process robustness and control.

Results of experiments are reported in Table 1.4 for moldings of a microfluidic system with micro pillars (diameter of 500 μm) and a thin plate with micro ribs (width of 300 μm, thickness of 150 μm). As a result of the specific design, the micro injection molding machine exhibits higher molding mass repeatability, shorter cycle time, smaller part volume, and shorter filling time.

Table 1.4 Experimental Comparisons of Two Different Polymer-Based Micro Products with Micro Features Produced by a Conventional Injection Molding Machine and a Micro Injection Molding Machine

Machine characteristic	Conventional injection molding machine	Micro injection molding machine (2 pistons, 1 screw)
Molded polymer-based micro product	Microfluidic system with micro pillars 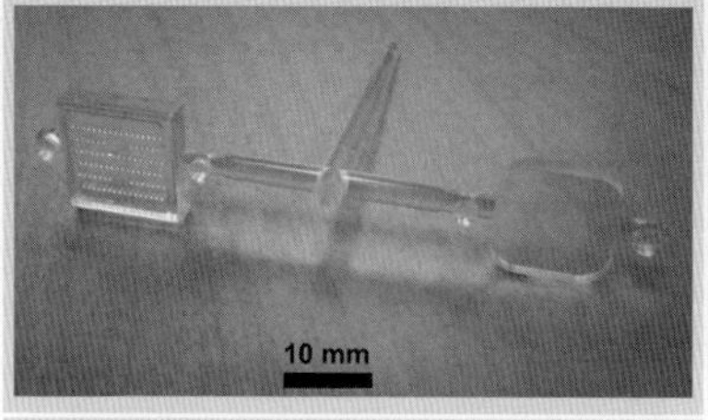	Thin plate with micro ribs 
Micro features	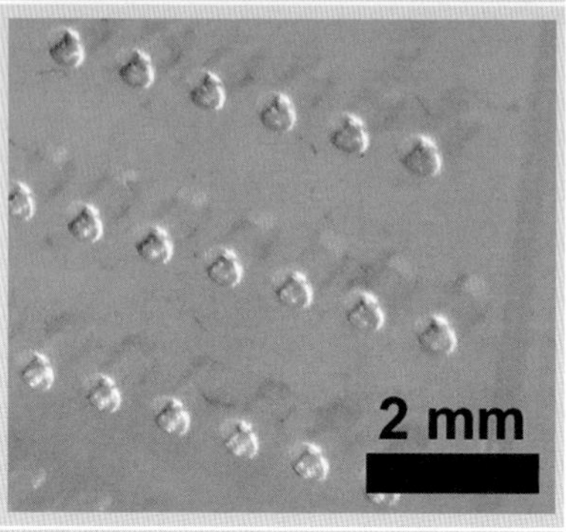	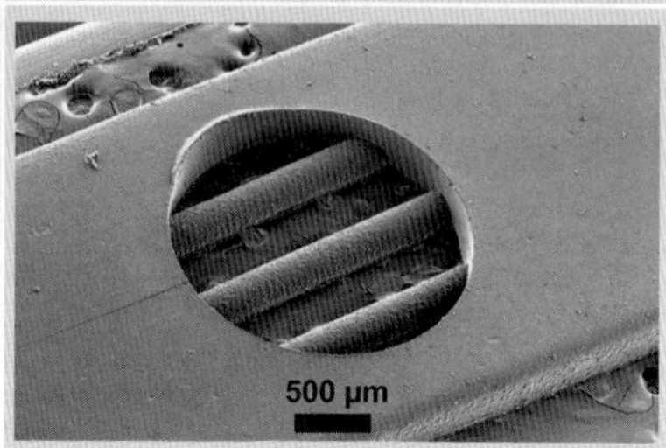
Material	Polystyrene	Polystyrene
Exp. average shot mass	2052.2 mg	118.8 mg
Exp. molding mass repeatability (1σ std. dev.)	4.6 mg	0.3 mg
Part mass/total mass ratio	33%	27%
Filling time	500 ms	80 ms
Cycle time	12–15 s	6–9 s

1.4 Appendix

Phase 1. Plastication

Feed 1

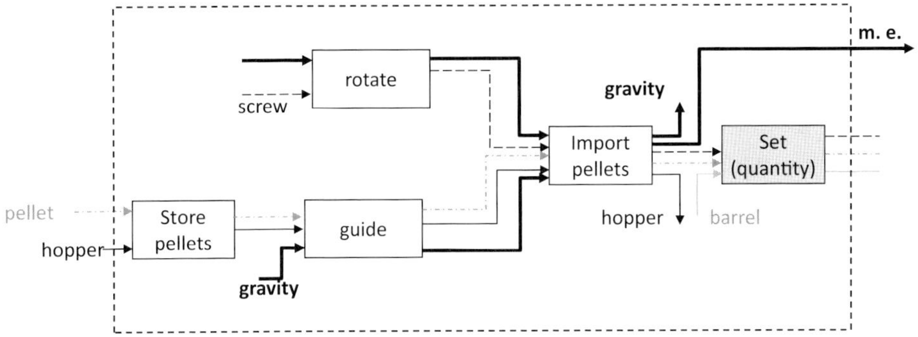

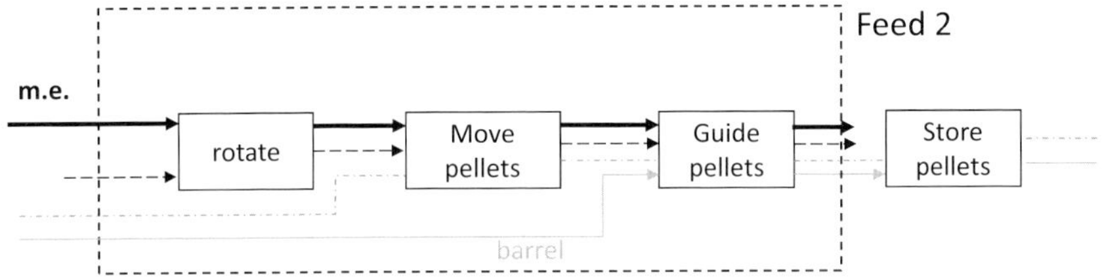

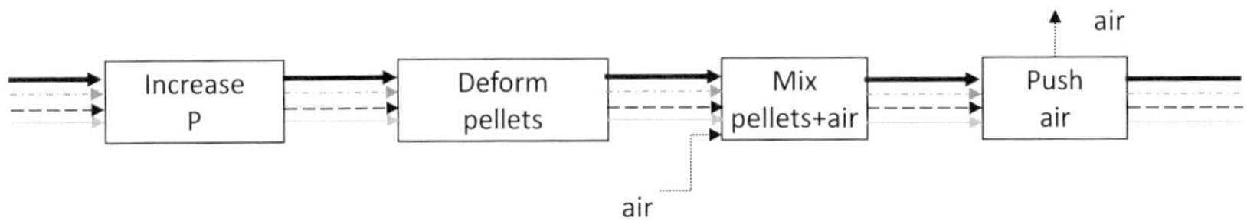

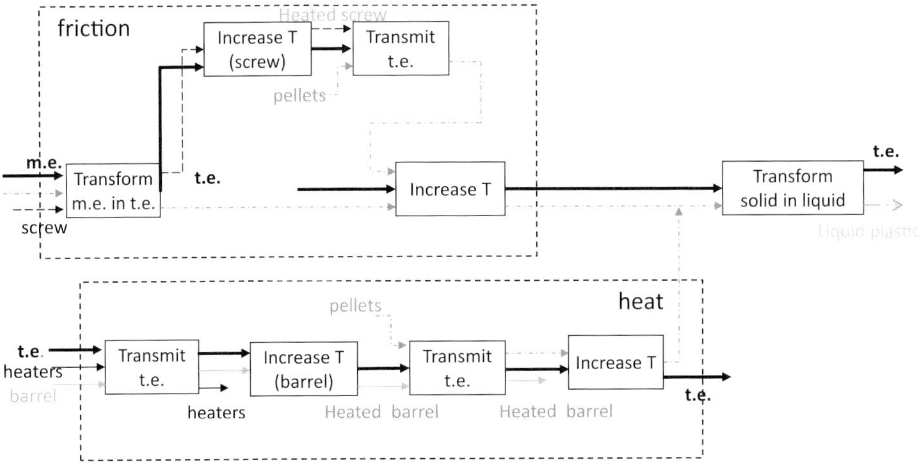

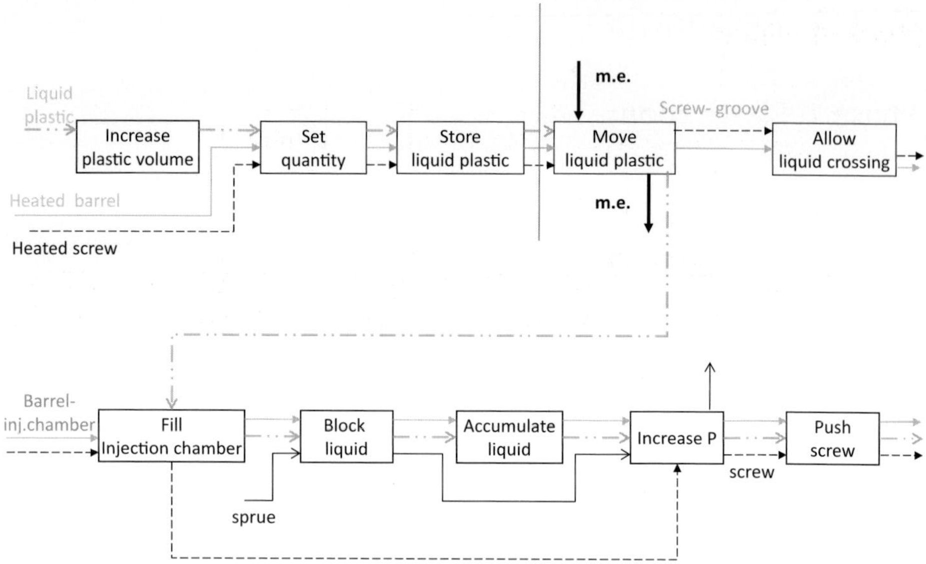

Phase 2. Ejection'

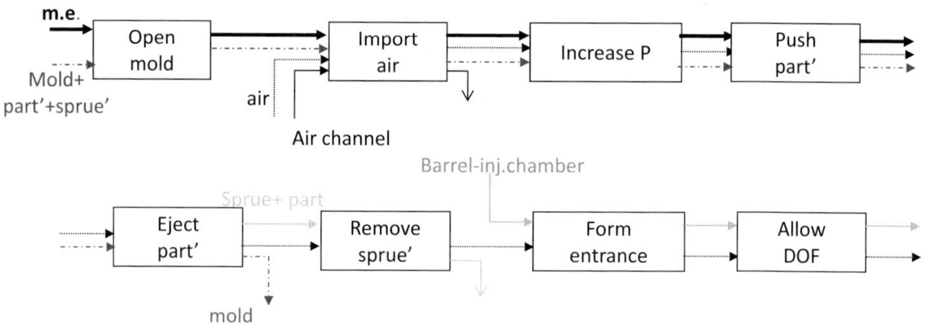

Phase 3. Injection

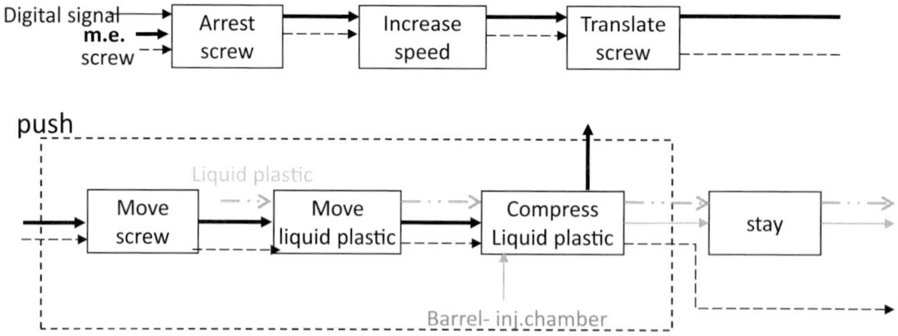

Phase 4. Filling

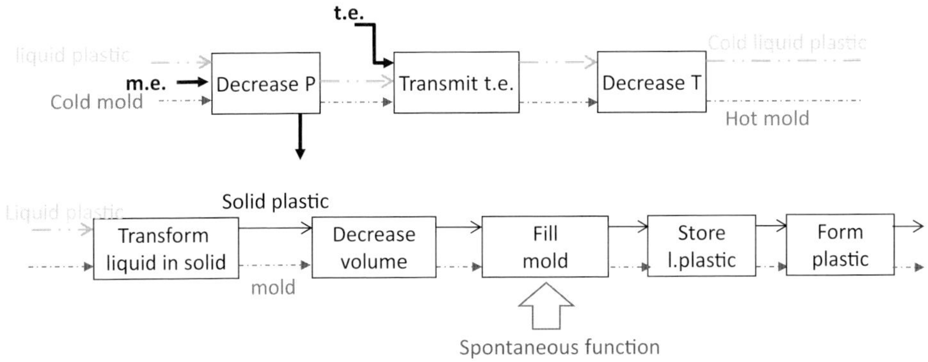

Phase 5. Packing

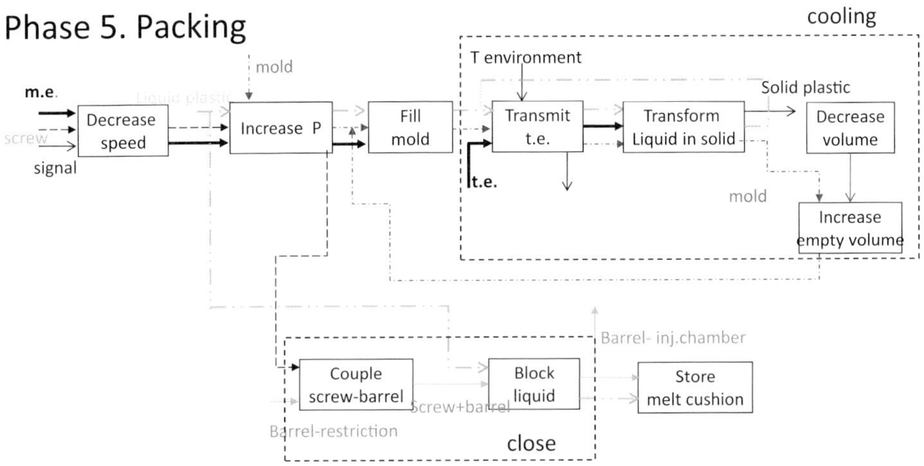

Phase 6. Cooling

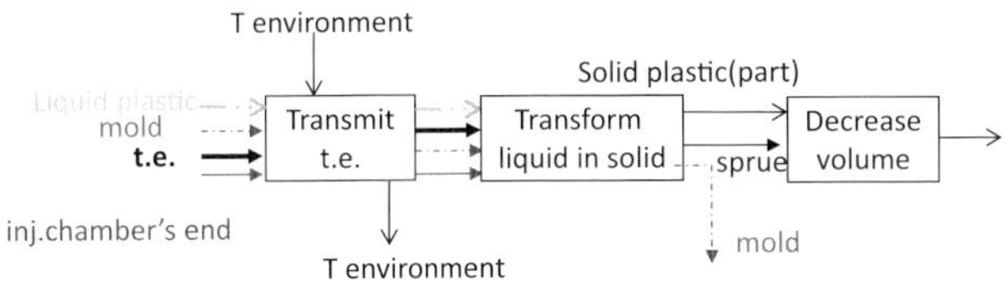

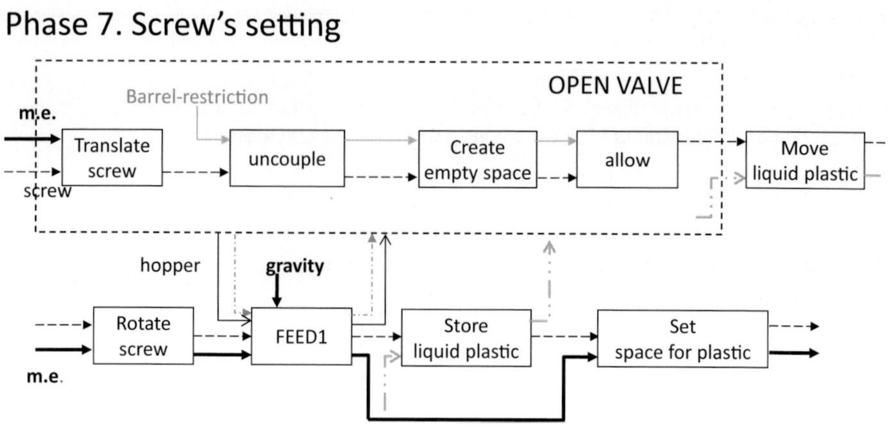

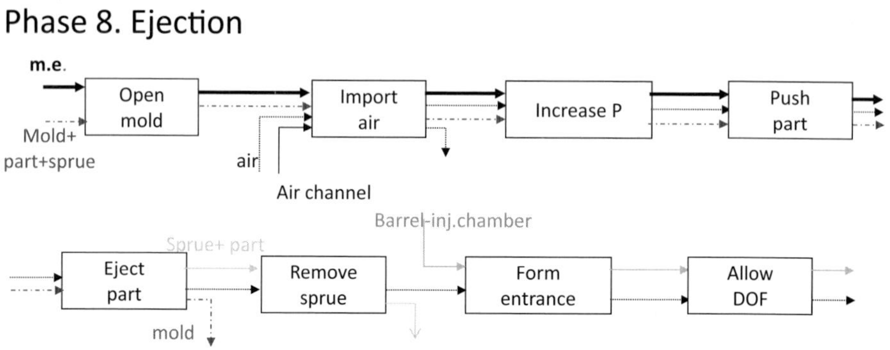

References

[1] Miles, L. D., *Technique of Value Analysis and Engineering*, McGraw-Hill Book Company, New York (1972)

[2] Pahl, G., Beitz, W., *Engineering Design: A Systematic Approach*, Design Council, London (1984)

[3] Stone, R., Wood, K., Development of a Functional Basis for Design, *Journal of Mechanical Design* 122(4) (2000), 359–370

[4] Hirtz, J., Stone, R., McAdams, D., Szykman, S., Wood, K., A Functional Basis for Engineering Design: Reconciling and Evolving Previous Efforts, *Research in Eng. Des.* 13 (2002), 65–82

[5] Szykman, S. R., Sriram, D., Smith, S. J., Proceedings of the NIST Design Repository Workshop, Gaithersburg, MD (November 1996)

[6] Cascini, G., Rotini, F., Russo, D., Functional modeling for TRIZ-based evolutionary analyses, DS 58-5: Proceedings of ICED 09, the 17th International Conference on Engineering Design (2009)

[7] Cascini, G., Rissone, P., Rotini, F., Russo, D., Systematic design through the integration of TRIZ and optimization tools, *Procedia Engineering* (2011), 674–679

[8] Umeda, Y., Tomiyama, T., Yoshikawa, H., FBS Modeling: Modeling scheme of function for conceptual design, in Proceedings of the 9th International Workshop on Qualitative Reasoning, Amsterdam, Netherlands, 11–19 May 1995, 271–8

[9] Suh, N. P., *Axiomatic design: advances and applications*, Oxford University Press, New York (2001)

[10] Dastagiri, M., Prasad, M.M., Application of Axiomatic Design in Injection Molding Process, Proceedings of Conference "Frontiers in Automobile and Mechanical Engineering" (FAME), 25–27 Nov. 2010, 222–228

[11] Kazmer, D., Axiomatic design of the injection molding process, ICAD2000 First International Conference on Axiomatic Design Cambridge, MA – June 21–23 (2000)

[12] Griffiths, C.A., Micro Injection Moulding: Tooling and Process Factors, Ph.D. Thesis, University of Cardiff, Institute of Mechanical and Manufacturing Engineering, Cardiff School of Engineering (2008), i–xxi, 1–217

[13] Tosello, G., Precision Moulding of Polymer Micro Components – Optimization, Simulation, Tooling, Quality Control and Multi-Material Application, Ph.D. Thesis, Technical University of Denmark, Department of Mechanical Engineering, ISBN: 978-87-89502-77-9 (2008), 1–242

[14] Scholz, S.G., Micro Injection Moulding: Process Monitoring and Optimization, Ph.D. Thesis, University of Cardiff, Institute of Mechanical and Manufacturing Engineering, Cardiff School of Engineering (2011), 1–226

[15] Islam, A., Hansen, H.N., Marhöfer, M., Breusch, A., Dormann, B., Bondo, M., Two-component micro injection moulding for hearing aid applications, *International Journal of Advanced Manufacturing Technology* 62, DOI 10.1007/s00170-011-3841-1 (2012), 605–615

[16] Michaeli, W., Opfermann, D., Kamps, T., Advances in micro assembly injection moulding for use in medical systems, *Int. J. Adv. Manuf. Technol.* 33, DOI 10.1007/s00170-007-0951-x (2007), 206–211

[17] Michaeli, W., Kamps, T., Micro assembly injection moulding with plasma treated inserts, *Microsystem Technologies* 14, DOI 10.1007/s00542-008-0684-2 (2008), 1903–1907

[18] Piotter, V., A review of the current status of microPIM: materials, processing, microspecific considerations, and applications, *Powder Injection Moulding International* Vol. 5, no.3 (September 2011), 27–36

[19] Wang, M.-W., Microceramic injection molding of a multilayer micropatterned micropart, *Int. J. Adv. Manuf. Technol.* 51, DOI 10.1007/s00170-010-2597-3 (2010), 145–153

[20] Piotter, V., Bauer, W., Knitter, R., Mueller, M., Mueller, T., Plewa, K., Powder injection moulding of metallic and ceramic micro parts, *Microsyst. Technol.* 17, DOI 10.1007/s00542-011-1274-2 (2011), 251–263

[21] Müller, T., Piotter, V., Plewa, K., Guttmann, M.,. Ritzhaupt-Kleissl, H.-J, Hausselt, J., Ceramic micro parts produced by micro injection molding: latest developments, *Microsystem Technologies* 16, DOI 10.1007/s00542-009-0992-1 (2010), 1419–1423

[22] Attia, U.M., Alcock, J.R., A review of micro-powder injection moulding as a microfabrication technique, *J. Micromech. Microeng.* 21, 043001 (22pp), doi:10.1088/0960-1317/21/4/043001 (2011)

[23] Prokop, J., Eberl, C., Funk, M., Prüfe, P., Lorenz, J., Piotter, V., Welz, M., Ritzhaupt-Kleissl, H.J., Mechanical testing of micro samples produced by a novel LIGA related process chain, *Microsystem Technologies* 17, DOI 10.1007/s00542-011-1256-4 (2011), 281–288

[24] Ruh, A., Klimscha, K., Piotter, V., Plewa, K., Ritzhaupt-Kleissl, H.-J., Fleischer, J., The development of two-component micro powder injection, moulding and sinter joining, *Microsystem Technologies* 17, DOI 10.1007/s00542-011-1326-7 (2011), 1547–1556

[25] Fantoni, G., De Grave, A., Hansen, H.N., Functional analysis for micro mechanical systems, *International Journal of Design Engineering* Vol. 3, No.4 (2010), 355–373

[26] Gabelloni, D., Fantoni, G., Apreda, R., Bonaccorsi, A., On the link between features and functions, Proceeding of the International Conference on Engineering Design, ICED11, 15–18 August 2011, Technical University of Denmark, Copenhagen

[27] Fantoni, G., Tosello, G., Gabelloni, D., Hansen, H.N., Modelling injection moulding machines for micro manufacture applications through functional analysis, 1st CIRP Global Web Conference on Interdisciplinary Research in Production Engineering (2012)

[28] Rosato, D., Rosato, M., Rosato, D., *Injection Molding Handbook, 3rd Ed.*, Kluwer Academic Publishers (2000)

[29] Whiteside, B.R., Martyn, M.T., Coates, P.D, In-process monitoring of micromoulding – Assessment of process variation, *International Journal of Polymer Processing* Vol. XXI, Issue 5 (2005), 162–169

[30] Griffiths, C.A., Dimov, S.S., Scholz, S., Hirshy, H., Tosello, G., Process factors influence on cavity pressure behaviour in micro-injection moulding, Journal of Manufacturing Science and Engineering, *Transactions of the ASME* (American Society of Mechanical Engineers) Volume 133, Issue 3, (June 2011) 031007 (10 pages), doi:10.1115/1.4003953

[31] Tosello, G., Hansen, H.N., Schoth, A., Process and part filling control in micro injection molding, Society of Plastics Engineers (SPE) – 66th Annual Technical Conference (ANTEC) (2008), 2529–2533

[32] Ganz, M., Micro Injection Moulding and Compression Moulding, Workshop at the 1st International Conference on Multi-Material Micro Manufacture (4M): Polymer Technology toward Nano, future technology for Europe, Karslruhe (Germany), 29th June–1st July 2005

[33] Altshuller, G., Shulyak, L., 40 Principles: TRIZ keys to technical innovation, Technical Innovation Centre, Worcester, Massachusetts (2001)

[34] Baruffi, F., Charalambis, A., Calaon, M., Elsborg, R., Tosello, G., Comparison of micro and conventional injection moulding based on process precision and accuracy, Proceedings of 15th CIRP Conference on Computer Aided Tolerancing – CIRP CAT 2018, Procedia CIRP 2018, 6 pages

2 Micro Molding Process Monitoring and Control

Ben Whiteside, Maksims Babenko, Elaine Brown

2.1 The Need for Process Monitoring in Micro Molding

It is important to be able to measure the operating characteristics of any manufacturing process in order to be able to detect any drift in the process performance, which may influence the quality of the resulting product. However, such measurements are particularly important for the micro molding process [1–4], where small variations in the material properties, process conditions, and hardware condition can have significant impacts on the final product properties.

In the process, injection rates are high in order to fill the cavity as rapidly as possible, to avoid premature solidification of material due to the rapid heat loss and large thermal gradients associated with the high surface area to volume ratio of the micro-molded products. The current industry standard uses a target injection pressure to trigger the switchover from injection velocity control to hold pressure control to provide the best repeatability. However, this places great demands on machine designers as each component in the injection system needs to respond extremely quickly to process parameters and be able to accelerate and decelerate to tightly defined velocities in very short periods of time. In addition, temperature control of the plasticizing unit and mold surfaces should be accurate to less than a single degree, and should be highly stable during the process.

These equipment demands have the consequence that there will always be some deviation between actual performance and target performance for micro molding processes, and it is important that these are understood to be able to determine the limitations of the equipment. Process monitoring systems allow detailed evaluation of the input material condition, machine dynamics, polymer shaping/structuring, and cooling behavior. This level of information provides an extremely powerful tool for optimizing any micro molding process, ensuring high productivity and improved product performance.

■ 2.2 Micro Molding Sensor Technologies

All process monitoring systems require sensors, which is any device that can convert a physical property or stimulus into a measurable property. Typically this will involve conversion into an electrical signal, which has a voltage or current that can be captured by a suitable system to provide a digital equivalent value. If the relationship between the physical property and the resulting electrical signal is understood a priori, then a simple conversion can be performed to calculate the actual physical property. However, it is much more typical that this relationship is determined experimentally using measurements of properties with a known value in a process called calibration. The calibration is typically performed by the manufacturer and the information supplied with the sensor usually in the form of a sensor calibration certificate. The calibration can be affected by environmental parameters, such as temperature, so it is important to ensure that the sensor is being used in the defined window of operation where the calibration is shown to be accurate.

Sensor integration in micro molding processes can be particularly challenging due to the requirement for sensors to have very fast response times, and to be capable of tolerating harsh environments in terms of high temperatures and pressures that are typically encountered during a typical micro molding injection cycle. In addition, simply finding a free volume within the tool itself for the sensor to occupy poses difficulties, as micro molding tools tend to be much smaller in size than their conventional injection molding counterparts, yet they can house a range of features for mold heating and cooling, vacuum venting, guide and alignment systems, moving cores, and ejection assemblies. The projected area of the cavity itself can also in some cases be much smaller than the area of the smallest available commercial sensor.

This section highlights some of the more suitable commercially available sensors for characterizing micro molding processes and discusses promising emerging technologies which may be beneficial.

2.2.1 Volumetric Flow Rate

It is important to know the flow rate of material into the mold cavity during injection as this will have a significant influence on the shear rate, corresponding stresses, and the resulting viscosity of the polymer melt during the filling phase of the process. All of these factors will influence the flow front profile, the cooling behavior, and the resulting morphology and microstructure of the material within the micro-molded component.

The majority of industrial micro molding machines offer flow rate measurements, which are typically calculated using periodic displacement measurements recorded from the encoder on the main injection drive motor. Such measurements will only provide accurate data if there is a direct, inelastic linkage between the drive and the injection mechanism (screw, plunger, etc.), but belt drives are commonly used in injection machines which will have an element of compliance and therefore will cause deviation between the encoder displacement and the injection system displacement in the event of high forces on the belt caused by high accelerations or pressures, for example.

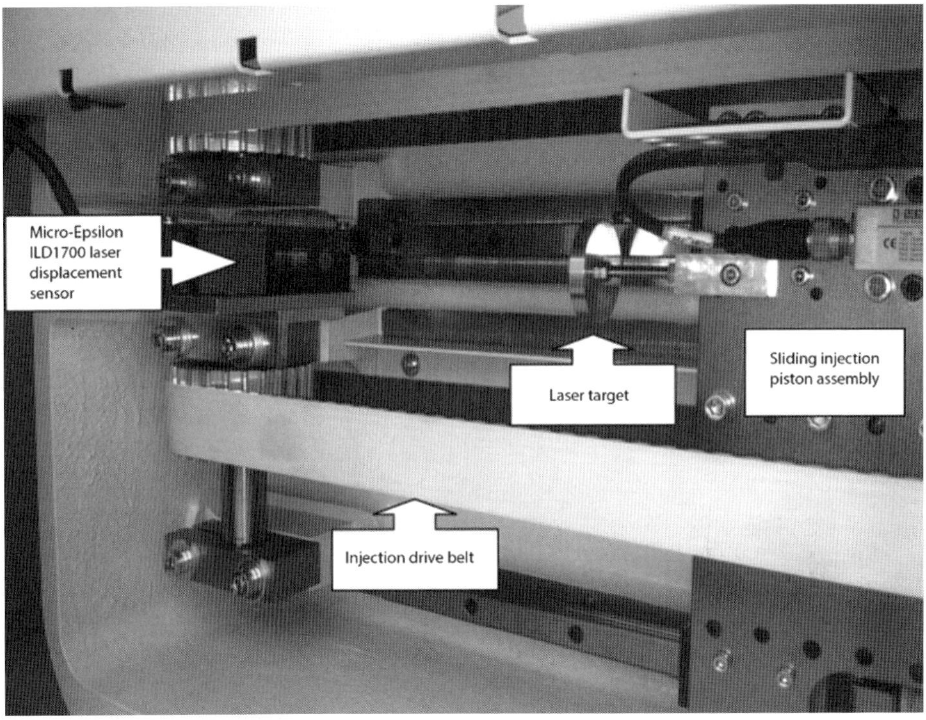

Figure 2.1 Laser displacement sensor installation on the drive unit of a Wittmann Battenfeld Micropower 15 micro molding machine

In order to account for this, it is useful to measure the position of the injection piston or screw directly using a secondary displacement sensor. Such an example is shown in Figure 2.1, where a non-contact laser displacement sensor (Micro-Epsilon ILD1700) is used to sense the position of a target directly attached to the injection piston of a Battenfeld Micropower 15 micro molding machine [5]. This sensor records data with a frequency of 2500 Hz and an accuracy of 4 μm. It can accurately measure the position of the injection piston and is unaffected by vibrations in the drive unit. The displacement values can also be differentiated in the software to provide a velocity measurement for the whole injection cycle. The volumetric flow rate at any given time can then be easily calculated by multiplying the velocity value by the projected area of the injection piston.

2.2.2 Temperature Sensors

Thermocouples are the most preferred method for characterizing the thermal behavior of the key heating components in injection molding machines. However, the recorded temperatures of the heated elements do not provide a truly accurate method of quantifying the temperature of the actual polymer during plasticizing, mold filling, and cooling. It is better to measure the material temperature directly, and for this purpose contact thermocouples and infrared (IR) sensors are typically used.

Contact thermocouples are typically embedded in elements that are flush mounted in the surface of the screw barrel, nozzle, or mold tool and are in contact with the melt directly. These provide a robust and accurate measurement of the temperature of the polymer but have a major disadvantage, in that they do not respond very quickly to changes in temperature due to the thermal inertia of the metallic element in which they are embedded.

Infrared sensors consist of optical elements that are sensitive to radiation of materials in the mid infrared range (having wavelengths typically in the 1.5 to 5 μm range). They are usually sealed units with the sensitive optical element situated behind a sapphire window that is flush mounted with the cavity surface [6–8]. Such sensors provide a near instant response to changes in polymer temperature, which makes them well suited for highly dynamic processes such as micro molding, but they also have some caveats that should be considered before they can be used. Different materials emit different intensities of IR radiation in the active range, and the effectiveness of the material to emit is termed emissivity. Perfect black bodies have an emissivity of 1, and perfect reflectors have a value of 0. This value should be known to correct the sensor output to provide accurate temperature measurement. The second main issue is that many polymers are quite transparent in the IR region and so the sensor detects emission from a range of depths in the near wall region, so it is difficult to determine the temperature at a given location. Solutions to these issues include the use of a fractional amount of carbon black material or similar pigment in the polymer, which ensures an emissivity value very close to 1 and that the temperature recorded is at the sapphire window/polymer interface.

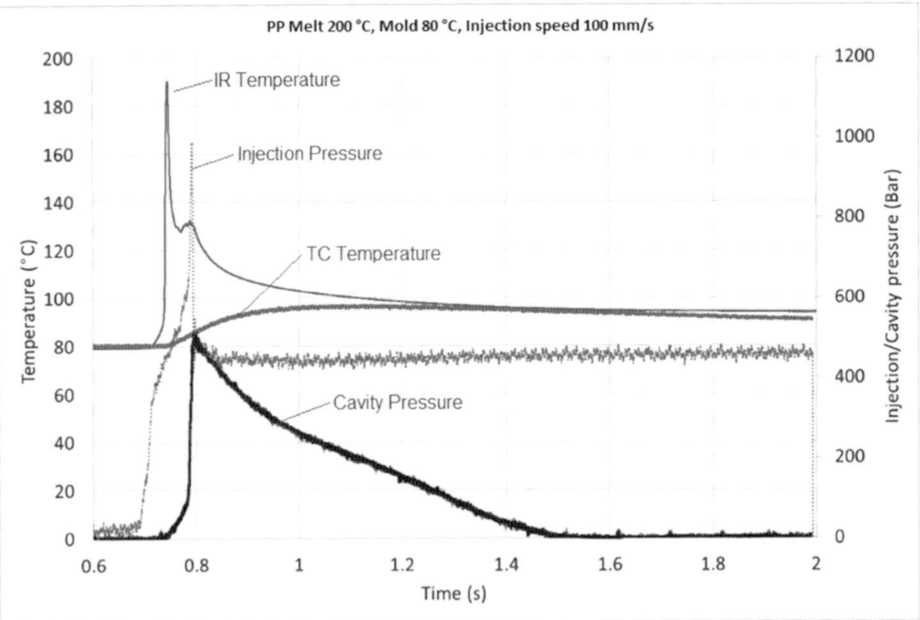

Figure 2.2 Typical process data collected during molding of a 20 mg micro-molded component

A comparison of IR and thermocouple measurements recorded during a single micro molding cycle is shown in Figure 2.2. Here it can be seen that the IR sensor reacts almost instantly to the arrival of polymer melt and shows a sharp, high peak corresponding with the 200 °C melt temperature. The thermocouple trace is much slower to respond and is only able to accurately record the polymer temperature after about 1.5 seconds, when the two traces converge.

2.2.3 Pressure Sensors

Pressure sensors are a key part of any process monitoring and control system as they provide insight into the quality of the polymer melt, the viscosity, and the quality of the filling process. There are a number of commercial sensors available for studying the micro injection process, which are built on two fundamental technologies—piezoelectric and piezoresistive. Piezoelectric sensors function by containing crystals capable of producing small amounts of electrical charge when subjected to a force, which is captured using a charge amplifier and passed on as a standard voltage signal. They can be made very compact, offer extremely fast response times, and are highly repeatable. The main drawback with such sensors is that they are not well suited for sustained measurements, as the charge will decay with time and cause an apparent drop in the signal even though the force remains constant. Piezoresistive sensors avoid this problem by exhibiting a change in resistivity when a force is supplied, which can be recorded using a very simple electronic circuit. They are similarly fast to respond too, but are generally a little larger (typically 4–10 mm diameter) than their piezoelectric counterparts, which makes them less well suited for micro molding applications. In practice, the typically short cycle times of micro molding processes (typically sub-5 s) make them an ideal process for the adoption of piezoelectric sensors anyway.

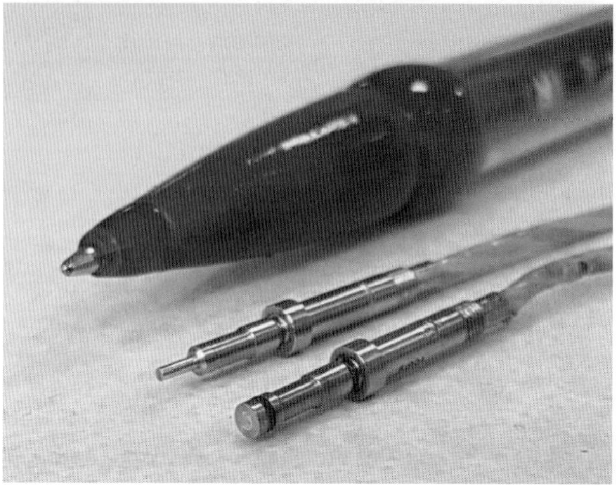

Figure 2.3 Kistler 6182 and 6183 piezoelectric pressure sensors

There are a number of ways in which piezoelectric sensors can be integrated into a micro molding tool. The simplest (and most common) is to mount the sensor flush with the cavity mold surface, where it can directly record the pressure encountered at the front face during cavity filling [9–11]. This is typically referred to as a direct measurement and two examples can be seen in Figure 2.3. However, sometimes the physical size of the sensor can restrict the number of locations where a sensor can be placed and an alternative strategy needs to be sought. One solution is to simply measure the force behind one of the ejector pins in the cavity; so special piezoelectric measuring sensors exist for this purpose. They simply measure the force transmitted by the pin, so it is trivial to convert this to an actual pressure measurement if the surface area of the tip of the pin is known. This is called an indirect measurement.

In some rare cases, neither of these techniques may be possible and more imaginative solutions are required. In some cases, a soft sensor (or virtual sensor) approach may be employed, where a measurement could be recorded away from the area of interest, but can be related to the required measurement through the use of a mathematical model or numerical simulation of the system. An example would be the use of a piezoelectric sensor to measure the tool surface strain to infer a cavity pressure measurement by calculating the deformation of the external surfaces of the tool when the cavity pressure is supplied.

2.2.4 Ultrasonic Sensors

Ultrasonic sensors can provide a means to measure changes both at the polymer/mold interface and in the consistency of the polymer simultaneously. Such systems use ultrasonic transducers that can both emit and sense ultrasonic waves. A transducer may be mounted either side of the polymer to allow transmission through the material during cavity filling. The velocity and attenuation of the signal can be monitored to provide information about the polymer consistency, level of crystallization, and filler content and dispersion, providing a very powerful, non-destructive method for quality control. Such transducers tend to be reasonably low cost and are packaged with customized waveguides for easy integration into polymer processing systems.

These systems are typically difficult to integrate into micro molding processes for reasons–stated before–relating to their physical size. However, compact reflective systems using film-based transducers [12, 13] offer significant promise for such applications, and are shown in Figure 2.4. The sensors consist of a thin (20 µm) piezoelectric film that is deposited on the rear of a cavity plate insert. An electrode is deposited on the surface of the film to create a sensor that can be excited by applying a sharply rising voltage pulse across the electrode and cavity plate, which will cause the sensor to resonate with a frequency of approximately 20 MHz. These waves are easily conducted by the steel of the cavity plate to the steel/polymer interface, where a first reflection will occur. The non-reflected portion of the signal then continues through the polymer melt until it reaches the far cavity wall, where a second reflection will occur. During transit, the sensor can then be switched into a listening mode that can detect the returning signals, providing a measurement system that requires just a single transducer. The high frequency ultrasonic waves are attenuated quickly and will only penetrate a few millimeters of polymer (which is quite acceptable for a micro injection process with typical wall thicknesses of 0.5 mm), and the high frequency provides very high measurement resolution, able to detect small changes in melt consistency. Another benefit is the fact that as the sensors are mounted on the rear of the cavity plate, no sensor witness marks are seen on the cavity surface, which can be an important consideration for functional surfaces or optical moldings. As in a transmission system, the reflection through the polymer can be used to verify polymer quality, and the first reflection from the cavity surface can be used to monitor shrinkage and detachment of the polymer during cooling.

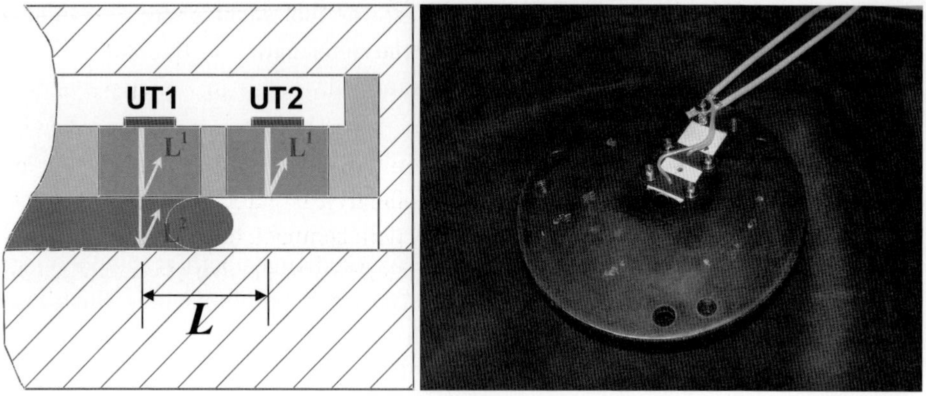

Figure 2.4 Left: ultrasonic sensors can be mounted on the rear of the mold cavity plate, where they can record the reflections from the mold/polymer interfaces at the near and far wall; right: piezoelectric film sensor installation on the rear of a mold cavity plate

■ 2.3 Visualization Systems

Sensors in the tool provide valuable information about the process behavior but only offer single point measurements at defined locations where a sensor can be installed. Ideally, it would be useful to record data at points throughout the cavity, to build a full picture of the process to use for control applications. One solution to this problem is to use imaging devices (which contain optical sensor arrays) to directly visualize what is happening in the mold cavity. This effectively allows millions of measurements to be recorded in parallel and provides a very powerful tool for understanding mold-filling phenomena.

2.3.1 Tool Design for Visualization

There are a number of considerations to be taken into account when designing a tool for micro molding visualization studies [13, 14], to ensure that the system will be effective.

The material used for the window should be transparent over the range of wavelengths of light that will be used for the imaging system, robust enough to tolerate very high pressures, and should not perturb the environment that is to be studied. An excellent candidate material is sapphire, which is transparent in the visible and near to mid IR ranges, is robust, and has very similar thermal behavior (heat capacity and heat transfer coefficient) to standard tool steels. It is not too expensive, but it can be brittle, so care has to be taken to avoid stress concentrations which can cause it to shatter.

For the design of the tool itself, care needs to be taken to ensure that the optical path is not longer than the maximum working distance of the lens, or the cavity will not be able to be focused. The illumination path should also be taken into account, and systems may use transmission, reflective, or lateral lighting systems. Mirror systems can be used to keep the optical path compact, but should be isolated from areas of the tool where contamination with fluids or particulates could occur.

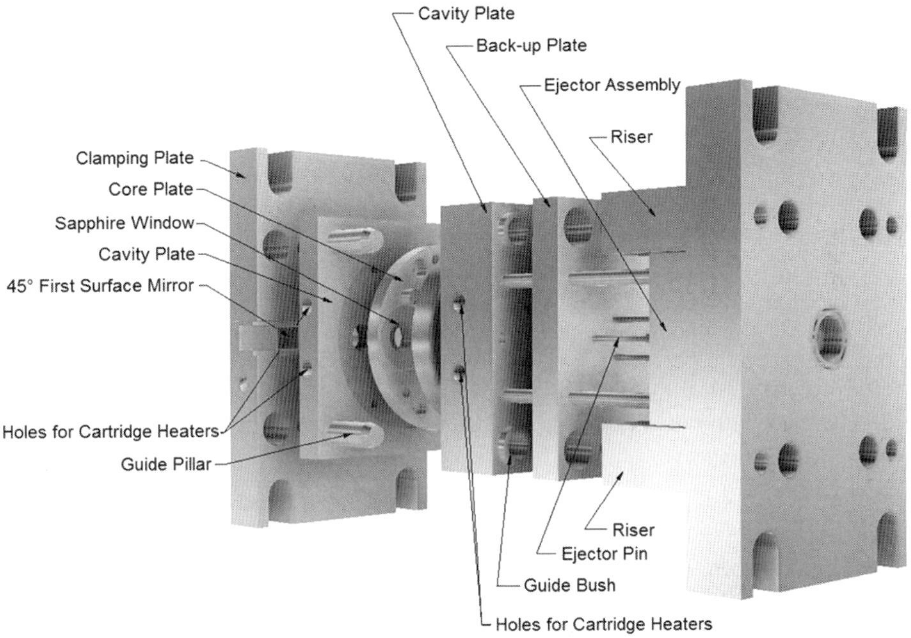

Figure 2.5 Mold design for in-process visualization

Figure 2.5 shows such a tool design, which is illuminated using a coaxial (through the lens) system and therefore has a very simple optical path. The design consists of replaceable cavity inserts that allow a number of cavity geometries to be studied, and the window can be quickly and easily replaced in case of breakage.

2.3.2 High-Speed Imaging

Micro molding cavities will typically fill in just a few milliseconds, so in order to record filling phenomena with an adequate number of frames, very high-speed acquisitions are required. High-speed camera technologies are becoming mature and there are a number of well-known applications worldwide, including studies of the impact on human occupants during vehicular car accidents and ballistics, to name but a few.

There are three main elements that require consideration for the optical train for any visualization system, namely: the camera, the lens, and the illumination.

When selecting a camera, consideration should be given to:

- The number of pixels on the sensor, which will influence the resolution and detail in acquired images
- The sensitivity of the sensor, which will determine the minimum exposure time and therefore the level of motion blur during highly transient motion
- The speed of the interface and the memory location of the stored images, which will determine the maximum bandwidth of data transfer (and therefore the maximum number of frames per second) and the convenience of data processing. There are two main architectures for data transfer. The highest specification cameras usually utilize very fast memory modules mounted in the camera with very high bandwidth buses linking them to the sensor. These provide the highest possible frame rates, but can be inconvenient as the data needs to be subsequently downloaded to a computer for processing. The alternative is to use a camera that streams directly to the system memory of an attached PC using an appropriate data link such as Camera Link, USB3, or gigabit Ethernet. These systems are much more convenient to use, but lack the raw performance of the onboard systems.

The choice of lens is important as this will influence the overall level of magnification of the image, the location of the camera relative to the cavity in order to achieve focus, and the overall level of light that is recorded by the sensor. These are all governed by key lens specifications including:

- The focal length (e.g., 80 mm) or magnification value (e.g., 5×) for macro/microscopy lenses, which will determine the viewable area
- The working distance, which determines the distance between the lens and object to be imaged
- The aperture of the lens (f-number), which determines the amount of light that will pass through the lens and reach the sensor

An ideal lens will provide an ideal level of magnification to form a full image of the area of interest, with a suitable working distance for the visualization mold tool arrangement, and the largest aperture for the maximum level of light to reach the sensor. Telecentric lenses are particularly advantageous for process imaging as they can be easily calibrated to allow direct measurement of feature sizes in images, suffer from no perspective distortion, and capture parallel light paths, which is useful for birefringence measurements to study residual stress formation during filling and cooling.

Illumination strategies are often considered as an afterthought, but should be integrated into the optical design as early as possible. High-speed imaging methods require high intensities of light to ensure minimum exposure times with corresponding maximum frame rates. The simplest method of illumination is to use a coaxial ("through the lens") technique similar to the system shown in Figure 2.6. This method uses an optical element which reflects light from a perpendicular source (commonly, a fiber optic link from a high intensity illuminator) at 90°, so it passes through the lens. The light from the object is then reflected back through the lens, where it passes back through the optical element to the camera. Of course, for this to work the element needs to be able to both reflect and transmit light, so a half-silvered mirror is commonly used and this is referred to as a beam splitter, as it allows 50% reflection and 50% transmission. The main drawback of such systems is that the light that takes the unwanted path at each incidence is effectively lost, so the intensity that reaches the camera is attenuated by a factor of 4.

Figure 2.6 NAC HX-3 high-speed camera apparatus and coaxial illumination set up on a Wittmann Battenfeld Micropower 15 micro molding machine

Transmission lighting systems separate the illumination and imaging paths completely, to avoid such issues, but require transparent windows on either side of the cavity. The design of such molds can be troublesome due to the presence of ejector assemblies and the actual cavity forms, so reflective systems tend to be far more common in use.

2.3.3 Thermal Imaging Methods

Thermal imaging systems have very similar considerations to the visible high-speed imaging systems discussed in the previous section, so they will not be discussed again here. The main differences for such systems are that no illumination is required, as the material itself is the emitter of the IR photons that are recorded. For accurate measurements using IR systems, careful consideration has to be given to the materials used in the optical train. Sapphire windows are very useful for mid-IR imaging devices, but many lower specification camera systems use longer wavelengths, typically 14 μm, where sapphire is completely opaque. It is also important to realize that sapphire itself is not completely transparent even at lower wavelengths, so radiometrically calibrated IR cameras will not be able to provide correct temperature data. A separate calibration procedure should be followed, which will image a surface at a known temperature through the entire optical system to allow a calibration factor between the photon flux and actual surface temperatures to be determined. This is typically performed by heating the target to a range of known temperatures and recording the resulting values measured by the camera to create a calibration file.

Once calibrated, thermography systems provide very powerful tools for monitoring polymer melt consistency, shear heating effects, and cooling behavior in the cavity [15].

■ 2.4 Data Acquisition and Archiving Systems

2.4.1 Data Acquisition Hardware

All data acquisition (DAQ) systems need some form of digitizer to convert analog sensor signals into digital values for computer processing and displays. There is a large variety of different systems in the marketplace with different capabilities and costs, so it is important to choose the optimal system for the task in hand. Key considerations include the number of parallel measurements to be performed (described as the number of channels), the accuracy of the conversion of analog to digital signals (the resolution), and the number of measurements that can be performed by each channel every second (the sampling rate). The number of channels required is simply equal to the number of sensors you have, or think you may have in the system in the future. The resolution of the system is given in terms of the number of bits that the analog to digital converter (ADC) will assign to the input signal. Typical values include 12 bit or 16 bit, which corresponds with either 4096 or 65536 individual values after conversion. So, for a pressure sensor that can measure a pressure range of 0–2000 bar and provides a 10 V output at the maximum value, a 10 V 12-bit DAQ system would have an ultimate resolution of 0.48 bar, while the 16 bit equivalent offers a resolution of 0.03 bar. The sampling rate determines the sensitivity of the system to highly dynamic events and should be fast enough to accurately describe the state of the system. This is particularly important for rapid transitions and peak detection. If the sampling speed is too low, then highly transient events can be missed, which can have a significant effect on the quality of data.

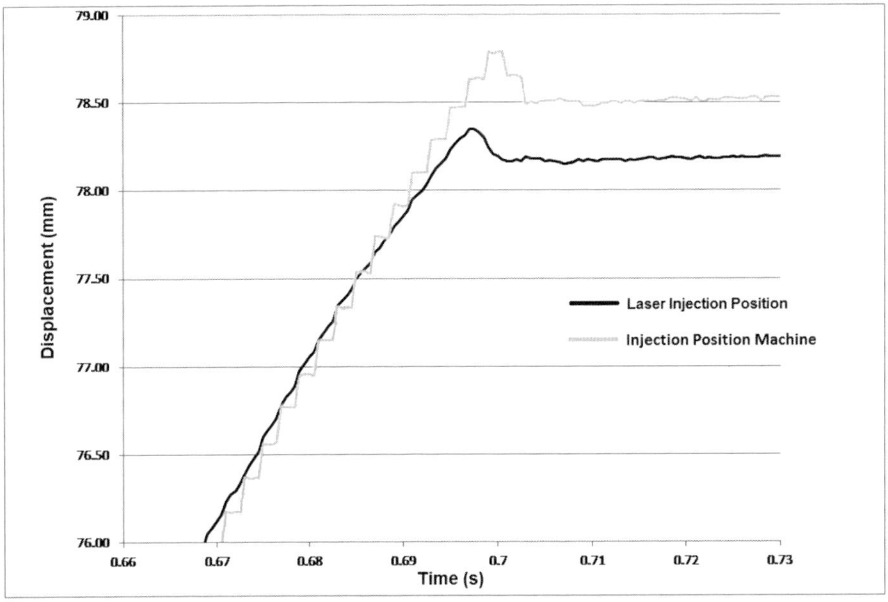

Figure 2.7 Comparison of data recorded through the Wittmann Battenfeld Micropower 15 and the National Instruments' Compact DAQ cDAQ 9181 systems

Figure 2.7 highlights the difference in quality of data for injection piston position between a standard DAQ system on a micro molding machine from 2009, and a bespoke data acquisition system. The standard system has a much lower sampling rate than the machine data acquisition, as it is restricted by the CAN bus frequency (which is the main communication bus between all the elements of the machine). This has the consequence that the position curve is poorly defined, and also introduces a slight delay in the acquisition, causing a misalignment of the peaks.

2.4.2 Synchronization of DAQ Systems

For many applications, it is important to synchronize systems to a standard time base to provide a means to compare results between cycles or different DAQ techniques. The standard way to do this is to ensure that all systems are triggered simultaneously from a machine control signal (typically the start of injection digital signal). This will define the initial time for all acquisitions, but each DAQ device may still run at a different sampling rate, so although the first data point will be aligned, the others may not be. This often is not a problem, as the data curves generated with high sampling rates can be directly compared; however, there are occasions where a direct synchronization between measurements is required (for verifying a sensor value corresponding with a particular frame in a video acquisition, for example) and, in such cases, a common pulse train will be used to externally trigger the acquisition of each individual value on each device in the system. This can be supplied by a separate device, but is normally generated by one of the DAQ devices in the system and shared with the other devices using coaxial or optical connections.

2.4.3 Communication and Storage Strategies

For research applications, in many cases data is stored locally on DAQ machines attached to the manufacturing process, and then processed and archived manually by the researcher. This is a highly inefficient process and not relevant for industrial process monitoring and quality control applications, where the data has significant value for the monitoring of production and quality trends on a factory-wide basis. There is now a shift toward integrating process monitoring systems with other business monitoring systems. Increased levels of communication using strategies such as the Internet of Things (IoT) and Industry 4.0 are pushing manufacturers to constantly monitor process data, and archive it in fast searchable databases where it can be recalled in an instant. This poses challenges in terms of network capacity to handle the high volume of data, but the demands can be reduced by performing data compression at the point of collection before the data is transmitted. One example of this would be to reduce a full cavity pressure data curve containing thousands of values to just one peak and one integral values instead.

Figure 2.8 shows a possible framework for a process monitoring system that uses a Service Oriented Architecture (SOA) approach, which provides compatibility with existing business monitoring systems [16]. Data is collected from all aspects of the manufacturing process and published to the SOA bus as it becomes available. The data can be stored directly in the archive database, but other software agents can monitor the system to determine the quality of production, or to identify a potential issue and put in place a remedial action. Such frameworks are also compatible with Big Data approaches that can intelligently search through the vast amount of information and identify key relationships that would not have been immediately apparent and can be exploited for further process improvements.

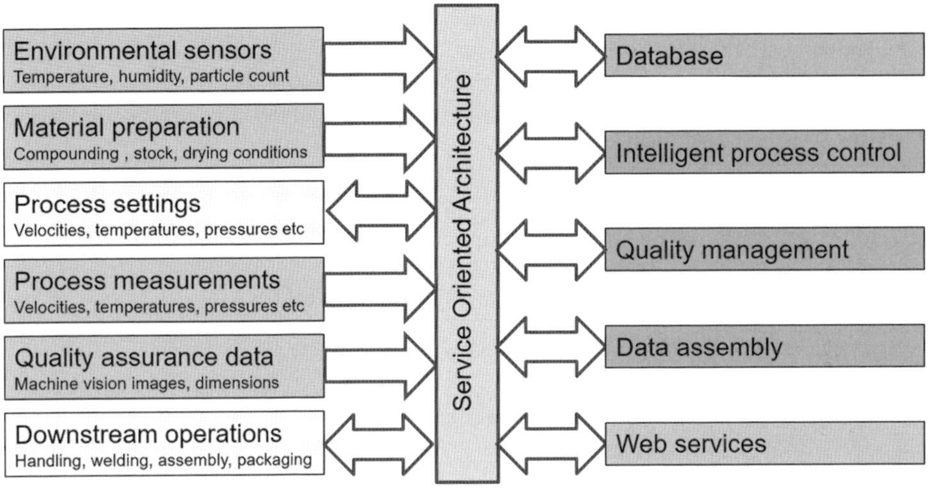

Figure 2.8 A process monitoring framework compatible with SOA methods

2.5 Applications for Process Monitoring Systems

2.5.1 Qualification of Machine Performance

Machine qualification is an important part of the quality management system for any large scale and complex manufacturing process. For a machine to be certified for use in the process chain, a sequence of performance tests are performed to ensure that it can run to specification. Additional process monitoring sensors can significantly aid this task by verifying machine actual performance relative to specification and highlighting areas where the machine may have drifted from target values and requires service or recalibration.

Figure 2.9 illustrates the value of extra sensors, where the displacement of the injection piston is compared with the value recorded by the encoder of the injection servomotor drive system using the experimental arrangement described in Section 2.2.1. There is a clear difference between the actual position and the position recorded by the machine during the hold phase of the process, due to the sustained pressure causing a slight elastic stretching of the drive belts.

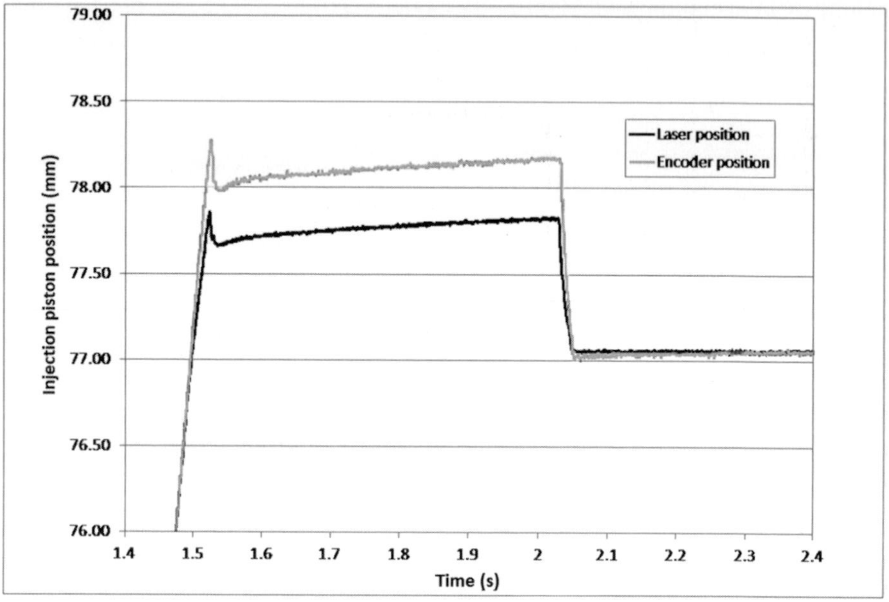

Figure 2.9 Comparison of the deviation between the injection piston position recorded by the drive encoder (gray) and the actual position (black) measured using a Micro-Epsilon ILD1700 laser displacement sensor

Figure 2.10 shows pressure sensor data recorded during four micro molding cycles with different injection velocities but a set switchover pressure of 600 bar and a hold pressure of 400 bar. It can clearly be seen that, at high injection velocities, the momentum of the injection assembly causes it to overshoot at the switchover position, resulting in a peak injection pressure that exceeds the set value. The data can be used to correct switchover values to take this into account and provide greater control of pressure during the molding process.

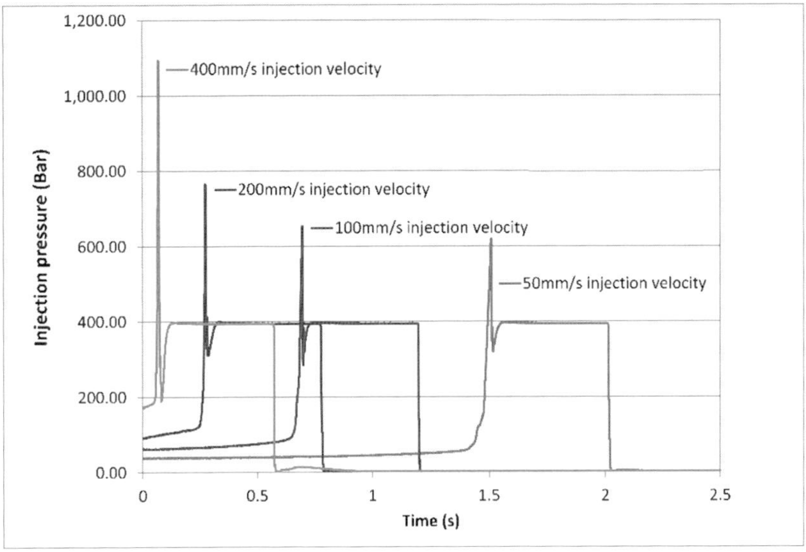

Figure 2.10 Injection pressure measurements for four different injection speeds with switch-over set at 600 bar and 400 bar hold pressure

2.5.2 Material Quality Assessment

Process monitoring systems have often been used for rheological characterization of polymer melts and there are many examples of studies [17, 18] using pressure sensors in rheological nozzles, hot and cold runner systems, and in cavities to sense viscosity changes associated with different batches of material or thermal degradation. These systems typically relate the pressure gradient during filling to the viscosity of the material and monitor the parameter for deviation from the specified window. Such systems provide a useful way of verifying material quality but usually need to be supported by other measurements to ensure that the viscosity changes are purely the result of a material change, rather than a change in the thermal history of the polymer (caused by a heater band failure, for example) or in the injection behavior of the machine.

Ultrasonic systems can also be employed for effective material quality measurement. Figure 2.11 shows the results of an experiment that adopted piezoelectric film sensors as described in Section 2.2.4 to monitor the thermal degradation of a POM copolymer. There was a very clear increase in the ultrasonic velocity through the material in the cavity as the material became more degraded, as was clearly indicated by the color of the parts.

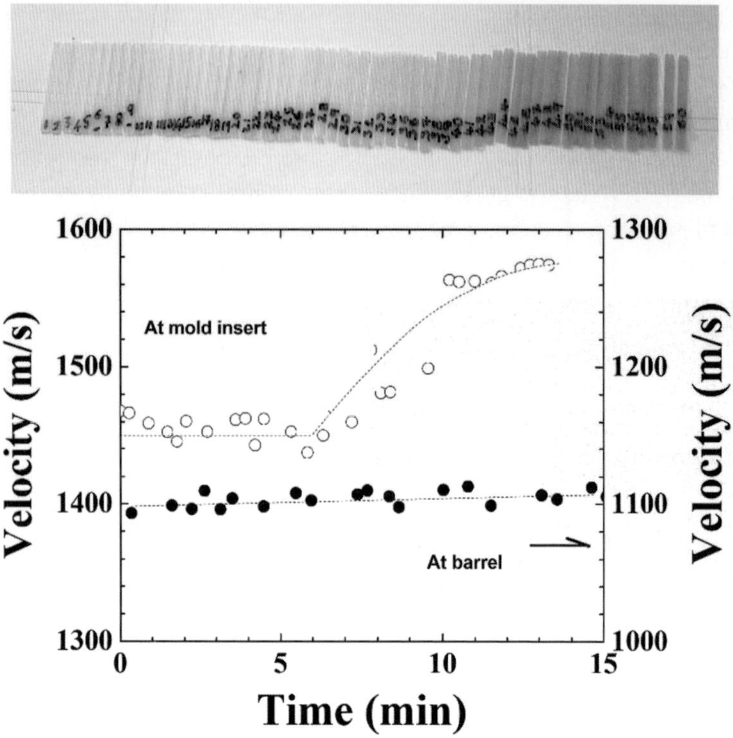

Figure 2.11 Ultrasonic velocity values measured in the cavity during POM degradation; the velocity increases with increasing degradation of material

2.5.3 Process Window Evaluation

Process monitoring tools are an invaluable asset for commissioning a new micro molding process. In conventional injection molding, injection pressure is a very useful measurement for optimizing part quality, but the sensitivity of the technique is significantly diminished for micro molding, since the runner and gate form a significantly higher proportion of the total shot size for such components. This has the consequence that injection pressure curves are mainly describing the filling of the runner and cavity, which is not a key concern for the manufacturer. When optimizing the process, the machine operator is seeking the ideal environment in the mold cavity itself and, with a process monitoring system in place, the feedback from a range of additional sensors placed within the mold cavity can significantly aid him or her in this task.

Figure 2.12 shows cavity pressure curves generated using a Kistler 9211 piezo-electric indirect measurement sensor mounted behind a 0.5 mm ejector pin in the cavity. This was part of a process to manufacture a component with a mass of 0.3 mg at three different mold temperatures. The hold pressure is set at the same value of 80 MPa for all cases, but, at lower mold temperatures, the pressure drop through the feed system is greater due to the increased viscosity of the polymer melt and to more rapid cooling creating a thicker frozen layer.

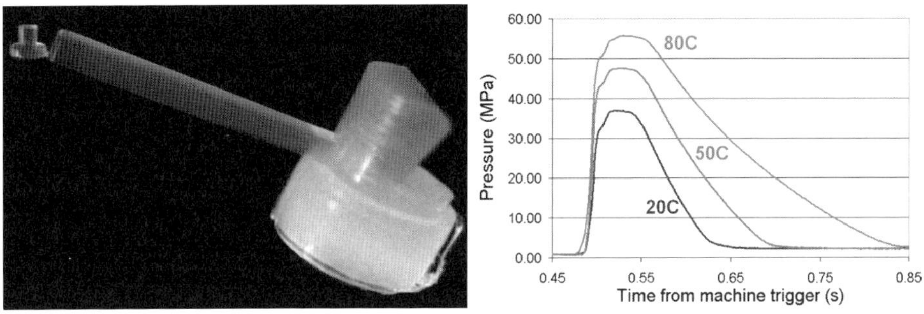

Figure 2.12 Left: 0.3 mg micro-molded component and runner system; right: recorded cavity pressure for three mold temperatures

Figure 2.13 shows the data generated from a FLIR X6540sc high-speed thermal camera during cavity filling of a polypropylene disc component with a thickness of 500 μm. Here the technique is being used to monitor the temperature at the flow front as the cavity fills and the data shows that temperature is significantly lower toward the end of the cavity. The machine operator can use this information to optimize the temperature across the whole component to produce better dimensional stability, improved surface finish, and ideal mechanical properties resulting from a better internal microstructure resulting, in turn, from the crystallization kinetics within the material.

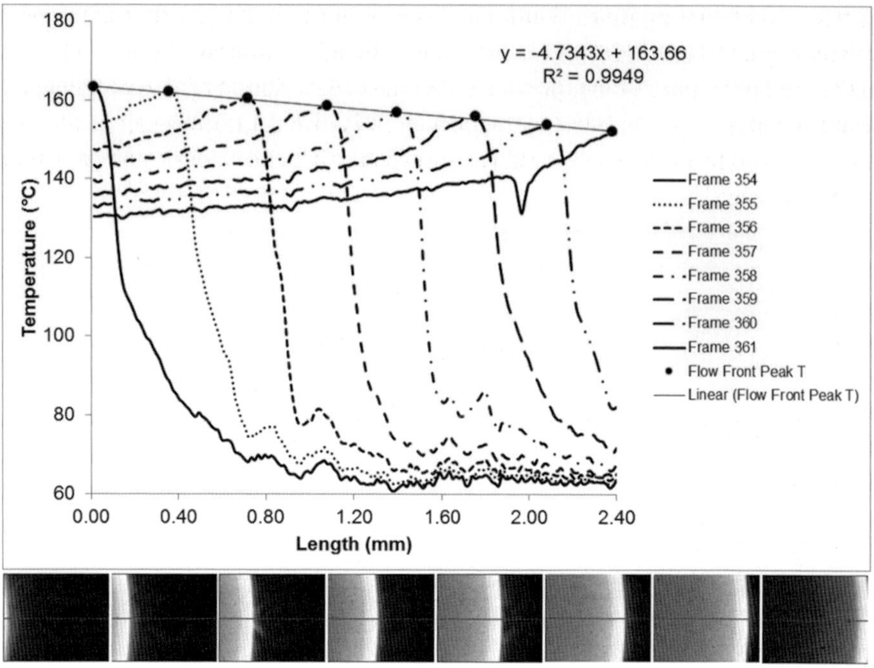

Figure 2.13 Flow front temperature measurements recorded during the filling of a disc component with a thickness of 500 µm

2.5.4 Simulation Boundary Conditions and Validation of Results

It is common to see cases where simulation software has been used to model a process and has failed to provide an accurate prediction of the mold filling behavior simply because incorrect boundary condition data has been used in the analysis. Common mistakes that are made are the use of melt and mold set temperature values (rather than those actually measured on the machine) and the use of default injection rates. Process monitoring systems allow these values to be measured directly to feed into the simulation model in order to create a more accurate model of the filling performance [19, 20]. The use of process monitoring tools allows understanding of systematic errors between set and actual values that can be used to improve the accuracy of future simulations.

These techniques can also be adopted to significantly improve less well-understood boundary conditions. A key parameter for determining the cooling behavior of a molding process is the heat transfer coefficient (HTC), which determines the flow of heat at the interface between the polymer and the mold tool. Small changes to this value can cause very large differences in simulation outcomes for micro molding processes, due to the very high surface to volume ratio of the components and the very rapid cooling times. Work has been performed using a thermal imaging system to quantify the HTC value for a micro molding process and it can be seen that the default values do not provide an accurate description of the cooling behavior of the polymer at the interface, as shown in Figure 2.14. However, by applying values derived from the cooling curves recorded using the camera system, a much better fit is seen.

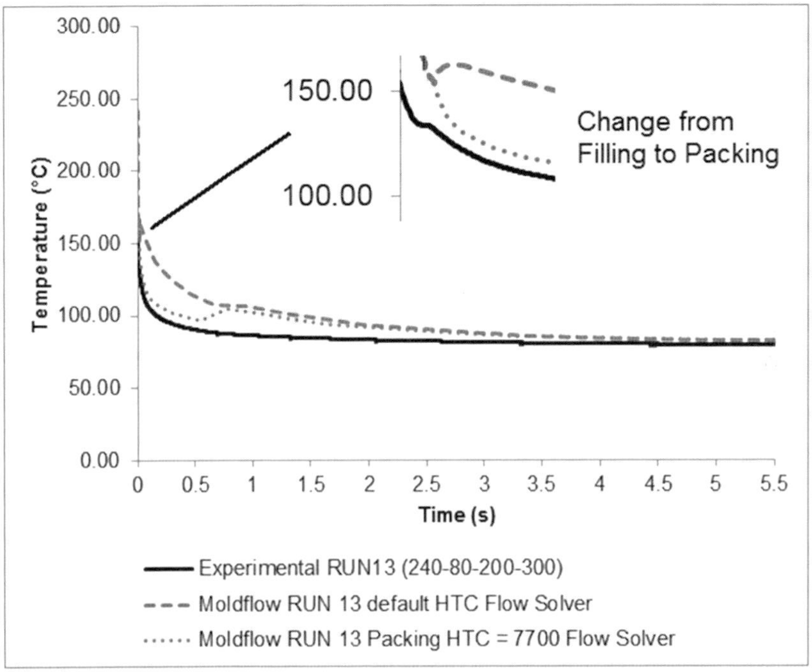

Figure 2.14 Comparison of simulated and experimental cooling curves describing component surface temperature

2.5.5 Sensor Development and Validation

Multiple sensor-based systems can be used to benchmark development sensor performance and also provide further understanding of the process environment [21]. The use of a common external clock signal to trigger the acquisition of each measurement sequentially allows data to be collected from multiple sources at the same point in time, allowing them to be directly compared. To illustrate this, Figure 2.15 shows the data collected using a Kistler 9211 sensor for injection pressure measurement, combined with a sequence of images acquired with a Mikrotron MC1310 high-speed camera, which were acquired synchronously. The increase in the pressure recorded by the sensor can be directly compared with the flow front position at that time, providing greater insight into the process and providing useful information that could be used to directly determine flow behavior from pressure curves alone.

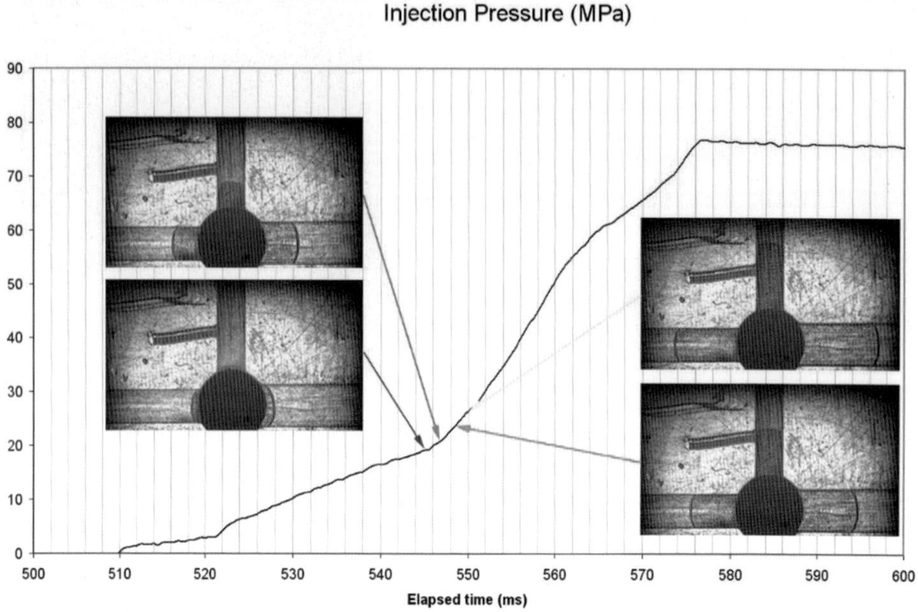

Figure 2.15 Synchronous injection pressure data and high-speed camera images of filling of a component with three channels of width 1.0, 1.5, and 2.0 mm

A second example is shown in Figure 2.16, where the recorded values of the amplitude for the reflection of ultrasonic waves generated using the piezoelectric film sensors described in Section 2.2.4 are synchronized with another high-speed camera visualization using a Mikrotron MC1310 camera. Here it can be seen that the amplitude of the reflection drops when the polymer enters the cavity, as some of the energy is transmitted into the polymer. As the part cools, the polymer detaches from the wall, causing the amplitude to return to the same level as when the cavity was empty.

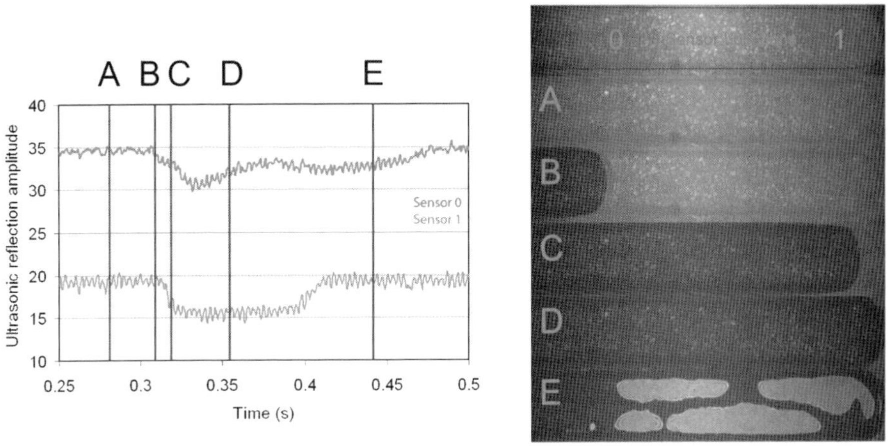

Figure 2.16 Synchronized ultrasound and high-speed visualization of cavity filling of a flat plaque component of width 2.0 mm and depth 250 μm. The figure is presented in the color supplement

2.5.6 Intelligent Process Control

A key focus for micro molding equipment providers is to be able to develop a system that would allow intelligent parameter adjustment to account for environmental or material changes and keep the production quality within specification. There have been many studies in the literature to attempt to develop such systems [22–24] but a working solution is still not available commercially due to the lack of data available from standard injection molding machines to ensure that a control strategy can be implemented.

There is a significant opportunity to implement additional process monitoring systems that can be combined with product quality systems (such as machine vision inspection stations) and Big Data analytics to be able to develop robust algorithms that would allow a degree of autonomous process control, resulting in fewer products that are out of specification and a corresponding increase in productivity. Such a system would have the ability to configure itself automatically for the molding of a new component and would have the ability to detect the failure of a critical component, safely pause the process, and alert the technical staff. This would have seemed like the stuff of science fiction just 20 years ago, but recent developments in sensor technologies, data management systems, and computer modelling are coming together to such an extent that a truly intelligent process is just around the corner.

References

[1] Attia, U. M., Marson, S., Alcock, J. R., Micro-injection moulding of polymer microfluidic devices. *Microfluidics and Nanofluidics* (2009) 7(1), pp. 1–28

[2] Giboz, J., Copponnex, T., Mélé, P., Microinjection molding of thermoplastic polymers: a review. *J. Micromech. Microeng.* (2007) 17(6), pp. R96–R109

[3] Whiteside, B. R., Martyn, M. T., Coates, P. D., Allan, P. S., Hornsby, P. R., Greenway, G., Micromoulding: process characteristics and product properties. *Plastics, rubber and composites* (2003) 32(6), pp. 231–239

[4] Griffiths, C. A., Dimov, S. S., Scholz, S., Tosello, G., Rees, A., Influence of injection and cavity pressure on the demoulding force in micro-injection moulding, *Journal of Manufacturing Science and Engineering,* Transactions of the ASME (American Society of Mechanical Engineers) (2014) 136, 031014, DOI: 10.1115/1.4026983, 10 pages

[5] Whiteside, B. R., Tuinea-Bobe, C. L., Coates, P. D., Characterising the Injection Kinetics of a Micro-injection Moulding Machine. Proceedings of 9th International Conference on Multi-Material Micro Manufacture, Vienna, Austria (2012) doi:10.3850/978-981-07-3353-7_330

[6] Bendada, A., Cole, K., Lamontagne, M., Simard, Y., A hollow waveguide infrared thermometer for polymer temperature measurement during injection moulding. *J. Opt. A: Pure Appl. Opt.* (2004) 5, pp. 464–470

[7] Bendada, A., Cole, K., Lamontagne, M., Simard, Y., Analysis of thermal contact resistance between polymer and mold in injection molding. *Appl. Therm. Eng.* (2004) 24, pp. 2029–2040

[8] Abeykoon, C., Martin, P. J., Kelly, A. L., Brown, E. C., A review and evaluation of melt temperature sensors for polymer extrusion. *Sensors and Actuators, A: Physical* (2012) Volume 182, pp. 16–27

[9] Zhang, N., Su, Q., Choi, S. Y., Gilchrist, M. D., Effects of gate design and cavity thickness on filling, morphology and mechanical properties of microinjection mouldings. *Materials & Design* (2015) 83, pp. 835–847

[10] Griffiths, C. A., Dimov, S. S., Scholz, S., Hirshy, H., Tosello, G., Process Factors Influence on Cavity Pressure Behavior in Microinjection Moulding. *J. Manuf. Sci. Eng.* (2011) 133(3), p. 31007

[11] Masato, D., Sorgato, M., Lucchetta, G., Analysis of the influence of part thickness on the replication of micro-structured surfaces by injection molding. *Materials & Design* (2016) 95, pp. 219–224

[12] Cheng, C. C., Ono, Y., Whiteside, B. R., Brown, E. C., Jen, C.-K., Coates, P. D., Real-time Diagnosis of Micro Powder Injection Molding Using Integrated Ultrasonic Sensors. *International Polymer Processing* (2007) 22, 2, pp. 140–145

[13] Cheng, C. C., Ono, Y., Kobayashi, M., Whiteside, B.R, Brown, E. C., Jen, C.-K., Coates, P. D., Real-time Diagnosis of Micromoulding Using Integrated Ultrasonic Sensors. *Transactions of the Institute of Measurement and Control* (2007) 29(5), pp. 383–401

[14] Han, X., Yokoi, H., Visualization analysis of the filling behavior of melt into microscale V-grooves during the filling stage of injection molding. *Polymer Engineering & Science* (2006) 46 11, pp. 1590–1597

[15] Babenko, M., A Study of Heat Transfer at the Cavity-Polymer Interface in Microinjection Moulding. UK Doctoral Thesis, University of Bradford (2015)

[16] Raza, U., Whiteside, B. R., Hu, F., An enterprise service bus (ESB) and Google Gadgets based micro-injection moulding process monitoring system, IET Conference on Wireless Sensor Systems (WSS 2012), London (2012), pp. 1–6

[17] Zhang, N., Gilchrist, M. D., Characterization of thermo-rheological behavior of polymer melts during the micro injection moulding process. *Polymer Testing* (2012) 31, 6, pp. 748–758

[18] Qin, X., Thompson, M. R., Hrymak, A. N., Torres, A., Rheology studies of polyethylene/chemical blowing agent solutions within an injection molding machine. *Polym. Eng. Sci.* (2005) 45, pp. 1108–1118

[19] Tosello, G., Schoth, A., Hansen, H. N., Implementation strategies for the optimization of micro injection moulding simulations. Proceedings of the 4th International Conference on Multi-Material Micro Manufacture (4M 2008), S. S. Dimov and W. Menz, eds., Whittles Publishing, Cardiff, UK (2008)

[20] Tosello, G., Costa, F. S., Hansen, H. N., Micro Injection Moulding High Accuracy Three-Dimensional Simulations and Process Control. Proceedings of Polymer Process Engineering 11: Enhanced Polymer Processing, Bradford, UK (2011)

[21] Whiteside, B. R., Spares, R., Brown, E. C., Norris, K., Coates, P. D., Kobayashi, M., Cheng, C. C., Optical imaging metrology for micromoulding cavity flows and products. *Plastics, Rubber and Composites* (2008) 37(2-4), pp. 57–66

[22] Zhang, S., Dubay, R., Charest, M. A., Principal component analysis model-based predictive controller for controlling part warpage in plastic injection molding. *Expert Systems with Applications* (2015) 42, 6, pp. 2919–2927

[23] Yarlagadda, P. K.D.V., Khong, C. A.T., Development of a hybrid neural network system for prediction of process parameters in injection moulding. *Journal of Materials Processing Technology* (2001) 118, 1, pp. 109–115

[24] Shen, C., Wang, L., Li, Q., Optimization of injection molding process parameters using combination of artificial neural network and genetic algorithm method. *Journal of Materials Processing Technology* (2007) 183, 2, pp. 412–418

3 Polymer Materials Structure and Properties in Micro Injection Molding Parts

Paritat Muanchan, Ryo Kaneda, Hiroshi Ito

3.1 Introduction

Micro scale molded parts in micro scale injection molding have been classified in three categories: molded parts with micro-structured surfaces; molded parts with less than one milligram pellet weight; and parts with micro scale precision, such as micro thin-wall parts [1]. In either case, a polymer melt is injected in a cavity by high injection speed and pressure. In addition, the cooling rate becomes extremely high because of the very thin cavities employed in micro injection molding. Thereby, high molecular orientation occurs on the inside of the molded parts in comparison with conventional injection molded parts. This molecular orientation has to be reduced because birefringence occurs. Furthermore, a skin layer is formed on the surface layer of the molded parts due to the high cooling rate. This skin layer should be reduced to raise the replication of the mold surface structure. The birefringence and replication properties are important with a transparent amorphous polymer. Semi-crystalline polymers consist of a crystal phase and an amorphous phase, and are characterized by the formation of a skin-shear core structure due to crystallization.

The influence of molecular orientation and crystallization in micro injection molding becomes larger than in conventional injection molding, because the polymer melt is injected in a cavity by high injection speed and pressure and then is quenched. Molecular orientation and crystallization affect the mechanical, optical, and thermal properties of molded parts. In this chapter, polymers scaling effects such as flow behaviors, higher-order structure, and engineering properties of molded parts in micro injection molding are presented and clarified.

3.2 Specific Properties of Polymers for Micro Injection Molding Applications

As mentioned above, specific properties different from the ones in conventional injection molding develop in micro injection molding because of differences in higher-order structures such as molecular orientation and crystallization. Precise replication of mold surface and size dimension of tiny parts is very important in this process. One important behavior is the surface replication of the designed mold. Surface replication for micro and nano surface patterns is a complex phenomenon in relation to flow behavior in the filling stage, shrinkage in the cooling stage, deformation at the mold release, etc.

The process cycle of micro and nano surface patterns transfer in injection molding can be described as being formed by three main stages: filling, packing-holding, and cooling. During the initial filling stage, when the molten polymer fills the cavity, compressibility effects are small and the flow is usually considered incompressible because of its isothermal state. Once the entire fine surface pattern is filled, the process enters the packing-holding stage, during which a pressure is applied on the material. At this point a compression occurs and the solution behavior is driven by the polymer compressibility. In the injection molding process, the compression phase is very short but results in dramatic changes in the nature of the polymer flow in the cavity. Velocities may take values comparable with those during the filling stage, while the pressure increase propagates into the entire cavity. As soon as the pressure reaches its maximum value, the velocity amplitude falls. It is the beginning of the holding stage, during which compressibility effects are still important. The inflow of material must compensate for the decrease of the specific volume from cooling. The polymer in the gate then freezes and the mass of polymer in the cavity is set. The physics during the remaining cooling phase is governed by the energy and state equations. After compression, the resin cools down under the glass transition temperature of solidification temperature for the polymer. A series of injection molding processes ends by removing the molded parts from the mold. Note that this state is an early stage for the next molding, and the same processes are repeated for the continuous molding. If necessary, the molded parts are processed in a secondary process such as coloring and joining, and they become final parts.

As described here, the method of injection molding efficiently combines the processes of giving the shape to the molten polymer and the cooling of the molded material. Therefore, the injection molding process is recognized as a powerful mass production tool in which uniform quality can be realized over a wide range of sizes with a high degree of freedom in terms of shape. As generally known, this process is widely used to manufacture molded parts, from car bumpers (meter-size) to cellular phone connectors (millimeter-size). Based on the same idea, the development of mass production of the micro and nano structures (micro/nanometer size) in the injection molding process is being actively pursued. Specific properties of mass production of the micro and nano structures are presented in the following.

Micro optical components such as lenses and array devices are usually made from amorphous polymer materials. This means non-crystalline polymers. In these cases, the optical properties are affected by the refractive index, birefringence, and molecular orientation. Under high pressure, high shear, and the rapid cooling process the molecular orientation is extremely increased. On the other hand, micro scale mechanical components such as mechanical gears and thin plates with micro structures are usually made from semi-crystalline polymer materials. The specific properties of importance are mechanical properties (e. g., elastic modulus, strength, residual strain, hardness, and deformation) affected by crystallinity and molecular orientation. Under high pressure, high shear, and the rapid cooling process the crystallinity and molecular orientation is also extremely increased. With respect to surface replication of polymer materials, the formation of a skin-shear region of molded parts is important, which influences the changes in crystallinity and molecular orientation of polymers.

3.3 Materials Scaling Effects in Micro Injection Molding

Micro injection molding is not only a minimization of conventional injection molding; there are also some important changes in methods and operation. Table 3.1 presents a list of the differences between these two techniques. Therefore, in this part, the scaling down effect from macro scale to micro/nano scale and the changes in material behaviors and the properties of molded parts have been clarified including rheology, pressure-volume-temperature (pvT) properties, thermal properties, and mechanical properties.

Table 3.1 Comparison between Conventional Injection Molding and Micro Injection Molding [2, 3]

Process component	Conventional injection molding	Micro injection molding
Machine	▪ Hydraulic and electrical machines ▪ Clamping force > 15t	▪ Electrical or electro-pneumatic machines ▪ Clamping force < 15t
Flow simulation	▪ 1D, 2D, 3D calculation	▪ 3D calculation required
Mold development	▪ CAD rules for the molded part geometry ▪ Injection gate diameter > 1 mm	▪ Simulation for the dimensions and the positioning of feeding channel ▪ Injection gate diameter < 1 mm
Realization	▪ CNC machining, EDM (Electrical Discharge Machining), grinding for base mold manufacture ▪ EDM and CNC machining for mold cavity generation	▪ CNC machining, EDM, grinding for base mold manufacture ▪ LIGA, μEDM, ultra-short pulse, Electro Chemical Milling (ECM), laser ablation, Deep Reactive Ion Etching (DRIE), micro milling, micro grinding for micro mold cavity generation
Plasticization	▪ Plasticization screw (Ø > 20 mm) or plunger	▪ Plasticization screw (Ø < 20 mm) or plunger
Injection	▪ Injection screw (Ø > 20 mm) + thermal heating ▪ Shear rates < 10^4 s^{-1}	▪ Injection screw or plunger (Ø < 20 mm) ▪ Shear rates > 10^6 s^{-1}
Temperature	▪ Manufacturer's recommended	▪ Higher than manufacturer's recommended ▪ Variotherm process for the mold
Holding	▪ Switchover set as function of the pressure or the screw position	▪ Switchover based on the plunger/screw position
Cooling	▪ Generally few tenths of seconds	▪ Instantaneous part cooling and extremely rapid freezing of the injection gate

3.3.1 Rheology in the Micro and Nano Dimensional Ranges, at Low, High, and Ultra-High Shear Rates

Rheology is defined as the science of flow and deformation of materials that usually is used to describe the consistency of different molded parts in terms of viscosity and elasticity. Viscosity is meant as resistance to flow or thickness and elasticity is meant as stickiness or structure.

The flow behavior plays a key role in polymer processing technology, mainly in process design, process monitoring and control, quality control, and problem solving in micro injection molding [4]. The flow behavior of polymer melts can be characterized in terms of viscosity. Fluids can be divided into three groups according to their flow behavior, including (1) Newtonian fluids, (2) non-Newtonian fluids (time-independent), and (3) non-Newtonian fluids (time-dependent).

Shear viscosity is given as the ratio of shear stress to shear rate, which is strongly dependent on the shear rate. A typical shear viscosity curve of polymer melt has four important regions: (1) 1st Newtonian (zero-shear viscosity) plateau, (2) transition region, (3) shear thinning region, and (4) 2nd Newtonian plateau, when all the molecules have disentangled (see Figure 3.1).

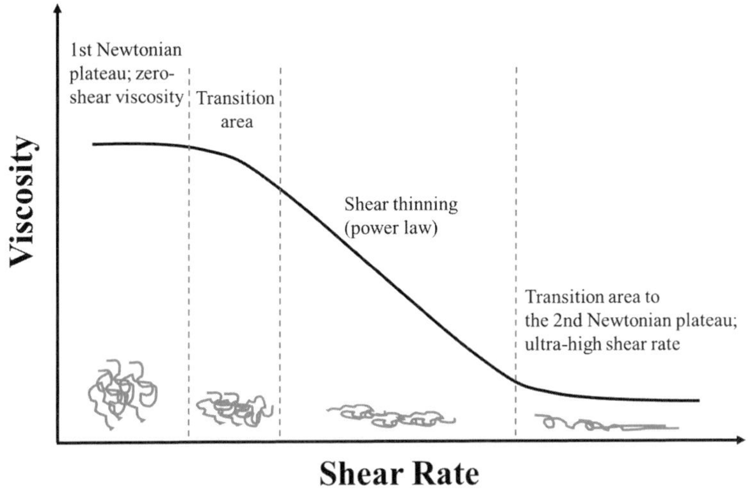

Figure 3.1 Typical viscosity versus shear rate curve

To understand the viscosities of polymer melts, different measurement instruments and calculation models are employed in order to obtain high accuracy data. At low shear rates, a rotational rheometer is the suitable instrument for measuring polymer melt viscosity at both macro and micro scale. With high shear rates, capillary viscometers and slit die viscometers are widely used to measure accurate viscosity data at the micro scale. However, when scaling down to the small channels, polymer melt viscosity is deviated from the capillary flow model and the slit flow model because of the surface interaction between polymer and wall channel, which assumes a dominant effect. In general, micro injection molding is usually operated at the high shear rate and ultra-high shear rate regions. However, it has not currently been reported on an appropriate model able to calculate the viscosity at ultra-high shear rates. For the viscosity at the nano scale, a precise instrument able to measure the real viscosity of the polymer melt is still not available. One approximation method to model the apparent viscosity taking into account the surface tension of the polymer melt and its contact angle on the wall of the flow channel as key factors has been proposed recently [5, 6].

At the high shear rate ranges of the shear thinning region, the scaling effect or size dependence on the common polymer melt viscosities for micro injection molding is usually reduced when the size is decreased at the micro scale [7]. C.-S. Chen reported [8] the reduction of viscosities of polyoxymethylene (POM) and polystyrene (PS) observed when the size of the micro channel was reduced in the range of 500–150 µm. For POM, the melt viscosity in the test range is approximately 5–12% and 29–35% for the channel sizes 300 µm and 150 µm, respectively, i. e., lower than values obtained with a traditional capillary rheometer. The rheological behavior of POM at the micro scale is different from the one at the macro scale, as it is also for PS, but displays a less significant reduction than PS. The result suggests that the wall slip occurs more easily for PS within micro channels than for POM due to the enlarged effect of molecular weight. Therefore, the micro viscosities and wall slip effects also must be considered when the channel size is smaller than or close to 100 µm. Based on the simulation, the calculations show that the surface tension effect is not important in polymer flow at micro scale, which is reported by D. Yao et al. [9]. In addition, polymers with long, complex, branched chains still need higher modelling accuracy for the determination of the calculated viscosities. For example, D. Zhao et al. [10] found that the viscosities of PS and poly(methyl methacrylate) (PMMA) with complex chain structures were increased, but those of polypropylene (PP) and high density polypropylene (HDPE) were reduced with the decrease of the die diameter from 1,500 to 500 µm using a capillary rheometer. The results suggest that the calculation of the viscosity of polymer melts in micro channels with the different wall-slip mechanisms should be considered based on the polymer molecular chain structure.

At ultra-high shear rates above 10^6 s^{-1}, polymer melts exhibit complicated rheology behaviors during micro injection molding. At this shear rate range, the viscosity is usually strongly dependent on the temperature. However, there is still a lack of knowledge on the rheology of polymer melts at this range. For example, A. Haddout and co-worker [11] reported that the PP and HDPE melts follow a pseudo-plastic shear thinning behavior at shear rates of up to 10^6 s^{-1}, above which their shear viscosity curves reach a quasi-Newtonian plateau. The second Newtonian plateau for PP and HDPE melts were observed at shear rates above 2×10^6 s^{-1} by M. Rides et al. [12]. A.L. Kelly et al. [13] also observed the shear thickening behavior for some grades of PP at shear rates above 10^6 s^{-1}. In addition, Takahashi et al. [14] observed a second shear thinning region of HDPE at shear rates above 7×10^6 s^{-1}.

3.3.2 Polymer pvT Properties at the Micro Scale

The pvT properties are significant for polymer engineering and polymer physics. For micro injection molding, pvT properties are very useful for processing design, processing and quality control, process simulation, and process optimization. For example, some defects such as shrinkage, warpage, sink marks, etc. can be solved by software programs which input pvT data for numerical simulations. There are many conventional measurement methods in polymer processing, including the piston-die technique, confined fluid technique, on-line technique, and modified techniques. To develop and create novel materials with high quality, evaluation methods within a short time using a small amount of samples become increasingly important for micro injection molding. However, it still remains difficult to evaluate the processability, structures, and physical properties of final molded parts simultaneously. Nevertheless, the evaluation of material characteristics using a small amount of samples has become possible. Furthermore, it is difficult to achieve an optimum molded parts design due to the lack of databases for optimizing material processing conditions for polymer processing at micro scale. Therefore, developments in pvT measurement are still being driven forward [15].

Measurement of Polymer pvT Properties and Rheology

H. Ito et al. [15] presented an intelligent analysis system for micro injection molding (see Figure 3.2), which combined the material characterization evaluation system for the measurement of material properties such as shear viscosity and pvT behavior. Using this system, the evaluation of a material's characteristics, processabilities, structures, and physical properties of all molded parts can be conducted using a small amount of samples (ca. 50 g), in a short time, and the flowing ability can be analyzed simultaneously.

The shear stress is obtained by measuring the pressure changes in a micro scale slit (400 µm) by three pressure gauges when the pistons are moved up or down simultaneously. The strain rate can be estimated via the flowing amount in the slit, and eventually the shear viscosity value is obtained. The specific volume is calculated based on the changes in position of the pistons due to the changes in temperature or pressure. This system confirms that the difference between the two viscosities of PP showed 30% at the low strain rate region and 10% at the high strain rate region as compared with the conventional capillary measurement.

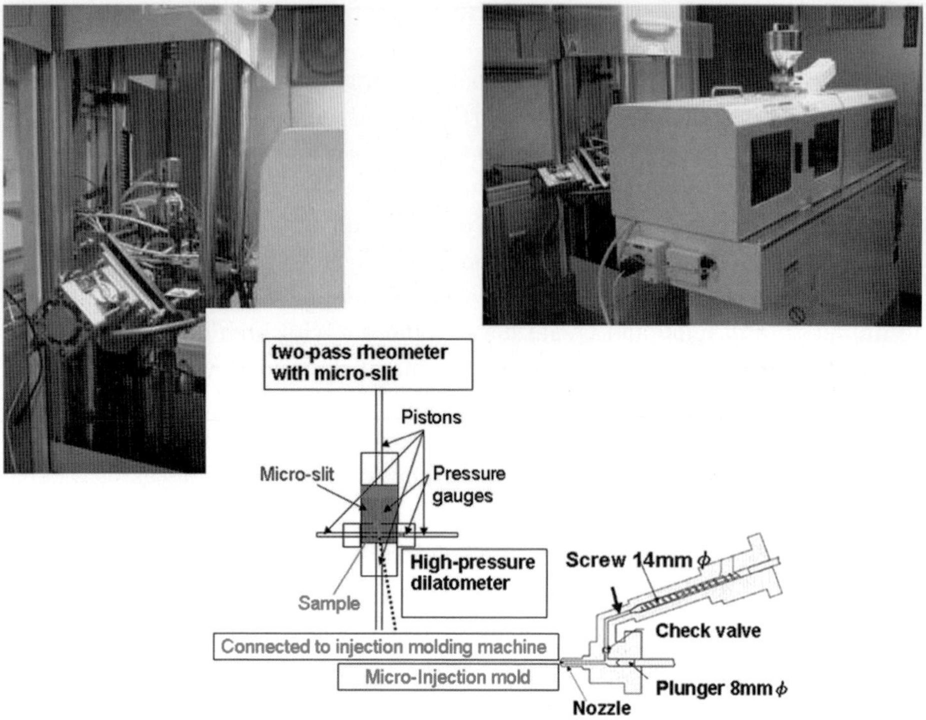

Figure 3.2 Schematic illustration and photograph of the intelligent analysis system for micro injection molding process [15]

Figure 3.3 shows the measured pvT of cyclic olefin copolymer (COC) and PP obtained by the intelligent analysis system. The pvT diagrams describe the specific volume as a function of pressure and temperature. For the amorphous polymer, the changed specific volume of COC indicated the onset of solidification at the glass transition temperature (T_g). The T_g shifted to the higher temperature region with increasing pressure, a trend that has been widely reported. For the semi-crystalline polymer, PP exhibits a different thermal response as compared with an amorphous polymer. The results revealed that the specific volume began to decrease at 130 °C (pressure: 10 MPa), which indicated that the density started to increase due to the crystallization effect. That temperature corresponds to the onset crystallization temperature and shifted to higher temperatures with increasing pressure.

In the case of the amorphous polymer, it has been noted that the starting of the transformation of the polymer from a glassy state to the rubbery state is not a sharp transition. On the other hand, the crystalline structure in the semi-crystalline polymer can maintain structural continuity above T_g. The specific volume of the semi-crystalline polymer is sharply increased where the crystal melts due to the difference in the specific volume between the glassy state and the melting state [15].

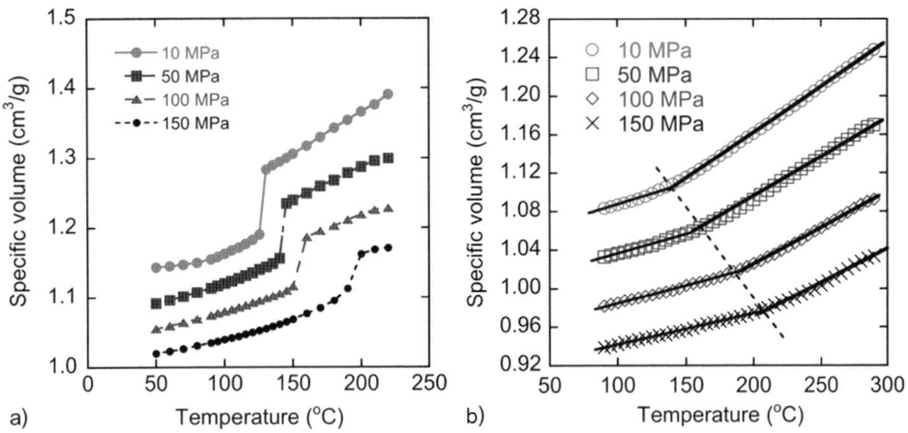

Figure 3.3 pvT behaviors measured by the intelligent analysis system of (a) COC and (b) PP [15]

3.3.3 Thermal Properties of Polymers at the Micro Scale

For the production of micro injection molding parts, it is essential to evaluate the scaling effect from the macro scale (conventional injection molding) to the micro scale (micro injection molding) on the thermal properties, which is significant for the usability of injection molded parts and processing design. Thermal properties such as thermal transition behavior, thermal stability, and thermal degradation play a key role in the development of miniature parts. The more widely used of the comprehensive suite of the thermal analysis techniques are differential scanning calorimetry (DSC), thermogravimetric analysis (TGA), thermomechanical analysis (TMA), and dynamic mechanical analysis (DMA).

The scaling down effect from the macro size to the micro size leads to the change in the volume of skin-core structures of the injection molded parts. In the case of semi-crystalline polymers at the micro scale, the crystalline entities present a lower size than those displayed at the macro scale due to faster cooling rates. Moreover, when the thickness of a micro mold cavity decreases to a few hundred micrometers, the molten polymer and micro mold cavity will encounter higher temperatures, speeds, and pressures, and the rheological behavior of the polymer flow in the micro cavity can differ from the one in a conventional cavity; the wall-slip effect occurs and extensional viscosity increases in the micro channels [16]. Therefore, the volume ratio of skin layer increases when the molded part thickness is reduced. The specific processing conditions can also affect the various stages of crystallization, i.e., nucleation and growth of crystalline lamellae. For amorphous polymers, H. Ito et al. [1] observed that the convex side on the surface of COC molded parts presents the change in molecular orientation at the micro scale.

Based on this behavior, the change in thermal properties due to the scaling down of the molded parts from the macro scale to the micro scale and/or nano scale will be observed. The details of molecular orientation will be discussed in Section 3.4 for both amorphous and semi-crystalline polymers.

Thermal Transition Behavior

For the thermal transition behavior of pellets and micro molded parts of HDPE, J. Giboz et al. [3] found that the DSC results revealed that the endotherm of the HDPE micro molded parts was shifted toward a lower temperature compared to the one obtained with HDPE pellets. The melting temperature for the HDPE micro molded parts was close to 129.5 °C, whereas the HDPE pellets presented a melting temperature close to 132.5 °C. For the scaling down effect, the melting temperature of HDPE micro molded parts with 150 and 200 μm of thickness prepared by micro injection molding and HDPE macro molded parts with 1.5 and 2.0 mm of thickness prepared by conventional injection molding were compared. The melting temperature of the HDPE micro molded parts was shifted toward a lower value when scaling down the size from the macro to the micro scale [17–19]. In addition, N. Zhang et al. [20] also observed the reduction of the melting temperature of Pebax (polyamide 12 (PA 12)/poly(tetra-methylene oxide) (PTMO) 80/20 wt %) when they reduced the size of molded parts from 500 to 100 μm.

Thermal Stability and Thermal Degradation

With the changes in thermal stability due to the scaling effect, the annealing method is a simple way to evaluate the deformation behavior of the observed structures with a nano scale precision. For example, the thermal stability of nano surface patterns on a micro molded surface of polycarbonate (PC) has been reported by H. Ito and co-workers [1]. The molded part's depths and heights of the surface patterns were analyzed by scanning probe microscope (SPM) with a long probe for nano surface patterns measurement. Figure 3.4 shows the stability of nano surface patterns as a function of time at different annealing temperatures. The height and area of the cross-sectional nano surface patterns were measured by atomic force microscopy (AFM) at different conditions. In the case of 120 °C, the results revealed that the thermal stability was obtained for about 2 h. On the other hand, the heights of the nano surface patterns were reduced with time at 130 and 140 °C. At an annealing time of 2 min, the pattern height increased because of thermal shrinkage. At 10 min, the pattern height showed a rapid decrease of over 0.2 as compared with the initial height. At below the T_g of PC (145 °C), the primary relaxation time and the activation energy for the flow of PC are long and small, respectively. In general, polymer molecular activity on a thin surface of less than 100 nm is higher than that for the bulk state. Therefore, it is difficult to explain the nano surface patterns variation over 500 nm with a short time during the annealing process. It was proposed that the thermal stability of the nano surface patterns would be affected by molecular activity, large surface area, and/or internal higher-order structure. J. Giboz et al. [3] compared the thermal degradation of an HDPE polymer in pellets, macro molded parts (1.5 mm-thickness), and micro molded parts (150 µm-thickness) using TGA measurement. The results revealed that the reduction of the size of the injection molded parts and the corresponding increase in the shear rate do not favor the material degradation in this case.

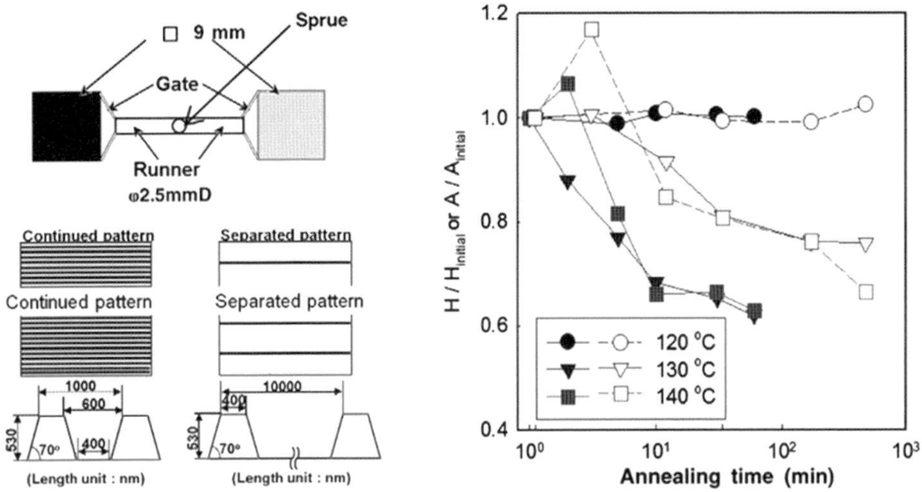

Figure 3.4 Schematic diagrams of mold with nano feature surfaces (left), and height (H) and area (A) of nano surface patterns as a function of annealing time (right) [1]

3.3.4 Mechanical Properties of Micro Molded Components (Micro Tensile Test and Nano Indentation)

Micro Tensile Test

The mechanical properties of micro molded parts are usually deviated from those of macro molded parts due to the change in free surface energy. However, this is not only due to the scaling effect on the mechanical properties but also depends on the processing conditions that influence the changes in the internal structures and molecular orientation.

The micro tensile test provides suitable equipment for the characterization of micro molded parts. H. Ito et al. [21] performed the mini tensile test to evaluate the mechanical properties of isotactic polypropylene (iPP) micro molded parts. The mold cavity thickness of the central region was varied to 0.3, 0.5, and 1.0 mm. Figure 3.5 presents the stress-strain curve of the iPP micro molded parts. In the case of 1.0 mm thickness, the elongation at break of molded parts prepared with the mold temperature of 100 °C was the lowest as compared with 40 and 70 °C. At 1.0 mm thickness, the molecular orientation of iPP micro molded parts at 100 °C was lower than at 40 and 70 °C. Moreover, the core region became thicker and a spherulite structure occurred in the region. In the case of 0.5 mm thickness, it shows an opposite behavior: the molded part prepared at 100 °C has the highest elongation at break as compared with 40 and 70 °C. At the high mold temperature of 100 °C, the Young's modulus and tensile strength of molded parts became lower as compared with those of lower molding temperatures. In the case of 0.3 mm thickness, the elongation at break of iPP micro molded parts at 40 °C became higher than at 70 and 100 °C. Furthermore, the Young's modulus of molded parts at the low mold temperature of 40 °C was slightly lower than at 70 and 100 °C. In the molded parts which have slight molecular orientation, these mechanical properties have a close relation to a high order structure, and the Young's modulus and the tensile strength increase with increasing molecular orientation and crystallinity. At 0.5 mm thickness, this is reasonable due to the relation between molecular orientation and a mechanical behavior. However, in the case of 0.3 mm thickness, the related behavior was due to the fact that a ductile brittleness transition had occurred. It was found that the involved mechanical characteristics were affected by the internal structure distribution of the molded parts.

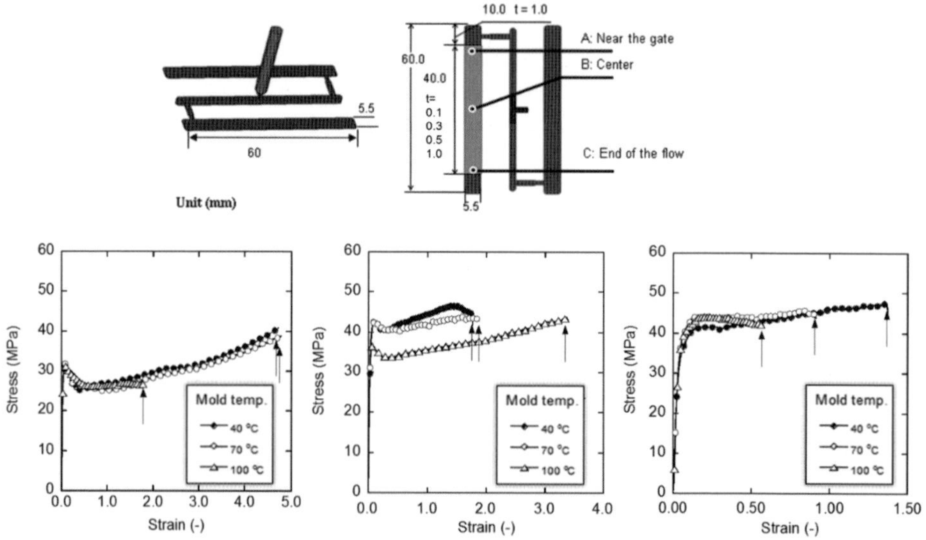

Figure 3.5 Shape of molded parts (top), and stress-strain curves of the molded parts (bottom) with cavity thickness: 1.0 mm (left), 0.5 mm (middle), and 0.3 mm (right) [21]

In addition, the scaling down effect from conventional injection molding to micro injection molding on the mechanical properties of the micro molded parts was found to be that these mechanical properties are significantly different from those of macro molded parts. By using micro injection molding, the Young's modulus reduces with an increase of cavity thickness from 200 to 500 µm due to the decrease in the skin ratio and molecular orientation factor in the case of Pebax [20]. Moreover, the fine-grained layer disappears when the cavity thickness is less than 200 µm. The skin area reduces to less than 20% when the cavity thickness is larger than 300 µm. Therefore, the yield stress broadly reduces with an increase in the cavity thickness owing to the skin ratio decrease. The yield stress increases because the fibrous crystals at the skin area have a high strength and a deeply penetrated texture. As compared with the standard (4.0 mm thickness) tensile specimens, miniaturized specimens exhibit a decreased Young's modulus and strain at break, while the yield stress increases. This result suggests that the miniaturization has a significant effect on the mechanical properties of the obtained molded parts.

Nano Indentation

Nano indentation is a precise method for the evaluation of micro scale and nano scale structures. The comparison between the mechanical properties of HDPE micro molded parts (150 µm thickness) and those of HDPE macro molded parts (1.5 mm thickness), using nano indentation measurements, showed the reduction in the elastic modulus and the hardness when scaling down from the macro scale to the micro scale of molded parts. This result suggests that the mechanical properties of micro molded parts are more homogeneous along the thickness than those of macro molded parts. Furthermore, mechanical properties measured by nano indentation are sensitive to changes in microstructures induced by the specific molding conditions. These measurements may be thus regarded as evidences of morphological changes and, in particular, they are able to account for oriented morphologies induced by the processing conditions [22]. H. Ito and co-workers performed the nano indentation test with the PP micro needle arrays, which were obtained by micro injection molding. It was found that the reduction in the loading force of the PP micro needle arrays when the weld line defect takes place on the micro molded part structures is as shown in Figure 3.6. Due to the scaling effect, the PP-micro needle shows the lower elastic modulus as compared with a PP flat film with 1.3 and 1.4 GPa, respectively.

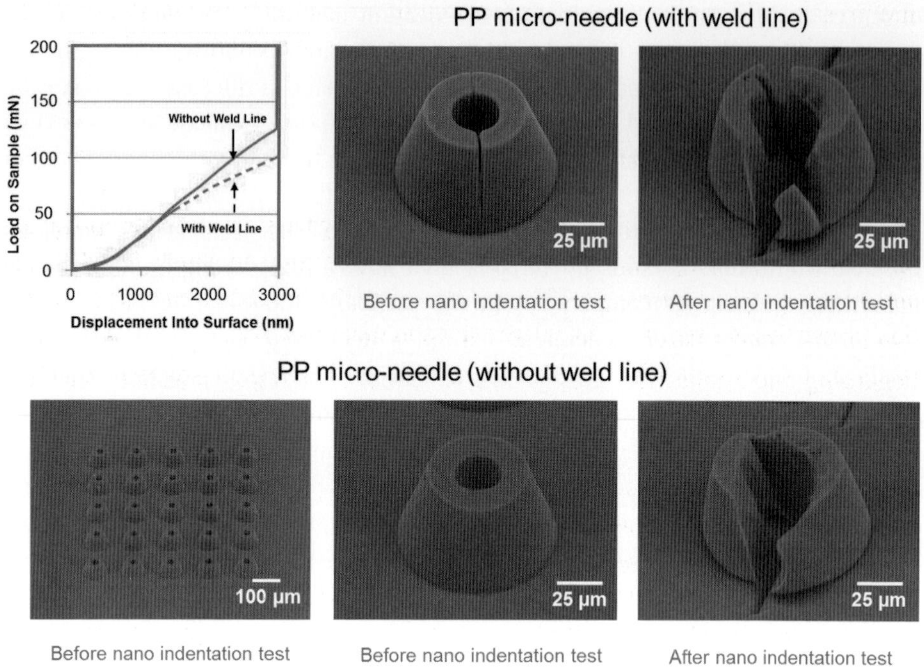

Figure 3.6 Nano indentation test results and PP micro needle before and after indentation test

■ 3.4 Molecular Orientation and Crystallinity in Micro Molded Parts

A multilayer so-called skin-core structure is formed inside of the injection molded parts. The skin layer and core layer are formed in the molded parts surface and the central portion in the thickness direction of the molded parts, respectively. The shear layer having molecular orientation caused by the flow of polymer melt is formed between the skin layer and the core layer. The crystallization of a semi-crystalline polymer is induced by this molecular orientation. In addition, a spherulite is generally formed in the core layer, which is a stress free area or relatively low stress area for a semi-crystalline polymer. The crystallinity of a semi-crystalline polymer is affected by the degree of molecular orientation and the size of the spherulite. This molecular orientation and crystallization vary according to polymer materials, molding conditions, cavity thickness, etc. In micro injection molding, the surface features and the thickness of the molded parts are at the micro/nano scale; the ratio of skin-core layers becomes higher than in conventional injection molding because a polymer melt is injected in a cavity by high injection speed and pressure. Therefore, the comprehension of the molecular orientation and the crystallinity are particularly important in micro injection molding.

3.4.1 Amorphous Polymers

The micro injection molding parts of amorphous polymers are used for optical applications, such as a pickup lens of CDs or DVDs, light-guide panels for LCDs, an optical filter, optical fibers, etc. [23–26]. Therefore, high transparency and replication properties are required for micro injection molding parts.

Regarding the replication of sub-micron features in micro injection molding, K. Mönkkönen et al. [27] carried out experimental investigations about the influence of molding conditions using amorphous polymers of five kinds: PC, styrene-butadiene block copolymer (SBS), impact modified PMMA, methyl methacrylate-acrylonitrile-butadiene-styrene polymer (MABS), and COC. They reported that the best replication results were achieved when melt and mold temperatures were higher than normal values. Furthermore, Y.K. Shen et al. [28] reported that a mold temperature is the most important process parameter in the results of a 3D numerical simulation of parts filling using PS.

To confirm the difference in replication described above from the distribution of the skin-core structure of molded parts, H. Ito et al. [1] observed the cross section of the micro injection molded COC using the mold with micro grid surface (see Figure 3.11(b)). An observed thin section was cut parallel to the melt flow direction (MD) from the molded parts. The cross section of the molded parts observed under a polarizing optical microscope is shown in Figure 3.7. The existence of a surface skin-shear core structure inside the molded parts was confirmed. They also observed that the convex parts on the molded surface showed lower molecular orientation and the shear region was exhibited below this feature. An inner core region with lower molecular orientation was formed. At a higher injection speed, the skin-shear core structure was also observed, and the thickness of the shear layer became greater, whereas the thickness of the skin decreased. At a higher mold temperature, the thickness of the shear layer became smaller. From these observations, it was revealed that the injection molding conditions affected the thickness of the skin-shear region. Moreover, the development of a skin-shear layer disturbed the polymer flow and the increase of internal pressure. The distribution of surface replication ratio was influenced by the thickness of the skin-shear layer. The optical birefringence varied with the skin-shear core structure, and the birefringence usually increased with increasing thickness of the skin-shear region.

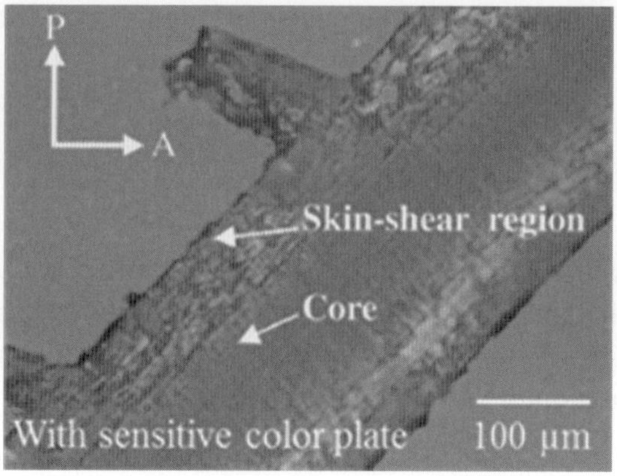

Figure 3.7 Polarized photograph of cross section of COC parts with micro feature surfaces [1]

To improve the surface replication by reducing the skin-shear layer thickness, H. Ito et al. [29] performed an experimental study using a zirconia ceramics insulated mold of 0.5 mm cavity thickness with nano surface patterns having anti-reflection function. A commercial PC was used for this study. The flow length and capability of surface replication can be improved by using the heat insulator mold of zirconia due to the decrease in cooling rate. From the intersectional polarized photograph of the molded parts, when the mold temperature is high, it was found that the core layer, having little influence on the shear, is formed. These results are consistent with the increase of the flow length. From the measurement results of retardation and birefringence, it was found that the molecular orientation near the gate decreased by using the heat insulator mold. Furthermore, due to the better surface replication, plates using the heat insulator mold show higher light transmission due to less light reflection at the surface of plates.

In the thin-wall injection molding of the 0.5 mm cavity thickness using the PC, D. Yao et al. [30] obtained better results of the flow length and birefringence by using a mold of high-frequency proximity heating. Also, T. Sato et al. [31] reported that the skin-layer disappeared by an injection molding process using infrared radiation heating and the birefringence decreased.

In addition, the injection compression molding is effective to improve the replication of sub-micron features and optical properties [32]. Thus, the replication of sub-micron features and optical properties of molded parts occurs depending on skin-core structure.

3.4.2 Semi-Crystalline Polymers

Semi-crystalline polymers consist of a crystal phase and an amorphous phase, which is a formed skin-core structure with crystallization. A semi-crystalline polymer filled in a cavity crystallizes in different states of the structure and crystallinity according to polymer materials, molding conditions, cavity thickness, etc. because the shear and cooling rates are different. The influence of molecular orientation and crystallinity in micro injection molding becomes larger than in conventional injection molding, because the polymer melt is injected in a cavity by high injection speed and pressure and then is quenched.

F. Liu et al. [33] reported that iPP micro molded parts with 200 µm thickness presented a large fraction of shear layer in comparison to the iPP macro molded parts with 2000 µm thickness, which presented a large fraction of core layer, and the crystallinity of the micro molded parts became higher than that of the macro molded parts by the shear layer having a highly oriented shish-kebab structure. In addition, they described that the large fraction of oriented region of the micro molded parts indicates a good mechanical performance.

Furthermore, J. Jiang et al. [34] reported that the obvious skin-core structure was observed in the micro gear of iPP molded parts with 450 µm thickness, and the nano indentation modulus decreases along the flow direction and increases with increasing mold temperature; the highly oriented shear layer has the highest nano indentation modulus compared with the skin and core layers.

H. Ito et al. performed thin-wall injection molding, and the structure and physical properties of iPP molded parts were investigated by using a cavity as shown in Figure 3.5 (top) [21]. The molded parts were produced at various mold temperatures and cavity thicknesses; the effects of the molding conditions and cavity thickness on the skin-core structure and physical properties were examined at the micro scale. When the cavity thickness decreased, the flow length decreased because of an increase in viscosity and the rapid solidification of the melted polymer, as shown in Figure 3.8. From these results of wide angle x-ray diffraction (WAXD) and birefringence measurements, the c-axis and a*-axis crystalline orientation at each point was observed, and it showed a similar tendency to that of the case of conventional injection molded iPP. The molecular orientation in the MD direction was particularly high at the vicinity of the gate because of the high shear stress in this region. The molecular orientation increased with a decrease of cavity thickness as shown in Figure 3.9. However, it showed the same high molecular orientation as high-speed spun fibers at cavity thickness 0.1 mm. Results of cross-sectional observation also revealed a skin-core structure inside the molded parts as shown in Figure 3.10. However, a spherulite structure at the core region was not observed at thin cavity thickness at low mold temperature. The thickness ratio of thickness of skin-shear layers to total cavity thickness increased at lower cavity thickness. The results of the tensile test showed that the elongation at break of the molded parts of cavity thickness of 0.3 mm was lower compared with those of 0.5 and 1.0 mm, and that it is closely related to the internal structure distribution.

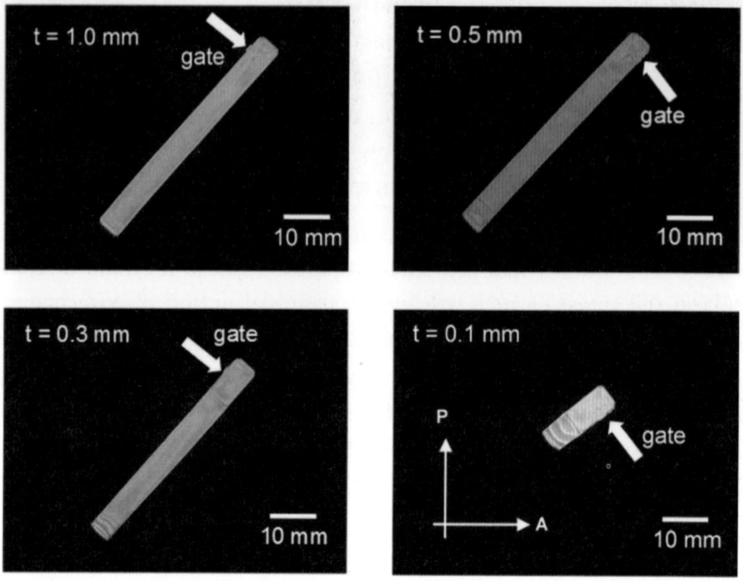

Figure 3.8 Photographs of molded parts for various cavity thicknesses; mold temperature: 40 °C [21]

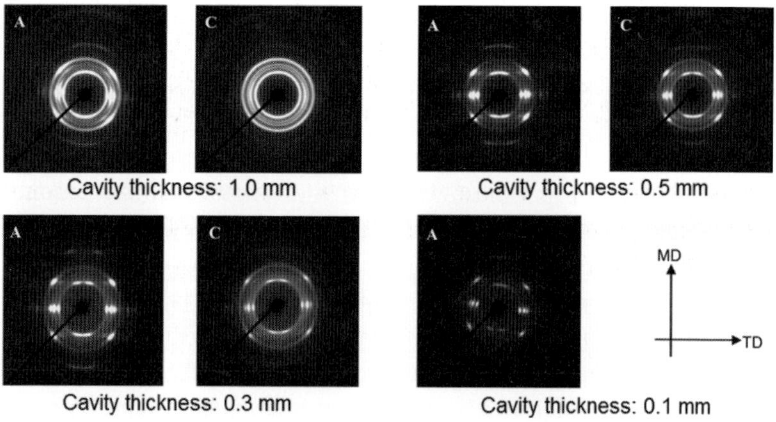

Figure 3.9 WAXD patterns of molded parts at positions A and C; mold temperature: 40 °C [21]

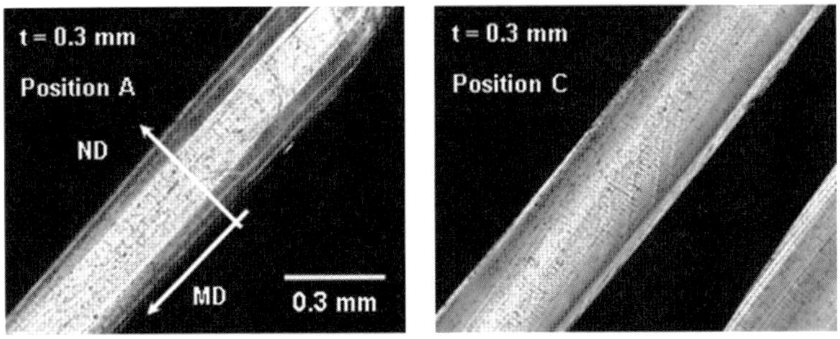

Figure 3.10 Cross-sectional views of the molded parts; cavity thickness: 0.3 mm [21]

C. Guo et al. [19] reported that HDPE micro molded parts (200 μm thickness) and macro molded parts (2000 μm thickness) exhibited a similar skin-core structure, but the micro molded parts showed a much larger fraction of oriented layer, and the crystallinity of micro molded parts was higher than that of the macro molded parts. In addition, they described that the large fraction of oriented region of the micro molded parts indicates a good mechanical performance. Regarding the micro injection molding for HDPE, X. Lin et al. [35] reported that the skin-core structure varied significantly through the plaque's cross section and the thickness of each layer changed with the process parameters and geometric position, and finally led to a variation of the mechanical performance.

Furthermore, H. Ito et al. [36] conducted an experimental investigation of thin-wall injection molding for HDPE to obtain the high infrared radiation (IR) transmittance with an opaque state (low visibility ray (VR) transmittance), because the HDPE is used as a condenser lens for some IR systems such as radiation thermometers to transmit IR of the 8–12 μm region. As a result, for a high mold temperature and a long cooling time, the increased core-layer thickness improved the IR transmittance. Simultaneously, the VR transmittance decreased because the crystallinity became higher.

As described above, the skin-core structure of a semi-crystalline polymer influences mechanical and optical properties; in particular, the influence becomes large in micro injection molding.

3.5 Micro/Nano Composites in Micro Injection Molding

In compounding a filler into a polymer, the material properties such as mechanical properties, thermal properties, and electrical properties can be enhanced [24]. The kinds of filler used are different depending on the required performance of the molded parts. In addition, the molded parts performance of polymer composites with compounded filler is also affected by the shape, size, content, and surface treatment of the filler. Furthermore, the molded parts performance depends on the dispersion of fillers in the composites [37–45].

It is necessary to use fillers having dimensions smaller than the surface structure size in micro injection molding, particularly in the case of thin-wall parts with micro/nano surface patterns. G. L. Liou et al. [46] investigated the filling efficiency in micro features for the reinforcement of different kinds and sizes. As a result, the filling efficiency in the micro feature was almost the same as in the main body of the component for the reinforcement with nano size.

Furthermore, L. Xie et al. [40] investigated mechanical properties using micro tensile test samples with weld lines of PP/carbon nano fibers (CNFs) and PP/titanium dioxide (TiO_2) composites with various loading fractions (10, 20, 30, and 35 wt %). The results show that when the CNFs content is higher than 10 wt %, the tensile strength of samples with weld lines made of nano composites becomes lower than neat PP, while the raising CNFs content contributes to the improved E modulus of micro injection molded weld lines. Additionally, with the increasing fraction of CNFs in PP, the weld line area's elongation percentage is decreased, whereas for the case of TiO_2, the 10 wt % is the threshold for the micro injection molded weld line tensile strength turning from a decrease trend to an increase. The same as CNFs, the elongation of micro weld line samples were in general lower than those of neat PP as well, due to the addition of TiO_2 nano particles.

H. Ito et al. [47] investigated the processability, structure, and properties of PP/hexagonal boron nitride (h-BN) composites using two types of thin-wall molded parts as shown in Figure 3.11. The shear viscosity decreased by compounding h-BN and the composite flowed easily into the cavity. Consequently, the polymer orientation decreased (see Figure 3.12) and the replication ratio of the micro surface patterns increased (see Figure 3.13(b)). The strength and elongation at break of molded parts were lowered, but the modulus was greatly improved. As a result, deformation at demolding was suppressed. The thermal diffusivity was increased with increasing h-BN content. However, the effects of the h-BN filler size were small. It is necessary to form a network using fillers to improve the thermal diffusivity of composites by compounding fillers. H. Ito et al. [47] used an h-BN filler in this composite system. The layer of h-BN fillers in the PP matrix was structured in the injection molding process, as shown in the SEM images of Figure 3.13(a). Therefore, the PP layer blocked the heat flow along the thickness direction and the thin-wall molded plates seemed to reveal anisotropic thermal diffusivity. The thermal diffusivity along the thickness direction is greatly increased if isotropic shapes of fillers are used. Results of WAXD patterns show that the polymer orientation became higher near the gate than at the flow end, although thermal diffusivity increased with increasing distance from the gate. This tendency indicates that not only the filler orientation but also the polymer orientation affects thermal diffusivity. Heat transfer needs a path to move; therefore, heat transfer also occurs along the polymer chain. The thermal diffusivity is measured along the normal direction (ND) axis. Therefore, heat passes along the ND axis, increasing the distance from the gate. The highest thermal diffusivity is achieved at the flow end position.

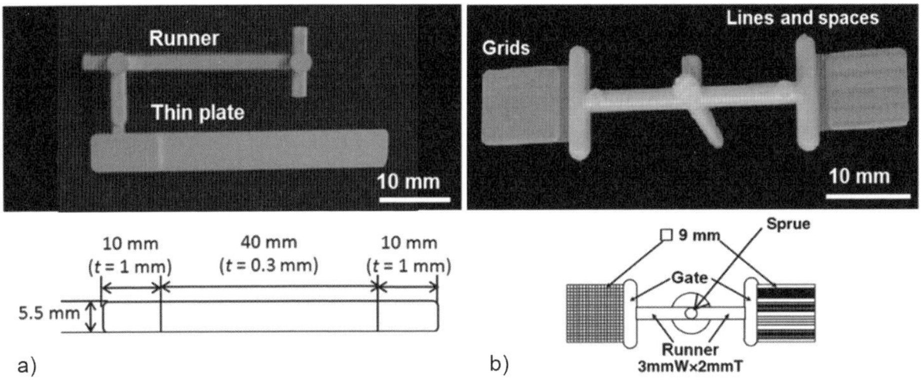

Figure 3.11 Schematic diagrams and photographs of molded parts: (a) thin plate, (b) lines and spaces and grids [47]

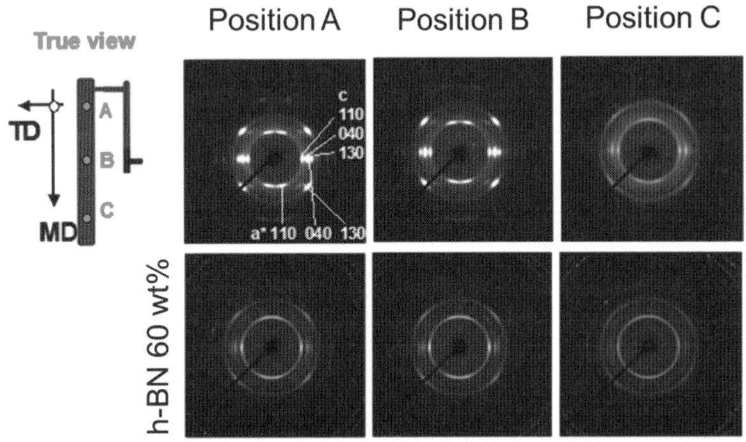

Figure 3.12 WAXD patterns of different positions with or without h-BN (PT 110 (60 wt %)) [47]

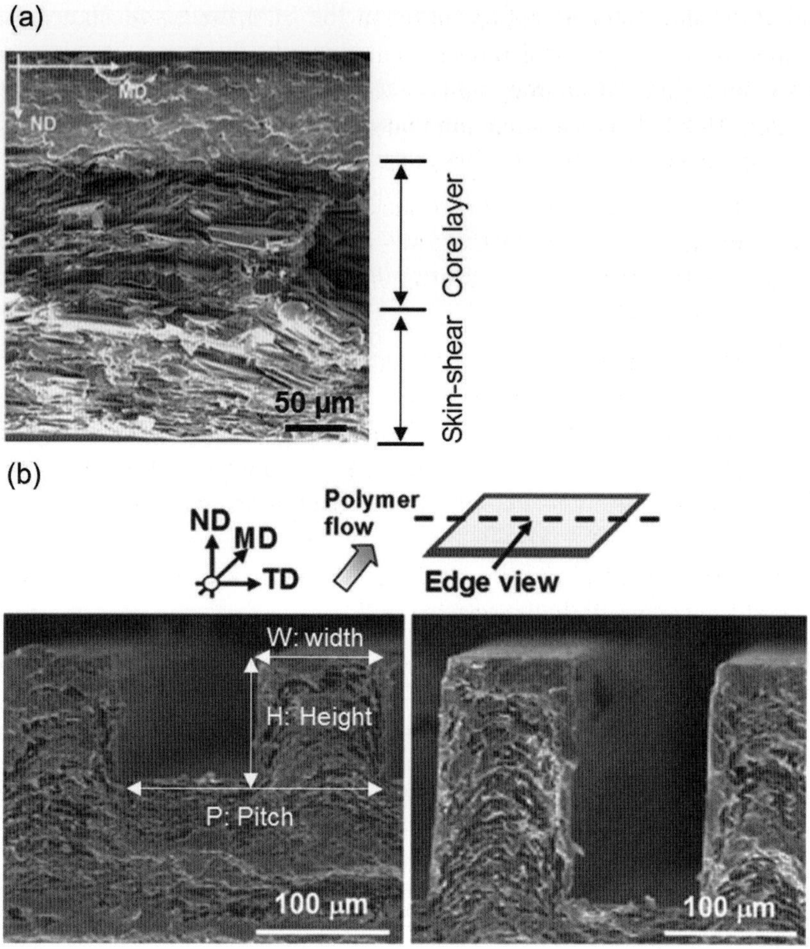

Figure 3.13 SEM images of cross section of molded parts: (a) thin plate (PT 110 60 wt %, inj. speed = 150 mm/s, near the gate position), (b) lines and spaces: (W = 100 µm) × (H = 100 µm) × (P = 200 µm) (left) and (W = 100 µm) × (H = 200 µm) × (P = 200 µm) [47]

Regarding the effects of filler surface treatment, particle size, and injection molding conditions on the mechanical properties of silica-filled PMMA, composites were investigated by Y. Kobayashi et al. [37]. Micro/nano silica particles filled PMMA composites were prepared using high shear compounding and thin-wall micro injection molding. In order to modify the surface of these fillers, the polymer grafting method was employed. Three types of silica, i.e., micro-sized untreated, micro-sized PMMA-grafted, and nano-sized PMMA-grafted silica, were incorporated into the PMMA matrix. The particle size presents the biggest effect on the mechanical properties, since it affects the distribution as well as the dispersion of the fillers in the matrix. Distinct morphological differences could also be observed when different silica sizes were incorporated, especially the state of plastic deformation at the skin and core regions of the moldings. Filler treatment contributed minimally toward mechanical property enhancement, although there were evidences pointing toward improvements in filler dispersion as well as filler-matrix interaction. This led to an assumption that there was better filler-matrix adhesion in nano filler composites than in micro filler composites.

Powder injection molding using micro metal particles or nano silica particles is extremely promising as a production process for micro molded parts because the technique enables near-net-shape fabrication of micro-structured parts with nearly no post-processing [48].

In powder injection molding for metal injection, the particle size and distribution of metal powder are also important factors affecting the properties of green compacts and sintered molded products. K. Okubo et al. [49–51] have investigated the miniaturization of metal injection molding parts and have realized mass production of micro metal injection molding parts with a fine structure. In metal injection molding processes, metal powder that is mixed and kneaded with an organic binder is plasticized and injection molded in the same way as plastics to produce near-net-shape parts with high precision [52].

In addition, H. Ito et al. [53] investigated the processability, structure, and properties of quartz glass/polymer composites used to fabricate a new microfluidic plate with glass. The molded parts length increased concomitantly with increasing injection speed as a general trend. The shrinkage between the mold and green parts became low at both the flow and transverse directions. On the other hand, the thickness of molded parts increased, becoming greater than the designed value. The internal morphology affected the shrinkage of green molded composites. The surface replication ratio of molded and sintered parts showed high values, and sintered molded parts having a high aspect ratio of 3.4 with a micro line width of 10 µm were obtained.

References

[1] Ito, H., Suzuki, H., Kazama, K., Kikutani, T., *Curr. Appl. Phys.* (2009) 9, pp. 19–24

[2] Giboz, J., Copponnex, T., Mélé, P., *J. Micromech. Microeng.* (2007) 17, pp. 96–109

[3] Giboz, J., Copponnex, T., Mélé, P., Proceedings of the 7th International Conference on Multi-Material Micro Manufacture (4M) (2007), pp. 281–284

[4] Zhang, N., Gilchrist, M.D., *Polym. Test.* (2012) 31, pp. 748–758

[5] Muanchan, P., Suzuki, S., Kyotani, T., Ito, H., *Polym. Eng. Sci.* (2017) 57, pp. 214–223

[6] Shao, Z., Gopinadhan, M., Kumar, G., Mukherjee, S., Liu, Y., O'Hern, C.S., Schroers, J., Osuji, C.O., *Appl. Phys. Lett.* (2013) 102, p. 221901, doi: 10.1063/1.4808342

[7] Chen, S.C., Tsai, R.I., Chien, R.D., Lin, T.K., *Int. Commun. Heat and Mass Transfer* (2005) 32, pp. 501–510

[8] Chen, C.-S., Chen, S.-C., Liaw, W.-L., Chien, R.-D., *Eur. Polym. J.* (2008) 44, pp. 1891–1898

[9] Yao, D., Kim, B., *J. Micromech. Microeng.* (2002) 12, pp. 604–610

[10] Zhao, D., Jin, Y., Wang, M., *Polym. Eng. Sci.* (2017) 52, pp. 1806–1814

[11] Haddout, A., Villoutreix, G., *Inter. Polym. Proc.* (2000) 15, pp. 291–296

[12] Rides, M., Kelly, A. L., Allen, C. R. G., *Polym. Test.* (2011) 30, pp. 916–924

[13] Kelly, A. L., Gough, T., Whiteside, B. R., Coates, P. D., *J. Appl. Polym. Sci.* (2009) 114, pp. 864–873

[14] Takahashi, H., Matsuoka, T., Kurauchi, T., *J. Appl. Polym. Sci.* (1985) 30, pp. 4669–4684

[15] Ito, H., Saito, T., Yasuhara, T., Kikutani, T., SPE ANTEC Tech. Papers (2005), p. 693

[16] Su, Q., Zhang, N., Gilchrist, M. D., *J. Appl. Polym. Sci.* (2016) 133, p. 42962

[17] Giboz, J., Copponnex, T., Mélé, P., *J. Micromech. Microeng.* (2009) 19, p. 25023

[18] Giboz, J., Copponnex, T., Mélé, P., *J. Appl. Polym. Sci.* (2017) 134, p. 44239

[19] Guo, C., Liu, F. H., Wu, X., Liu, H., Zhang, J., *J. Appl. Polym. Sci.* (2012) 126, pp. 452–462

[20] Zhang, N., Su, Q., Choi, S. Y., Gilchrist, M. D., *Mater. Design* (2015) 83, pp. 835–847

[21] Ito, H., Yagisawa, Y., Saito, T., Yasuhara, T., Kikutani, T., Yamagiwa, Y., *Theoretical and Applied Mechanics Japan* (2005) 54, pp. 263–268

[22] Giboz, J., Vite, M., Bec, S., Loubet, J. L., Copponnex, T., Mélé, P., Proceedings of the 7th International conference on Multi-Material Micro Manufacture (4M) (2010), pp. 103–106

[23] Hansen, H. N., Hocken, R. J., Tosello, G., *CIRP Ann. Manuf. Technol.* (2011) 69, pp. 695–714

[24] Attia, U. M., Marson, S., Alcock, J. R., *Microfluid. Nanofluid.* (2009) 7, pp. 1–28

[25] Yang, C., Yin, X. H., Cheng, G. M., *J. Micromech. Microeng.* (2013) 23, pp. 1–21

[26] Griffiths, C. A., Dimov, S. S., Brousseau, E. B., Hoyle, R. T., *J. Mater. Process. Technol.* (2007) 198, pp. 418–427

[27] Mönkkönen, K., Hietala, J., Pääkönen, P., Pääkönen, E. J., Kaikuranta, T., Pakkanen, T. T., Jääskeläinen, T., *Polym. Eng. Sci.* (2002) 42, pp. 1600–1608

[28] Shen, Y. K., Yeh, S. L., Chen, S. H., *Int. Comm. Heat Mass Transfer* (2002) 29, pp. 643–652

[29] Kayano, Y., Zouta, K., Takahagi, S., Ito, H., *Inter. Polym. Proc.* (2011) 26, pp. 304–312

[30] Yao, D., Kimerling, T. E., Kim, B., *Polym. Eng. Sci.* (2006) 46, pp. 938–945

[31] Saito, T., Satoh, I., *Polym. Eng. Sci.* (2002) 42, pp. 2418–2429

[32] Rytka, C., Kristiansen, P. M., Neyer, A., *J. Micromech. Microeng.* (2015) 25, pp. 1–16

[33] Liu, F., Guo, C., Wu, X., Qian, X., Liu, H., Zhang, J., *Polym. Adv. Technol.* (2012) 23, pp. 686–694

[34] Jiang, J., Wang, S., Sun, B., Ma, S., Zhang, J., Li, Q., Hu, G. H., *Mater. Des.* (2015) 88, pp. 245–251

[35] Lin, X., Caton-Rose, F., Ren, D., Wang, K., Coates, P., *J. Polym. Res.* (2013) 20, pp. 122–133

[36] Kaneda, R., Takahashi, T., Takiguchi, M., Hijikata, M., Ito, H., *Inter. Polym. Proc.* (2016) 31, pp. 385–392

[37] Kobayashi, Y., Leong, Y. W., Sarata, M., Hamada, H., Yamada, K., Kotaki, M., Yoshinaga, K., Karakawa, H., Ito, H., Kazama, K., Kikutani, T., ANTEC Tech. Papers (2007), pp. 2000–2004

[38] Hwang, S., Hsu, P. P., *Ind. Eng. Chem.* (2013) 19, pp. 1377–1383

[39] Xie, L., Ziegmann, G., *J. Alloys. Compd.* (2011) 509, pp. 226–233

[40] Xie, L., Kirchberg, S., Steuernagel, L., Ziegmann, G., *Microsyst. Technol.* (2010) 16, pp. 1855–1859

[41] Alateyah, A. I., Dhakal, H. N., Zhang, Z. Y., *Adv. Polym. Technol.* (2013) 32, p. 21368

[42] Huang, C. K., Chiu, S. W., *J. Appl. Polym. Sci.* (2005) 98, pp. 1865–1874

[43] Wang, X. C., Jing, X., Peng,Y. Y., Ma, Z. K., Liu, C. T., Turng, L. S., Shen, C. Y., *Polym. Eng. Sci.* (2016) 56, pp. 319–327

[44] Yuan, M., Winardi, A., Gong, S., Turng, L. S., *Polym. Eng. Sci.* (2005) 45, pp. 773–788

[45] Zhang, M.Q., Rong, M.Z., Zhang, H.B., Friedrich, K., *Polym. Eng. Sci.* (2003) 43, pp. 490–500

[46] Liou, G.L., Young, W.B., *Inter. Polym. Proc.* (2010) 25, pp. 264–269

[47] Ito, H., Kazama, K., Kikutani, T., Proceedings of the 2007 International Manufacturing Science and Engineering Conference (MSEC2007) (2007), pp. 133–140, doi: 10.1115/MSEC2007-31035

[48] German, R.M., *Powder Injection Molding* (1990) MPIF, Princeton, New Jersey

[49] Okubo, K., Tanaka, S., Hamada, H., Ito, H., Asia-Pac. *J. Chem. Eng.* (2009) 4, pp. 133–139

[50] Okubo, K., Tanaka, S., Ito, H., *Microsyst. Technol.* (2009) 15, pp. 887–892

[51] Okubo, K., Tanaka, S., Ito, H. SPE ANTEC Tech. Papers (2009), pp. 2592–2597

[52] German, R.M., *Powder Metallurgy Science* (1994), MPIF, Princeton, New Jersey

[53] Ito, H., Honta, K., Ikari, T., Ito, S., SPE ANTEC Tech. Papers (2010), pp. 2172–2176

4 Surface Replication in Micro Injection Molding

Matteo Calaon, Danilo Quagliotti, Guido Tosello

Transferring features from one substrate to another implies replication of geometries (deterministic and random), which is a well-established method within various manufacturing technologies. If the replication concerns micro- and nano-sized surface features, the replication fidelity between different process chain steps is highly dependent on process variation, process conditions, tool accuracy, material behavior, and feature geometries.

In the field of micro and nanotechnology, where development and fabrication value are generally very high, solutions to decrease final product costs are oriented toward economy of scale to remove barriers between the introduction of new technologies and the use of controllable manufacturing processes.

Polymers have largely been employed in micro and nanotechnology applications due to their attractive physical properties (i.e., chemical resistance, low water absorption, optical transparency in the UV range) and for the capability of being processed with technologies enabling large volume production. Micro injection molding (μIM) is recognized as one of the most important manufacturing processes enabling mass fabrication of polymer components. Replication of produced master geometries materializes thanks to the filling of mold cavities with a polymer material in its liquid/viscous state with high injection speed and pressure. Due to the accuracy reached and the precision of the manufacturing processes involved, methods and approaches to process chain validation for the fabrication of both deterministic and random structures are required.

4.1 Replication of Micro and Nano Structures

The replication of micro- and nano-sized surface features is highly dependent on the replication technology involved [1, 2]. Transferring geometries from one substrate to another by means of different energy inputs is a well-established method within various manufacturing technologies, which can be grouped into three main categories: replication from a fluid state, replication from a solid/viscous state, and finally by material deposition.

The degree of replication of produced master geometries into the functional or end-use substrate is evaluated by dimensional comparison between the mold and workpiece geometries. The dimensions of surface micro and nano structures are typically orders of magnitude smaller than the structured surface area. In the lateral direction, the ratio of the micro and nano geometry dimension to the workpiece dimension is typically of the order of $(10^{-1}-10^{-6})$ m. In the vertical direction, the ratio between the structure height and the substrate thickness is typically of the order of $(10^{-1}-10^{-4})$ m [3].

Typical replicated surfaces in the micro and sub-micrometer range include both well-defined geometries like grooves, holes, pillars, etc. and random topography usually characterized by low/very low surface roughness values. Deterministic well-defined geometries are grouped in most of the cases in periodical arrays. The graph in Figure 4.1 assesses the current polymer replication capability, both at the research and application level [4–14], of the micro injection molding process in terms of micro/nano features' vertical replication at different lateral dimensions.

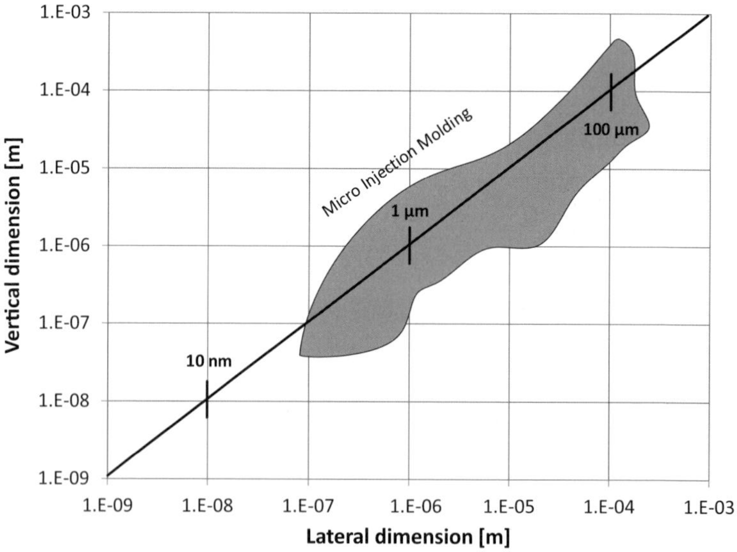

Figure 4.1 Micro injection molding of thermoplastic polymer processing capabilities in terms of micro/nano features' vertical replication, inspired by [3]

The industrial attention in using replication concepts among different substrates at all scales (from the micro- to the nanometer range) as manufacturing technologies able to bring specific products with integrated nano features into the consumer market is enormous. Nevertheless, the combination of several replication steps in the same process chain with the intent of enhancing productivity and lowering single product costs, faces the challenges of reaching the physical limits of manufacturing processes (reproducibility, minimum target dimensions) and measuring instruments (measurement resolution and uncertainty). Current research related to precision molding and micro injection molding is addressing these challenges with a focus on tooling, processes, quality control, and materials.

Due to the increasing demand of combining functional structures with different length scales, tooling technologies are facing new scenarios in which instrument alignment, resolution, and, most importantly, reproducibility need to be reconsidered [15].

In micro tooling manufacturing, new micro molds and insert designs, including surface topographic modifications, for over-molding, in-mold assembly, or packaging of multimaterial micro components are under major development.

In the micro replication assembly of multimaterial micro components, the development of more suitable injection materials (advanced polymer compounds) and interface designs (surface functionalization) for a stronger bond and more adequate interaction replication between the selected materials/components are of major importance. Also, a further optimization of micro replication processes (enhanced process capabilities) for avoiding unresolved lingering issues in multimaterial micro component/systems replication, such as the appearance of non-acceptable flashes, flow lines and weld lines in the interface between the materials/components, which strongly decrease the quality of the product, have been studied [16]. Additionally, a new feeding and high-precision handling system for micro component manipulation will make possible the accurate allocation of the inlay parts (parts to be over-molded/assembled/packaged) without deformations and the accurate placement of the already over-molded parts at the next process stage. All of this, along with the possibility of avoiding, in most cases, the assembly step, will allow the achievement of multimaterial injected products with better quality (typical scrap rate decrease of 30–40%), in lower cycle times (typically, reductions in the order of 30–50% can be observed).

In process and part quality control, the definition of degree of replication is often related to the specific features under investigation [17, 18] (i.e., grooves, pillars, holes). Typically, absolute dimensions and the ratio between replicated and tool geometries are evaluated. If geometries differ from deterministic structures deviating to more complex shapes, the definition of the geometry to be measured is required [3]. Practical challenges are related to the definition of edges and measurement alignments [19]. Uncertainty associated with the measurand on the master and on the tool should be evaluated due to the reduced conformance zone of product quality acceptance, particularly in the case of nano geometries. Considering the mirroring of structures during the different replication phases, definitions of measuring strategy allowing for measurements relocation are required. Process variation, process conditions, tool accuracy, material behavior, and feature geometries at this dimensional scale are of paramount importance because the lower limit of the measurements' precision and accuracy is reached. Dimensional and geometrical metrology is required to establish traceability of measuring results as an integrated part of product characterization to enable definition of product tolerance ranges [20–22].

New stable materials/structures, design principles, and advanced materials technologies are required, leading to both new reduced cost substitutes to existing materials and to new higher added-value products and related functionalities [23].

4.2 Engineering of Micro- and Nanostructured Surfaces

Patterning large areas with micro- to nanostructured geometries yields different surface functionalities. Distributed nano patterns are used, for example, to control cell adhesion and position to monitor the status of cells' growth and to study their activation of the body's defense mechanism against foreign materials. Deterministic micro- to sub-micrometer structures, able to carry samples/reagents measured in the order of nano to picoliters, are employed in lab-on-a-chip micro-nanofluidic technologies. The need for practical low-cost and high-volume production for both distributed and deterministic nano geometries has shifted attention toward polymer replication. Transferring nano geometries fabricated on a nickel master into the polymer substrate with high accuracy is a key aspect. Origination of a master with the inverse structures of the final part is the underlying principle of any replication technology. Nickel electroplating is largely used to transfer manufactured nano geometries on a substrate able to withstand high temperature and pressure during injection molding.

In the present section, the top-down (in which structures are obtained by pattering the bulk material substrates) approaches used to produce the nickel shims used for polymer replication of deterministic and random sub-micrometer structures are described. First, an alternative fabrication technology to the most commonly used LIGA ((lithography), galvanoformung (galvano forming), (electroplating), abformung (molding)) process [24], on which micro channels are not defined on a thick photoresist but directly in silicon using dry etching, is illustrated. DEEMO (acronym for dry etching electroplating and molding) [25] makes fabrication of structures/channels for lab-on-a-chip applications in ranges from nanometers and micrometers possible. Secondly, an alternative method performed outside clean-room facilities with the potential to structure large areas with nano periodic semi-sphere-like structures with feature dimensions ranging from 500 nm down to 30 nm in diameter was employed [26]. Anodization of an aluminum substrate to create aluminum oxide connected membranes acting as a template for a later nickel electroplating procedure is reported.

4.2.1 Lithographic Techniques

Mostly based on developments obtained from so-called silicon technology, with the focus on techniques adopted to fabricate deterministic micro and sub-µm structures, lithography and etching play fundamental roles in the creation of different structures. Generally, an initial substrate (silicon in the present case) is coated with a thin photosensitive polymer (called a resist) by spinning procedures followed by gentle heating inside an apposite oven to enhance adhesion with the substrate and internal stress relaxation. Subsequent polymer radiation (photons, electrons, ions) initiates its physical and chemical properties, allowing pattern transfers with the same geometries as those previously manufactured on masks. The patterned resist is then used as a master-mask for subsequent etching operations of uncovered substrate areas. The technique used to manufacture structures within the production process chains adopted in the current project was based on electron beam lithography (EBL) and UV lithography. EBL allows the high resolution direct patterning of the resist thanks to high energy electron beam radiation. The critical parameter for resolution optimization is the beam dimension that can reach dimensions below 10 nm in width [27]. Moreover, the resolution of EBL is strictly linked to the scattering phenomena of interactions with materials' target atoms. Photolithographic techniques, instead, use an ultraviolet (UV) light source to activate a photosensitive polymer resist covered/uncovered with a pre-patterned mask. A final etching defines the depth of the designed structures by removing material from the silicon substrate. A mask material (silicon oxide for the proposed process chain) is used to selectively shadow the substrate material to be etched. Dry etching, providing highly directional (anisotropic) etching of deep structures, is employed in combination with the reaction of plasma etching in the proposed DEEMO process [28].

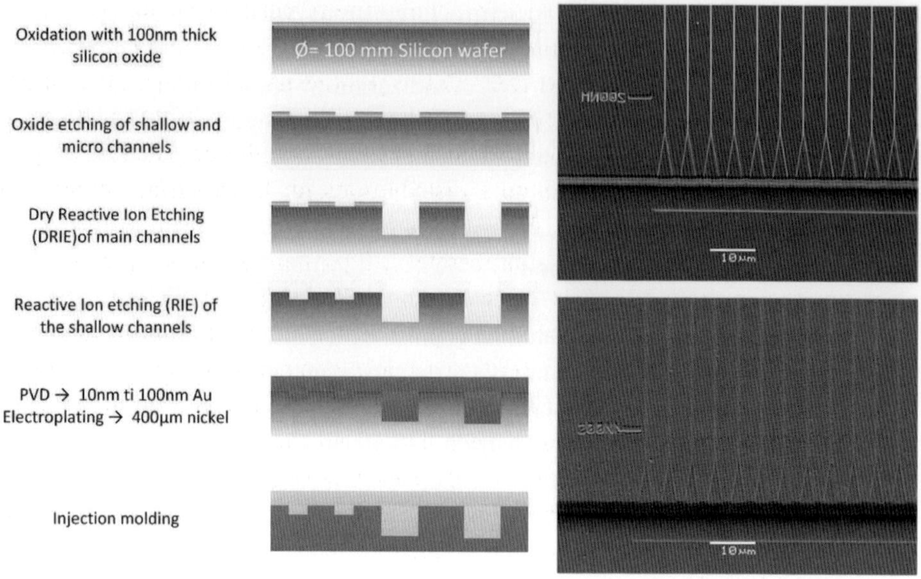

Figure 4.2 Left, replication processes within the DEEMO process chain. Right, top: SEM pictures of the nickel master; right, bottom: polymer replicated structures. Structures are representative of a critical dimension in a microfluidic design system that enables stretching of DNA macromolecules thanks to nanoconducts (nominal dimension width = 300 nm, height = 110 nm, pitch = 5 µm) intersecting micro-scale 5-µm-deep channels

The master geometries are fabricated on a single-side-polished silicon wafer. The channels were defined by 100 keV e-beam lithography. Prior to the e-beam lithography, the silicon was oxidized. The thickness of this oxide was approximately 100 nm. The channels defined by e-beam were etched by anisotropic selective reactive ion etching (RIE). Once the micro/sub-µm geometries are written on the silicon master, the second lithography step to create the deeper and larger features is performed. A first seed layer composed of Ti/Au, with a thickness of 10 nm/100 nm, respectively, was deposited by physical vapor deposition (PVD) in order to enhance the corrosion resistance of the obtained nano structures and to promote the deposition of a thick nickel layer. The nickel electroplating process parameters are optimized to avoid potential problems such as high surface roughness, pitting, poor adhesion, and high internal residual stresses.

4.2.2 Non-Lithographic Electrochemical-Based Techniques

The aim of developing master-making solutions with the capability of structuring large areas involving low-cost production and not necessarily confined to clean-room facilities has moved the focus of research to electrochemical-based procedures. In particular, so-called random structures are produced by a process chain integrating the electrochemical oxidation of aluminum, responsible for the self-organization of pore structures in the original aluminum substrate and later replication of the obtained periodical structures into a positive nickel pattern. With initial applications oriented toward proactive and decorative films on its surface, aluminum oxidation studies have started to focus in the last decades on potential properties enhanced by the porous alumina distributions on large surface areas and relatively narrow pore sizes [28, 29]. Applications such as template materials for the synthesis of nanowires or nanotubes [30], light emitters to support surface plasmon resonances (SPRs) [31], antireflection polymer films [32], and the capability of creating various configurations of pore arrangements in combination with lithography techniques [33] are in continuous development. Aluminum oxide films with regular pore distributions can be formed by electrooxidation of high-purity aluminum substrates in acidic electrolytes. The structure of the oxide film consists of a uniform array of parallel alumina cells packed hexagonally, each containing a nearly cylindrical pore. The uniform diameter of the pores is a consequence of the equilibrium between the growth rate of the alumina and its dissolution in the acidic electrolyte. The hexagonal self-order of the pores can be explained by the repulsive forces due to the volume expansion associated with the anodization process [34]. The pore formation is schematically shown in Figure 4.3, corresponding to the typical phases/regimes of oxide formation. After aluminum oxide dissolution, functional structures superimposed onto the original aluminum substrate by the oxide layer during its development are unmasked for the subsequent replication step involving nickel electroplating [35].

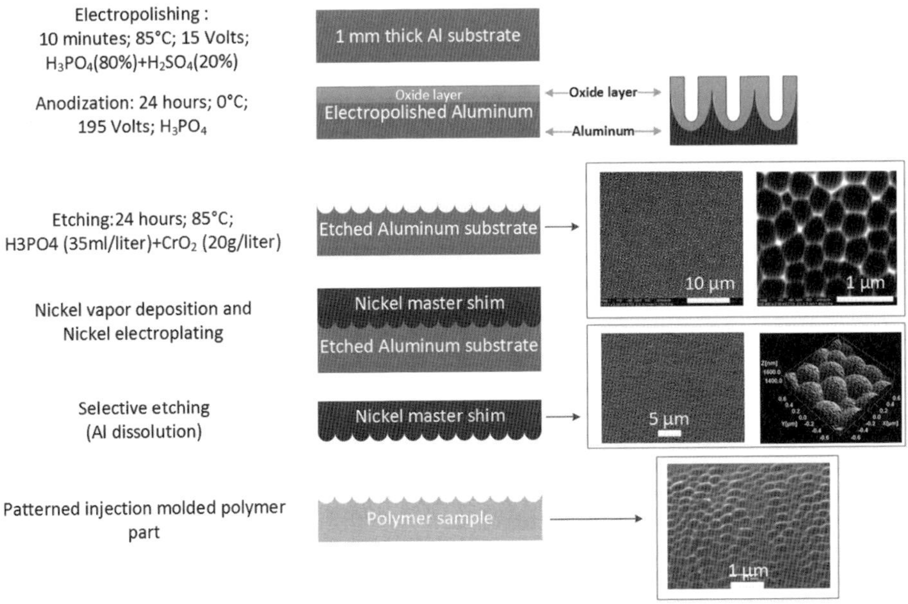

Figure 4.3 Production process chain to fabricate large nanostructured areas on polymers

The hard anodization process capability of creating nanoporous anodic aluminum oxide (AAO) with a hexagonal arrangement has become attractive because of its low-cost process chain. Its popularity has also grown because of its versatility to pattern 3D geometries while controlling the pore size, interpore distance, and aspect ratio of the nano pores of the resulting alumina membranes used for polymer replication based on the anodization of aluminum. The method consists of the electrochemical oxidation of aluminum substrate, which creates a nanoporous anodic aluminum oxide (AAO) film with a hexagonal arrangement. Surface pre-treatments of the 1 mm pure (99.999%) aluminum plate involved, as a first step, an annealing operation below the melting point (at 500 °C for three hours) to obtain a larger grain size and to eliminate residual stresses from previous manufacturing operations. As a second step, the aluminum substrate was electrochemically polished in an 80% H_3PO_4 and 20% H_2SO_4 electrolyte solution kept at 85 °C, which was vigorously stirred during the process. The anodizing process was carried out using an equipped cooling stage able to maintain and control the large evolution of heat due to the electric field of the barrier oxide. A phosphoric acid electrolyte was preserved at a constant temperature of 0 °C for 24 hours under high voltage (195 V). The formed oxide layer was then dissolved in an etchant solution containing H_3PO_4 (35 ml/liter) + CrO_3 (20 g/liter) at 85 °C to expose the pseudo hexagonal lattice structure. A nickel seed layer was deposited by physical vapor deposition (PVD) prior to the nickel electroplating deposition process in order to enhance the corrosion resistance of the obtained nano structures and to promote the deposition of a thick nickel layer. Nickel electroplating optimized process parameters reduced the risk of pitting, poor adhesion, and high internal residual stresses. Finally, after having achieved an approximately 600-µm thick nickel layer, the aluminum substrate was dissolved by chemical etching (NaOH at 65 °C for 24 h) to reveal the sub-nano regular pattern. The process chain also involved machining of rectangular and circular geometries by micro milling with an end mill with a diameter equal to 100 µm on which typical structures from the anodization of aluminum and subsequent nickel electroplating were produced (Figure 4.4).

Figure 4.4 SEM pictures of nickel patterned geometries; characteristic periodic nano structures on vertical walls are shown

■ 4.3 Replication Assessment of Polymer Surfaces at the Sub-Micrometer Scale

Quantification of the degree of replication becomes a key issue that drives the use of metrology as the enabling technology for product development and manufacture. Ultra-precision manufacturing requires instruments able to quantify quantities of selected geometries with a resolution in the sub-nanometer range, challenging the performance limit of existing instruments.

Technology is continuously evolving. In the last fifty years, the mechanical field, as well as electronics, medicine, telecommunications, and IT, has seen a continuous development. All these fields are correlated: firstly, the development of one is not independent of improvements in other fields; secondly, there is at least one field that can be considered both complementary and enabling to all fields: metrology. Metrology, in general, is traditionally regarded as the key discipline in making the industrial manufacture of components possible. In particular, it enables process control on the basis of measurands, either defined on the components or on some specific process characteristics. In this way, a product is described using absolute values combined with tolerances. Dimensional metrology covers the measurement of dimensions and, in principle, geometries based on distance measurements. In general, the importance of metrology is proportional to how far it helps other technologies to improve. Indeed, global competition, together with market needs, is the perfect engine able to push technology toward new frontiers.

As a consequence of the key role of metrology in precision manufacturing, improvements in metrological instrumentation have facilitated significant advances in micro- and nanotechnologies. However, the resolution and accuracy of current instruments are being stretched to their limits by the current needs of researchers as well as manufacturing processes in the immediate future. A natural trend is to focus the attention on surfaces. The interface of solid states at the micro and sub-micrometer dimensional scale plays a very important role, being one of the most significant aspects to affect surface properties (e.g., electrical, optical, mechanical properties, etc.). Understanding surfaces is a prerequisite for understanding their properties: for this reason, instruments allowing surface characterization play a fundamental role at this moment in the nanotechnology evolution [36].

The increasing complexity of parts and the number of critical features being integrated into the components being produced require 100%-part inspection with improved data fusion from different instruments [37]. Reference tactile methods (for example, atomic force microscopy) have to be interfaced with optical microscopy techniques (for example, optical interference, confocal microscopy, white light interferometry) and vice versa to overcome the limitations of both measuring principles. This enables the capability of performing traceable measurements of special regions of interest corresponding to inaccessible regions, in proximity to steep slopes, narrow cavities, undercuts, or hidden parts. The capability of performing traceable multiparametric metrology will finally convert experimental new technology, based on deterministic or functionalized surfaces, into controllable manufacturing production processes. Besides all these considerations, traceable measurements results and uncertainty estimation analyses should be directly linked to product dimensional miniaturization and proportionally developed with the production processes to assess product conformity within tolerance ranges.

All areas of measurement technology are being combined with computer integrated manufacturing and data management systems for information-based process control for which on-line software calibration is required. Although integrated metrology still needs a universal definition, it has become the term associated with the slow migration from off-line to in-line and in-situ measurements. The proper combination of off-line, in-line, and in-situ measurements will enable advanced process control and rapid yield learning [38].

4.3.1 Dimensional Evaluation of Replication Fidelity

The calibration of measuring instruments is indeed required to define measurements results as traceable measurements to the SI. Traceability is the property of the result of a measurement whereby it can be related to stated references, usually national or international standards, through a documented unbroken chain of comparisons all having stated uncertainties [39]. It is therefore of fundamental importance as stated by the definition of traceability that measuring uncertainty has to be expressed in order to allow comparison of different measuring results, ensuring the traceability to the SI. One of the main problems in dimensional nano metrology is derived from the fact that for undetermined or non-regular geometries, this traceability chain cannot be closed due to the lack of relevant measurands or transfer artifacts able to characterize the nano objects produced.

Measurements results are always subject to errors, hence, the expression of the measurements' true value is, in most of the cases, an estimation of an average value from which different considerations of the instrument and the measurand are made. Typical contributors to the resulting overall measurement uncertainty are:

- Accuracy: closeness of agreement between a measured quantity value and a true quantity value of a measurand [39]. Accuracy therefore correlates the mean of repeated measurements to the true value, closeness to the true value.
- Precision: closeness of agreement between indications or measured quantity values obtained by replicate measurements on the same or similar objects under specified conditions [39]. The precision is related to the measuring instruments

and its repeatability evaluated as the dispersion of results after repeated measurements.

- Resolution: smallest change in a quantity being measured that causes a perceptible change in the corresponding indication [39]. The resolution is related to the capability of distinguishing between adjacent values of the indicated quantity.
- Error: measured quantity value minus reference quantity value [39]. The error is quantified as the difference from the calibrated true value of the artifact and the measured one. Errors can be distinguished at random, caused mainly by environmental conditions, and systematic errors that give a constant (i. e., repeatable) deviation from the true value.

4.3.2 Applications of Profile Measurements

The replication quality of deterministic structures intended as features to which a standard definition of known geometrical shape can be applied, i. e., different aspect ratio rectangular shapes representing different channels of lab-on-a-chip design, is presented. Considering possible implementation in a mass production environment, the precision of measuring results and the accuracy of measurement relocation are very relevant. Focus is given to the quality control of the produced polymer parts to quantify the replication degree associated with the measurands on the master and on the replicated parts.

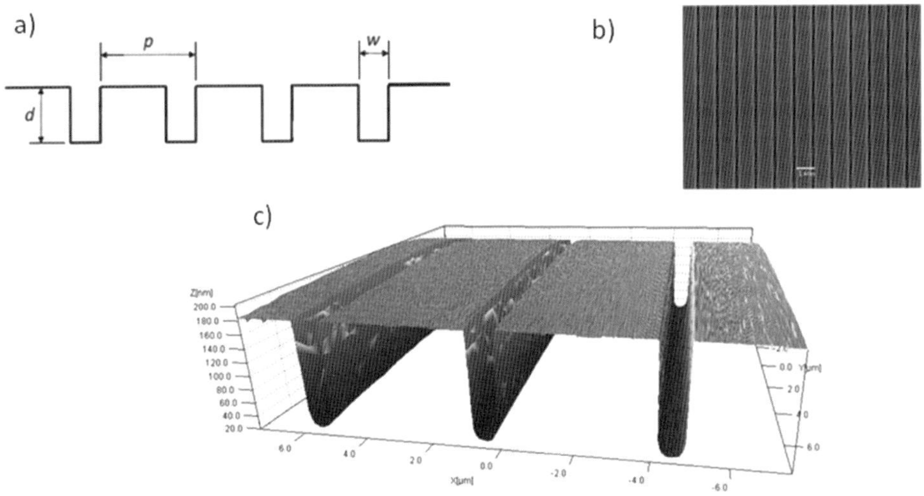

Figure 4.5 (a) This physical measurement standard repeats rectangular grooves along one direction. The shape is defined by the groove width *w*, the shape wavelength *p*, and the groove depth *d*; (b) manufactured silicon nano grating; (c) corresponding AFM scanned profile

Eq. (4.1) shows the H_{dev} of the produced polymer parts with respect to the nickel master protrusions. The analysis based on this geometrical dimension, H_{dev}, allowed the introduction of less uncertainty on the measuring strategy and a more realistic understanding of the replication insert-polymer quality.

$$H_{dev} = \left(|Polymer\,channel\,depth| - |Nickel\,channel\,height| \right) \tag{4.1}$$

The height, δ_z, of a groove or line is determined in the following way: let x_{lej} and x_{rej} be the x coordinate for the left and right edge of the profile with number j, ($j \in [1,m]$). This means that the width, w_j, of the step height (or groove) on line j is given by $w_j = (x_{rej} - x_{lej})$. On top of the step height (or in the bottom of the groove) and to the left and right of the step height, the flat portion A_j, B_j, and C_j of the profile with line number j can be defined as indicated in Figure 4.6 [40].

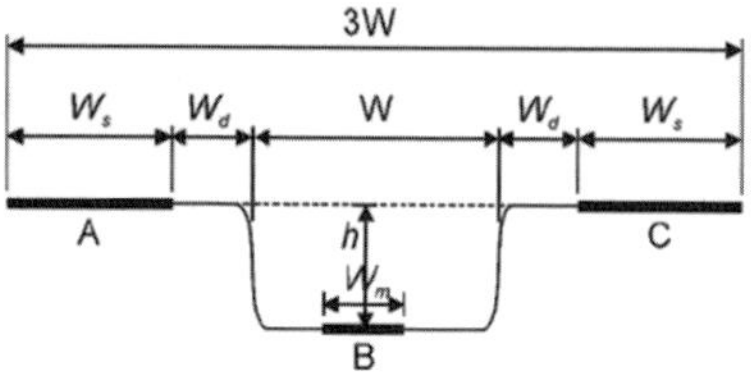

Figure 4.6 The definition of groove depth, *h*, in analogy to ISO standard 5436 [41]

The analysis based on Eq. (4.1) allows for product compliance assessment with specifications of dimensions on the molded components selected after the process optimization from the combination of the process parameters giving the highest replication conformance compared to the dimensions of the nickel master geometries. The quality control system capability in terms of uncertainty with respect to tolerance given as a limit value for functional design to ensure that the measurement uncertainty is sufficiently small to validate the tolerance limits should be verified beforehand.

4.3.3 Applications of Amplitude and Slope Replication of Polymer Surfaces

Surface replication of the tool insert component when molding the polymer melt is essential in advanced micro injection molding technology. The surface texture replication can be investigated using amplitude (Sa, Sq) and the slope (Sdq) surface roughness parameters [42], comparing the produced parts with the tool used to replicate the geometry and evaluating the measurement uncertainty.

In this context, the replication quality of molded surfaces using metal masters with different types of finish was investigated. Four types of surface finish were considered (Figure 4.7). Both master and replicated surfaces were measured using a laser scanning confocal microscope. Hence, the replication fidelity was evaluated comparing the measurements of the polymer surfaces against the ones of the masters. The amplitude and the slope replications were considered, calculating respectively Sq and Sdq areal surface texture parameters.

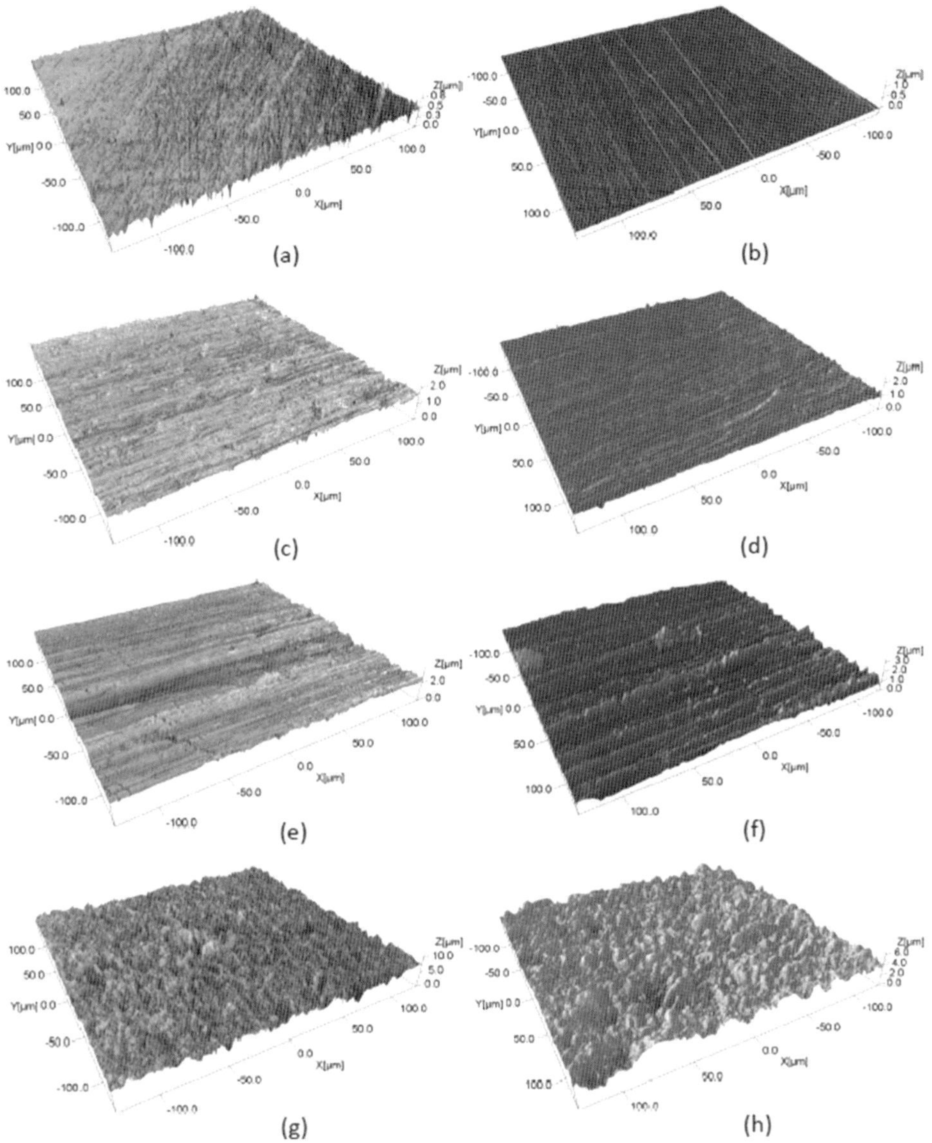

Figure 4.7 Examples of acquired surfaces. (a) Diamond buff polished steel master. (b) Injection molded replica of diamond buff polished steel master. (c) Grit paper polished steel master. (d) Injection molded replica of grit paper steel polished master. (e) Stone polished steel master. (f) Injection molded replica of stone polished steel master. (g) Dry blast polished steel master. (h) Injection molded replica of dry blast polished steel master. Molded material: ABS

The replication fidelity is considered for amplitude parameters as the complement to 100% of the relative deviation replicas/references and evaluated according to Eq. (4.2) and (4.3).

$$f_R = \frac{Sq_{poly}}{Sq_{ref}} * 100$$
$$f_R = \frac{Sdq_{poly}}{Sdq_{ref}} * 100 \quad (4.2)$$

Sq_{poly} and Sdq_{poly} are Sq and Sdq calculated on the replicated specimens (polymer specimens). Sq_{ref} and Sdq_{ref} are the ones calculated on the nickel standards (references).

The replication fidelity was calculated according to Eq. (4.2). The standard uncertainty of the replication fidelity was evaluated propagating the expanded uncertainties of masters and replicas, considering uncorrelated quantities, i.e., according to [39]:

$$u_{f_R} = \left(\sum_j \frac{\partial f_R^2}{\partial sx_j} u_{sx_j}^2 \right)^{\frac{1}{2}} \quad (4.3)$$

where f_R is given in Eq. (4.2), sx stands for Sq and Sdq, and sx_j stands for $sx_{replicas}$ and sx_{master}. Furthermore, the standard uncertainties of masters and replicas $u_{sx_j}^2$ were evaluated according to the procedure specified in Chapter 10, Section 10.5.1 (see also Figure 4.8). The method evaluates two groups of uncertainty contributors: the first one measuring calibrated artifacts and the second one measuring components under measurement.

A good amplitude replication was achieved for stone polished surfaces with a replication fidelity larger than 95%. The dry blast ones were evaluated with an amplitude replication fidelity of about 70%. The worst amplitude replication was achieved for both the diamond buff and grit paper polished surfaces with a replication fidelity of, respectively, 47% and 56%.

The tendency is almost the same for slope replication but the replication fidelity values are lower: 67% for stone polished surfaces, 50% for dry blast and grit paper polished surfaces, and 33% for diamond buff polished surfaces [43].

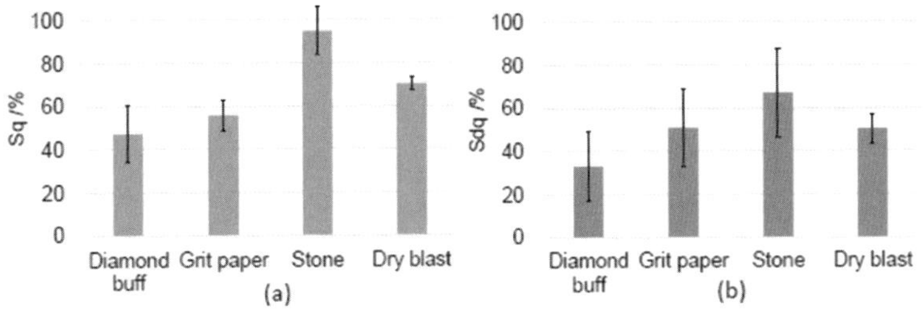

Figure 4.8 Replication fidelity of amplitude (a) and slope (b) for the different surfaces considered. Error bars indicate expanded uncertainty of the replication fidelity. Molded material: ABS

4.3.4 Applications of Areal Parameters

The highly complex scenario presented by the characterization, repositioning, and subsequent quantification of the measuring uncertainty of measurements of nano-patterned surfaces requires the correct implementation of measuring procedures. Two experimental approaches for the characterization of nanostructured surfaces with high geometrical complexity are described in the following.

The first approach [44] (Figure 4.9) is based on evaluation over an AFM-scanned area of 15 µm × 15 µm. The height variations of 50 randomly selected structures produced by aluminum anodization and subsequent nickel electroplating with respect to a reference plane obtained by correcting the global image with a first order least mean square fit algorithm were measured. Mean values of the average height were equal to (72 ± 40) nm. Moreover, apex curvatures of 50 more randomly chosen nickel bumps were fitted and the corresponding radii estimated. A coefficient of variation (standard deviation/average feature value) for an average radius value of 330 nm, describing the radius of curvature distribution, equal to 19%, was calculated.

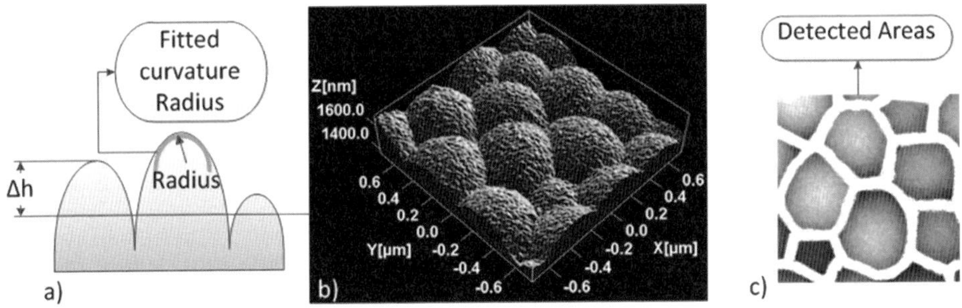

Figure 4.9 (a) Measurement technique to estimate the height variation and radius of the fitted curvature of the features; (b) example of a 1.5 µm × 1.5 µm nickel scanned surface; (c) texture cells detected on the nickel substrates used to calculate the average area [44]

Watershed segmentation allows the detection of the texture cells and their characterization through so-called features parameters [42]. In particular, an average area value of (0.27 ± 0.1) μm^2 was calculated above the same evaluation area of 15 µm × 15 µm on the nickel shim.

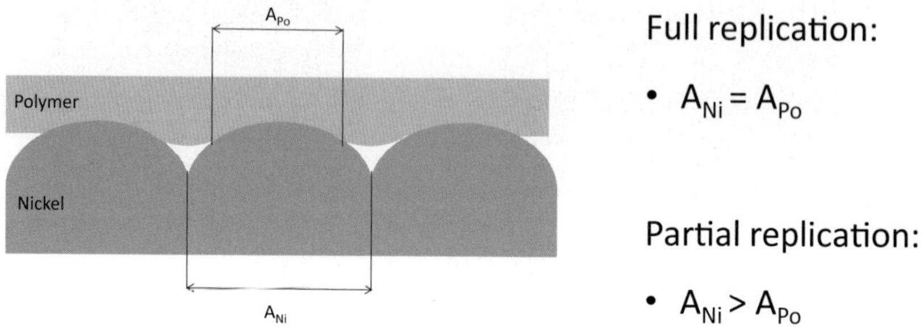

Figure 4.10 Degree of polymer replication, conceptual layout

Particle detection methods [45] can be used as measuring tools to detect and quantify the features' top surface area. Based on the hypothesis depicted in Figure 4.10 that full replication is given by $A_{Ni} = A_{Po}$ and partial features replication by $A_{Ni} > A_{po}$, where A_{Ni} is the detected top feature area of the nickel features and A_{Po} the corresponding replicated polymer area, different intercomparisons were carried out.

Finally, the graph in Figure 4.11 was used to compare AFM-calibrated measurements on nickel and the polymer to understand if the proposed method gives a graphical indication of the degree of replication between the master and replicated geometries for further implementation as a fast detection tool for compact nanostructured surfaces. The gap in between the two main populations of data indicates partial replication of polymer features compared to the original master geometries. The results confirm a good agreement in predicting the real physical behavior of replication.

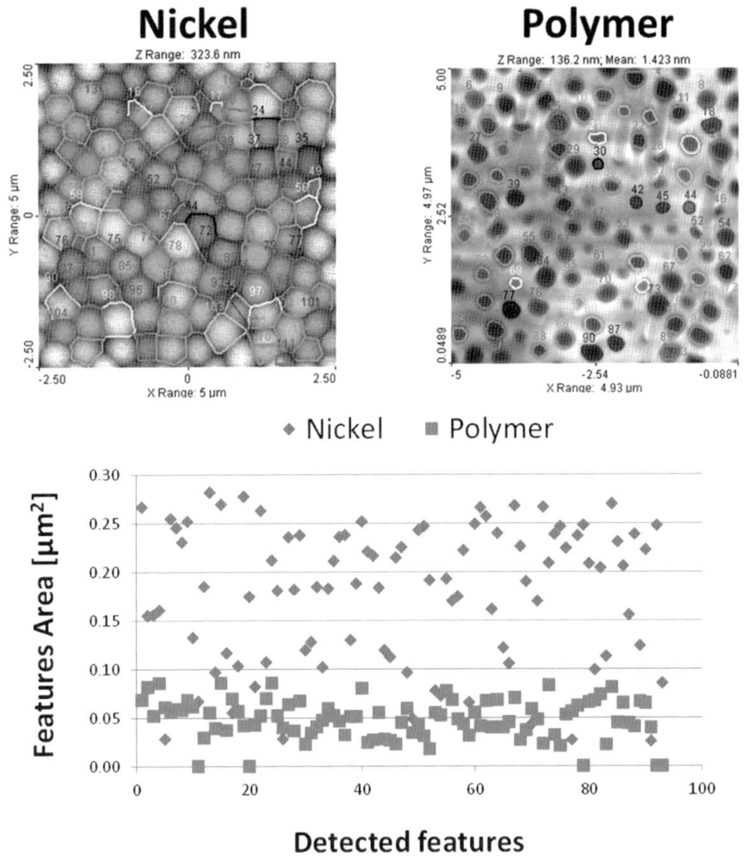

Figure 4.11 Detected nickel and polymer replicated features' dispersion using AFM. Evaluation area: 5 µm × 5 µm. The figure is presented in the color supplement

4.3.5 Application of Angular Intensity Distribution Measurements

This section introduces some of the latest developments of new techniques focused on in-line characterization of large nanostructured surfaces over large areas. Calibrated AFM measurements were used to develop a model based on scalar diffraction theory [46, 47], which is able to calculate the expected nickel and transparent polymer substrates' angular distribution of reflected and transmitted intensity, respectively (Figure 4.12). When light passes through the structure, its phase is modified. The far field is then calculated using this phase information together with a modified Rayleigh-Sommerfeld integral. This method requires only measured surface profiles, which are obtained from AFM scans, and the refractive indices of the materials as described in the literature [48].

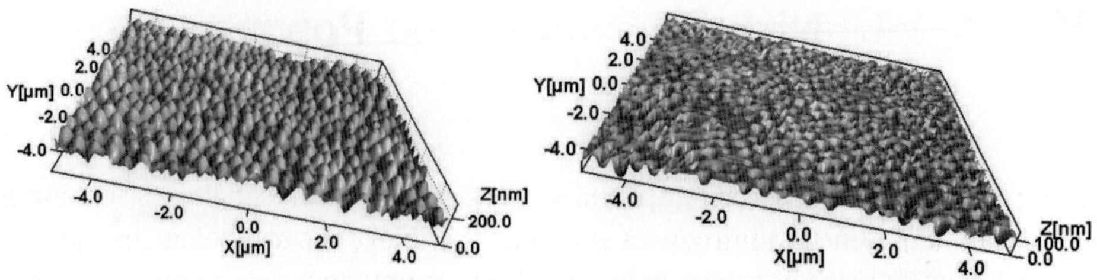

Figure 4.12 AFM imaging. Left: distributed pattern on the nickel tool. Right: distributed pattern molded on the polymer

The need to compare the scattered light of the distributed polymer features with a full ideal replication reference value made the analytical calculation of a calibrated pattern template necessary. The procedure developed involves carrying out an AFM-calibrated measurement on nickel master geometries. Then, the AFM image is inverted by the software and the surface pattern of the inverted image is calculated in order to obtain the reference value quantifying the ideal full replicated pattern. The different intensities of the two Gaussian curves (Figure 4.13) indicate the difference in optical responses between real and reference distributed structures. The difference is due to the different degree of replication of the compared structures that theoretically would allow quality assessment of produced structures [49].

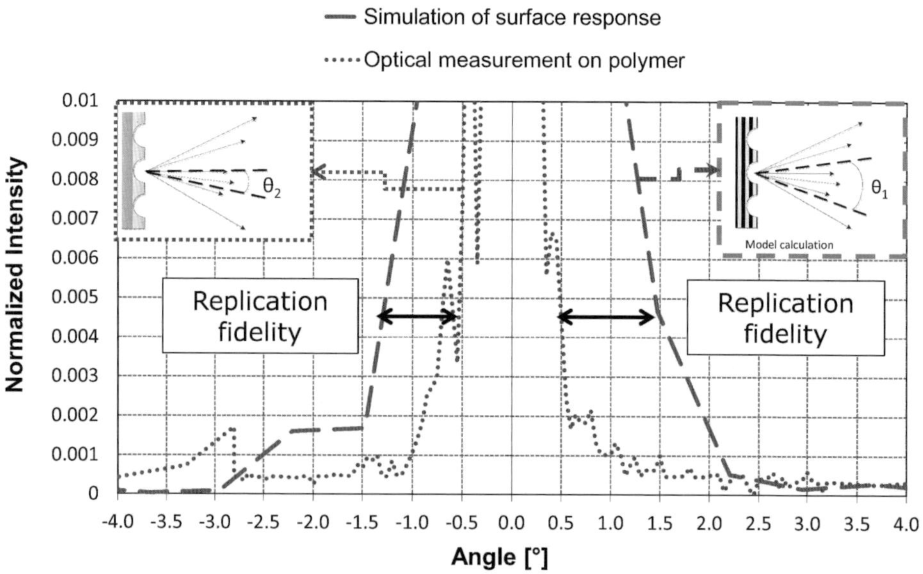

Figure 4.13 Inverted Ni structure from a simulated/measured (AFM) nickel structure obtained from anodization of Al 99.999% (theoretical full replication) compared with optical measurements carried out on a cyclic olefin copolymer (COC) substrate molded from the original nickel shim produced from a pure (99.999%) aluminum substrate

4.4 Influence of Tooling and Process Parameters on Replication of Microstructures

The possibility of processing polymer materials with technologies enabling large-volume production introduces solutions to remove the technology barrier between lab-scale proof-of-principle and the high-volume low-cost production of micro- and nano-technology-based products. In the previous sections, methods and approaches to process chain validation for final polymer micro and nano structure replication have been presented. In the following sections, the replication fidelity between different process chain steps and their dependency on process variation, process conditions, tool accuracy, material behavior, and features' geometries are considered and quantified.

4.4.1 Replication and Optimization of Deterministic Structures

The replication fidelity of deterministic geometries over the entire process chain can be quantified through traceable measurements on designed fingerprint test structures. To assess the critical factors affecting the replication quality throughout the different steps of the proposed process chain, test geometries are designed and produced on the side of the functional features. The so-called fingerprints of the lithography and molding processes are evaluated using atomic force microscopy. The entire process chain is therefore characterized and the degree of replication among the different replication steps quantified with precise measurements using a high accuracy relocation technique on the produced key test geometries [50].

The master geometries used in the presented study were fabricated as described in Section 4.2.1. Test geometries were composed of three crosses with a constant height (nominal design value equal to 62 nm) with horizontal and vertical wings with widths of 10 µm, 2 µm, and 500 nm, respectively. The crosses were positioned in the center of the produced 22 mm × 22 mm nickel insert, which was used for the injection molding of approximately 1000 parts at the end of the experimental test procedure.

Micro/nano crosses polymer replicas were produced by injection molding (Figure 4.14) using a commercially available cyclic olefin copolymer (COC), Topas 6013.

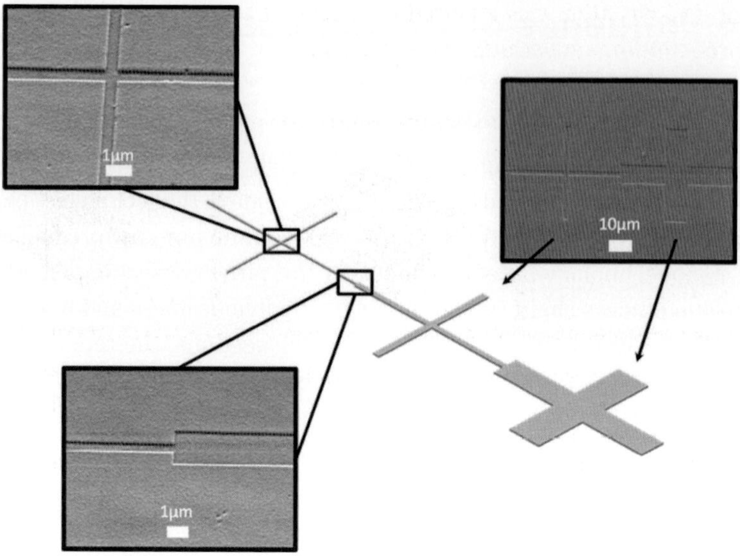

Figure 4.14 Scanning electron microscopy images of a COC molded part. The illustration shows the corresponding polymer replica details of the produced nickel micro/nano crosses

To evaluate the influence of process parameters on the replication of nano-scale structured surfaces, a statistically designed set of experiments was carried out [50]. The investigation was aimed at quantifying and optimizing the replication quality from the master geometry to the thermoplastic substrate on the parallel and perpendicular relative directions of the different cross wings with respect to the polymer flow.

At this dimensional scale, variations of the degree of replication due to replication process conditions, tool accuracy, material behavior, and feature geometry are in the order of a few micrometers down to a few tens of nanometers [3]. For this reason, high resolution AFM measurements are employed to assess and quantify the replication quality of the produced micro-nano crosses over the different substrates. Replication quality as a variation of H_{dev} (Eq. (4.1)) during the entire experimental plan is used as the output of the statistical analysis.

The results show the influence of both process parameters (mold temperature, packing time, packing pressure) and design parameters (channel width and direction with respect to the polymer flow) on the replication fidelity of the micro/sub-μm channels, providing effective guidelines for process and part design of injection molded microfluidic systems. It has been verified that a combination of high mold temperature and longer packing time improve the replication fidelity from the master to the polymer part. Moreover, results from the design of experiments have shown an improved process capability on replicating cross wings parallel to the polymer flow direction.

Measurement relocation was ensured by measuring with AFM the test structures on the three different substrates (Figure 4.15).

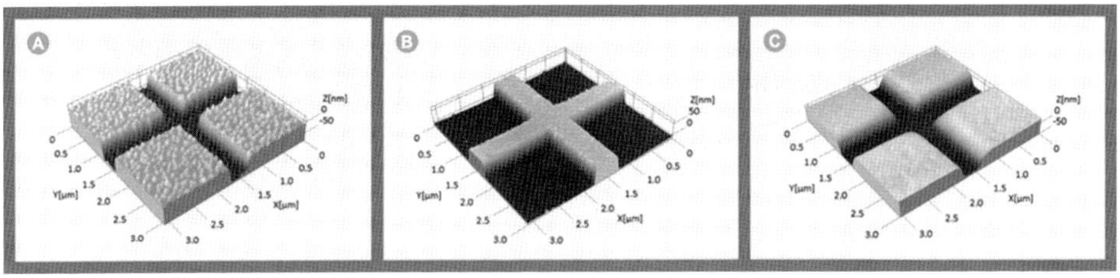

Figure 4.15 Process chain validation for sub-micro channel (width = 500 nm, depth = 60 nm) manufacture: (A) e-beam lithography on silicon, (B) nickel electroplating, (C) replication by micro injection molding (material: cyclic olefin copolymer, COC, TOPAS r 6013). Measurements by calibrated AFM measurements with uncertainty and relocation accuracy at nanometer level (magnification: X 1 ×, Y 1 ×, Z 5 ×)

A quantitative evaluation of the replication fidelity was performed considering the verification of dimensional tolerances with particular attention to the challenge of obtaining low measuring uncertainties [50].

An investigation into the crosses wings with different aspect ratios (1/161; 1/32; 1/8) shows the replication quality within the tolerance verification range. The average electroplated nickel tranches height among the three produced crosses is 98% of the average value of the corresponding etched grooves in the silicon wafer. The investigation helps, also, in understanding the process capability in terms of replication fidelity variation on the polymer substrate. The largest depth deviation was measured for cross number 3 with the smallest nominal width dimension of 500 nm. For an average measured polymer channel depth corresponding to 94% replication of the corresponding nickel trenches, a standard deviation of 4% among the smallest polymer crosses' depth was assessed.

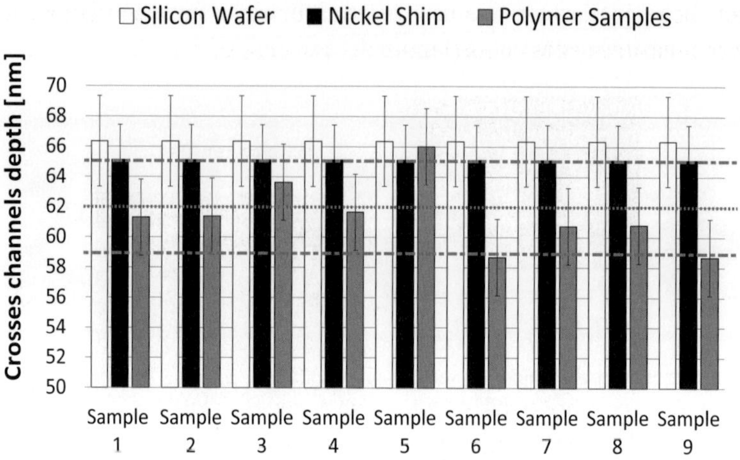

Figure 4.16 Quality control results of micro injection molded cross with nominal width of 500 nm. The square dot line represents the target design dimension of the crosses' height; the dash dot lines represent the tolerance range (as ±5% of the crosses' target design height). Histogram columns represent the channel depths measured on the silicon wafer (U = 3.0 nm), the nickel insert (U = 2.3 nm), and the polymer substrate (U = 2.5 nm). Error bars on the histogram columns indicate expanded combined uncertainty U (k = 2, conf. level 95%)

4.4.2 Replication Quality of Large Area Nano-Structured Surfaces

The establishment of periodic nanometer features can affect the physical (e.g., wetting behavior, hydrophobic/hydrophilic/self-cleaning properties) and optical properties (e.g., reflectivity) of a surface.

In order to understand the capability of micro injection molding on replicating periodical nanometer structures, the nickel insert produced as described in Section 4.2.2 was used. Three different thermoplastic materials were molded during the experimental phase (COC, PC, and PP).

In this example, the structured nickel insert area was 80 mm × 30 mm, and the typical single feature critical dimensions were, approximately, a diameter equal to 500 nm and height of 60 nm. Due to the extremely high number of nano features, a method based on the exact measuring repositioning able to compare measurements of corresponding features on the master and polymer surfaces could not be applied. Following the material supplier specifications, different process parameters were chosen to produce parts with different polymer materials. At the end of the experimental phase, in order to qualitatively characterize the achieved replication fidelity on the different polymer substrates, SEM pictures were captured. Images were taken in the same positions in an attempt to compare the polymer's properties with the replication obtained. On the sidewall and the bottom position of the fabricated micro-rectangular geometry, images were recorded to understand replication behaviors which are otherwise difficult to characterize with conventional AFM measurements or conventional optical systems.

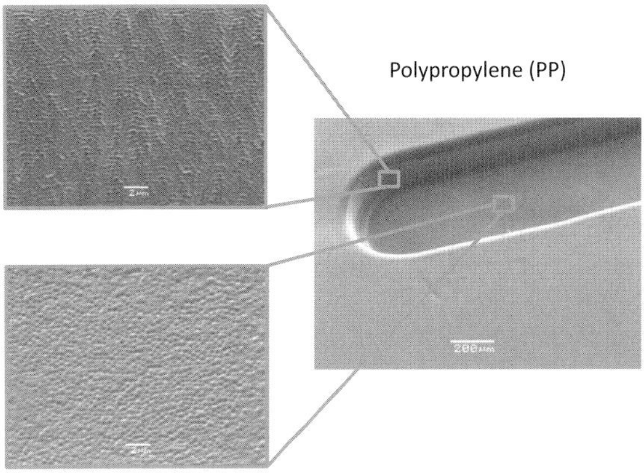

Figure 4.17 SEM pictures of replicated polymer surfaces

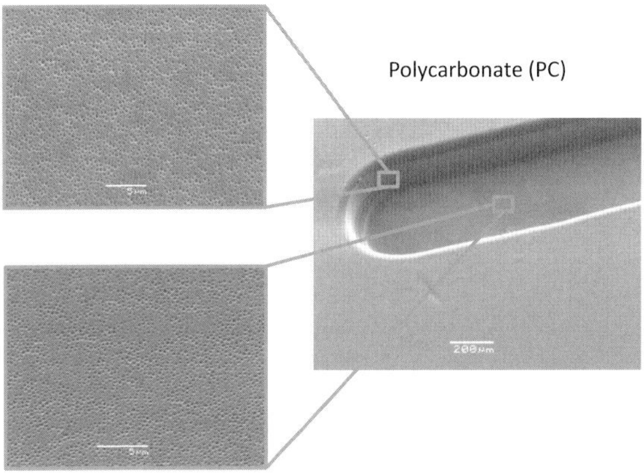

Figure 4.18 SEM pictures of replicated polymer surfaces

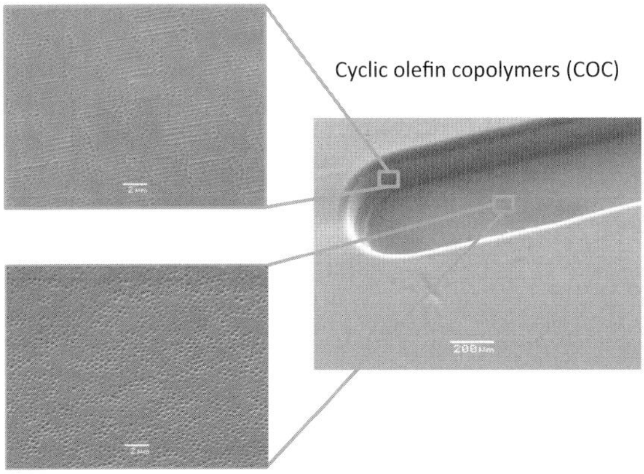

Figure 4.19 SEM pictures of replicated polymer surfaces

In Figure 4.17, the pictures show master structures density fully replicated in the polymer substrate after injection molding with a tendency of over-molding effects between pore boundaries. On the sidewall, high structure replication fidelity is observed with the effects of reciprocal sliding between the polymer part and nickel master geometries less clear for single features, but still present in the form of a wave effect along the entire wall.

Figure 4.18 shows a partial replication in terms of countable features on the bottom of the rectangular geometry. The effect of different processes is observed for PC on the structure's sidewall with a similar effect observed for COC. A small sliding effect of the features is visible on the vertical walls. The observed sliding effect is generated during the ejection phase when the polymer detaches from the nickel master stuctures.

Figure 4.19 shows the semi-sphere pattern replicated on the COC substrate. The images on the bottom of the structure indicate that full replication is not achieved. The image on the sidewall gives an impression of low polymer elasticity during part ejection indicated by scratch-like structures departing from every single replicated semi-sphere.

4.4.3 Influence of Process Parameters on Surface Replication

The present section highlights general guidelines to enhance surface replication. Influences of the main process parameters affecting replication of surface texture and geometries during injection molding of micro and nano components are described.

The state-of-the-art in replication of micro and nano surface geometries [3] has highlighted the importance of process monitoring, analysis, and optimization. By means of a specific set of activities, extensive research has been carried out to control, understand, and correlate the process conditions with respect to the material/part/process factor characteristics with the resulting properties of micro components. Different research groups have shown that, especially for micro- and nanostructures, mold temperature, melt temperature, injection velocity, injection pressure, and ejection temperature are important for the replication quality [51–54]. Temperature and pressure are such factors that can have a direct impact on dimensional stability. The injection pressure control, expressed in terms of both absolute value and repeatability (i.e., standard deviation), is fundamental for optimal part production and it is the critical process parameter for precision molding of high accuracy thermoplastic parts as well as for thermoplastic elastomers. An excessive value of injection pressure will lead to defects such as flash, whereas a large value of the standard deviation of the pressure indicates a poor cycle-to-cycle process repeatability (i.e., different filling conditions), causing different properties in the molded parts [55, 56].

Generally, parameters can be distinguished between: filling phase factors (melt temperature, mold temperature, injection speed) [57–59] and post-filling factors (packing pressure, packing time, cooling time, demolding forces) [60, 61].

In particular:

- High temperatures of melt, consistent with the polymer properties to avoid degradation, ensure lower viscosity, lower cavity pressure drop, and the ability of the melt to flow into micro to sub-micro geometries.
- High temperatures of the mold (T_{mold}), close to or above the glass transition temperature (T_g) of the polymer, are required to ensure high replication fidelity. Mold temperatures higher than T_g can ensure full replication because premature solidification is prevented. However, in this case, a variothermal process might be required in order to reduce the temperature after the cavity is filled in order to solidify the polymer so that the part can be demolded and dimensional stability is preserved. High T_{mold} is the main factor that allows minimization of the size of surface defects, such as weld lines, by enhanced surface replication. Mold temperatures above T_g (using thermoplastic cyclic olefin copolymer, COC) in combination with surface coating of the nickel molds greatly enhanced the quality of replicated features with 40-nm-wide and 107-nm-high pillars on a 200-nm pitch [62].
- High injection speeds reduce the injection time, preventing premature freezing of the polymer and also promote shear heating, which can locally decrease the viscosity and hence improve surface replication. On the other hand, high injection causes high shear stresses that can exceed the maximum shear stress allowed by the polymer before its degradation; it produces high injection pressure that can cause high internal stresses (detrimental for, e.g., micro optical parts) and could also damage high aspect ratio features of the mold itself.
- Higher packing pressures and longer packing times allow compensation for polymer shrinkage to preserve the geometries replicated during filling also after demolding. Shrinkage is significant at the micro- to nano-dimensional scale, especially for large injection molded components with structured surfaces. Especially at the sub-µm feature dimensional scale, non-fully-compensated shrinkage can compromise complete surface replication when the gate freezing time is shorter than the packing time needed. In these cases, an injection compression molding process should be employed instead of injection molding. The effect of injection molding pressure on the replication of a surface microstructure [63] has shown that a higher pressure forces the polymer to advance further into the microfeatures. For a low to moderate range of pressure, replication fidelity will increase. The deformation during the replication phase is partially viscous and partially elastic. As the holding pressure is released, the elastic deformation will recoil. The phenomenon is more evident if the holding pressure is set to high values. Replication as a function of holding pressures drops after an optimal

value for which the elastic shrinkage becomes significant with regards to the microstructure's critical dimensions. The side effect of applying high pressure during the replication process is reflected in mold deflection during injection molding [64]. Decreasing these deflections is essential for molding parts with tight tolerances and high aspect ratio geometries (e. g., thin walls).

Geometrical parameters are highly important critical factors for the optimized molding of micro/nano features:

- The spacing between features does not affect the surface replication fidelity in the sub-mm/100 µm range when the pitch equals the vertical dimension of features [58];
- The process capability is challenged for features with both vertical and lateral dimensions in the sub-µm/100 nm range;
- Further miniaturization in the sub-100 nm range for both vertical and lateral dimensions has not been reached with micro/nano injection molding processing, but rather only by nano imprint lithography [3] (see Figure 4.1).

The replication factor attainable by polymer injection molding depends also on the vertical dimension of the features to be replicated. Experiments carried out on injection molding of polypropylene showed that:

- It was possible to achieve 100% replication of features with vertical dimensions in the 100 µm range.
- On the other hand, with the same processing conditions, the replication factor decreased to 80% and 40% when the roughness of the master geometry had vertical dimensions in the range of 1 µm down to 100 nm, respectively [57].
- In addition, the inherent resolution of micro injection molding of polypropylene was demonstrated in [65], where an optical mold surface (Ra = 2 nm) could generate a polymer surface with a vertical range of 80 to 160 nm, depending on processing conditions, due to surface polymer crystallization effects.

The effects of these factors on the process replication capabilities are generally not consistent for different polymer materials, and could be adverse under specific conditions [58]. It is therefore not possible to define a unique process set-up that accommodates different materials and eventually different injection molding and micro injection molding machines.

A robust approach based on statistical analysis and dimensional metrology for different case studies should be performed. Optimization of the surfaces of micro/nano geometries is based on measurements on the µm/sub-µm scale. High accuracy measurements are needed for an effective process optimization: uncertainties in the range of 1–10% of the measurement results should be obtained. The uncertainty should be compared with the output variations caused by process parameter changes, also taking into account the process repeatability. The statistical significance of process parameters depends on the ratio between process-induced variation and the associated uncertainty, including both measurement calibration and process repeatability [3, 57, 66].

References

[1] Bigot, S., Minev, T., Dobrev, T., Dimov, S., Matthews, C., Function and length scale integration in micro/nano manufacturing. *Proceedings of the 2nd International Conference on Multi-Material Micro Manufacturing 4M* (2009), pp. 147–152

[2] Tosello, G., Calaon, M., Gavillet, J., Tang, P.T., Ravn, C., Hansen, H.N., Surface nano structures manufacture using batch chemical processing methods for tooling applications. *Proceedings of the 4th International Conference on Multi-Material Micro Manufacture 4M* (2011), p. 4

[3] Hansen, H.N., Hocken, R.J., Tosello, G., Replication of micro and nano surfaces technologies. *CIRP Ann.* (2011) 60(2), pp. 95–714

[4] Angelov, A.K., Coulter, J.-P., A Feasibility Study for Sub-100 nm Polymer Injection Molding. *Proceedings of the Society of Plastic Engineers (SPE)-65th Annual Technical Conference (ANTEC)* (2007), pp. 2987–2991

[5] Chang, H.C., Young, W.B., Experimental Study on the Filling of a Micro Injection Moulding with Cylindrical Dot Patterns. *Int. Polym. Process.* (2005) 20(3), pp. 245–249

[6] Gadegaard, N., Mosler, S., Larsen, N.B., Biomimetic Polymer Nanostructures by Injection Molding. *Macromol. Mater. Eng.* (2003) 288(1), pp. 76–83

[7] Griffiths, C.A., Dimov, S.S., Brousseau, E.B., Hoyle, R.T., The Effects of Tool Surface Quality in Micro-Injection Moulding. *J. Mater. Process. Technol.* (2007) 189, pp. 418–427

[8] Heckele, M., Schomburg, W.K., Review on Micro Molding of Thermoplastic Polymers. *J. Micromech. Microeng.* (2004) 14, pp. R1–R14

[9] Palanisamy, P., Yoon, S.-H., Lee, J.S., Mead, J., Barry, C.M.F., Lee, K.-H., Cha, N.G., Enhancement of Surface Replication by Micro-Gas-Injection Molding. *Society of Plastic Engineers (SPE)-67th Annual Technical Conference (ANTEC)* (2009) 3, pp. 1316–1322

[10] Scholz, S., Griffiths, C.A., Dimov, S.S., Brousseau, E.B., Lalev, G., Petkov, P., New Process Chains for Replicating Micro and Nano Structured Surfaces with Biomimetic Applications. *Society of Plastic Engineers (SPE)-67th Annual Technical Conference (ANTEC)* (2009), pp. 3021–3027

[11] Su, Y.-C., Shah, J., Lin, L., Implementation and Analysis of Polymeric Microstructure Replication by Micro Injection Molding. *J. Micromech. Microeng.* (2004) 14(3), pp. 415–422

[12] Yang, C., Li, L., Huang, H., Castro, J.M., Yi, A.Y., Replication Characterization of Microribs Fabricated by Combining Ultraprecision Machining and Microinjection Molding. *Polym. Eng. Sci.* (2010) 50(10), pp. 2021–2030

[13] Yoon, S.-H., Cha, N.-G., Lee, J.S., Park, J.G., Carter, D.J., Mead, J.L., Barry, C.M.F., Effect of Processing Parameters, Antistiction Coatings and Polymer Type When Injection Molding Microfeatures. *Polym. Eng. Sci.* (2010), pp. 411–419

[14] Giboz, U., Copponnex T., Mélé, P., Microinjection molding of thermoplastic polymers: a review. *Journal of Micromechanics and Microengineering* (2007) 17 (6), R96–R109

[15] Calaon, M., Hansen, H.N., Tosello, G., Garnaes, J., Norregard, J., Wei, L., Process variations in surface nano geometries manufacture on large area substrates. *Proceedings of the 14th euspen International Conference* (2014) 2, pp. 207–210

[16] Calaon, M., Tosello, G., Elsborg, R., Hansen, H.N., High accuracy and precision micro injection moulding of thermoplastic elastomers micro ring production. *Proceedings of the 4th International Conference on Multi-Material Micro Manufacture 4M* (2016), p. 4

[17] Schift, H., David, C., Gobrecht D´Amore, A., Simoneta, D., Kaiser, W., Gabriel, M., Quantitative analysis of the molding of nanostructures. *J. Vaccum Sci. Technol., B* (2000) 18, p. 3564

[18] Calaon, M., Hansen, H. N., Tosello, G., Garnaes, J., Ravn, C., Tang, P. T., Characterization of large area nanostructured surfaces using afm measurements. *Proceedings of the 11th euspen International Conference* (2011), pp. 37–40

[19] Calaon, M., Tosello, G., Hansen, H. N., Norregard J., Influence of process parameters on edge replication quality of lab-on-a-chip microfluidic systems geometries. *Proceedings of the ANTEC conference* (2013), p. 5

[20] Knapp, W., Tolerance and uncertainty. *Laser metrology and machine performance* (2001), pp. 357–366

[21] Kunzmann, H., Pfeifer, T., Schmitt, R., Schwenke, H., Weckenmann, A., Productive metrology – adding value to manufacture. *CIRP Ann.* (2006) 55(2), pp. 643–666

[22] Weckenman, A., Ernst, R., Studies on new tolerancing rules for micro and nanotechnology. *1st euspen conference* (2000) 1, pp. 214–221

[23] Commission of the European Communities, Preparing for our future: Developing a common strategy for key enabling technologies in the EU. Technical report, Commission Staff Working Document SEC (2009)

[24] Backer, E. W., Ehrfeld, W., Münchmeyer, D., Betz, H., Heuberger, A., Pongratz, S., Glashauser, W., Michel, H. J., Siemens, R. v., Production of separation-nozzle systems for uranium enrichment by a combination of x-ray-lithography and galvanoplastic. *Naturwissenschaften* (1982) 69

[25] Tanzi, S., Ostergaard, P. F., Matteucci, M., Christensen, T. L., Chech, J., Marie, R., Taboryski, R., Fabrication of combined-scale nano- and microfluidic polymer systems using a multilevel dry etching, electroplating and molding process. *J. Micromech. Microeng.* (2012) 22, p. 11

[26] Tosello, G., Calaon, M., Tang, P. T., Ravn, C., Hansen, H. N., Process chain for nano surface texturing of metal and polymer micro structures. *10th International Workshop on High Aspect Ratio Micro and Nano System Technology* (2013)

[27] Matteucci, M., *Biological and medical applications of deep x-ray lithography and liga techniques* (2007) Universita degli Studi di Trieste, March, PhD thesis

[28] Martin, C. R., Nanomaterials: a membrane-based synthetic approach. *Science* (1994) 266, pp. 1961–1966

[29] Lee, W., Ji, R., Gosele, U., Nielsch, K., Fast fabrication of long-range ordered porous alumina membranes by hard anodization. *Nat. Mater.* (2006) 5, pp. 741–747

[30] Steinhart, M., Wendorff, J. H., Greiner, A., Wehrspohn, R. B., Nielsch, K., Schilling, J., Choi, J., Gosele, U., Polymer nanotubes by wetting of ordered porous templates. *Science* (2002) 296(5575), p. 1997

[31] Nielsen, P., Morgen, P., Simonsen, A. C., Albrektsen, O., Hemispherical shell nanostructures from metal-stripped embossed alumina on aluminum template. *J. Phys. Chem.* (2011) 115, pp. 5552–5560

[32] Choi, K., Park, S. H., Song, Y. M., Lee, Y. T., Hwangbo, C. K., Yang, H., Lee, H. S., Nanotailoring the surface structure for monolithic high-performance antireflection polymer film. *J. Adv. Mater.* (2010) 22, pp. 3713–3718

[33] Choi, J., *Fabrication of monodomain porous alumina using nanoimprint lithography and its applications* (2004) Martin-Luther-Universität Halle-Wittemberg, PhD thesis

[34] Zoski, C. G., *Handbook of Electrochemistry* (2007) Elsevier

[35] Dylmer, H., Moller, P., Tang, P. T., *An electroplating method of forming platings of nickel, cobalt, nickel alloys or cobalt alloys* (1998), Technical report, PCT/DK1996/000270

[36] Marinello, F., *Atomic force microscopy in nanometrology: modelling and enhancement of the instrument* (2006) University of Padua (Italy) and Technical University of Denmark (Denmark), PhD thesis

[37] Weckenmann, A., Jiang, X., Sommer, K.D., Rube, U.N., Seewig, J., Shaw, L., Estler, T., Multisensor data fusion in dimensional metrology. *CIRP Ann.* (2009) 58, pp. 701–721

[38] ITRS 2009 Edition-Metrology. The international technology roadmap for semiconductors, Technical report, ITRS

[39] JCGM 200: 2012. International Vocabulary of Metrology – Basic and General Concepts and Associated Terms (Sèvres: Bureau International des Poids et Mesures, BIPM)

[40] ISO/DIS 25178-70:2011. Geometrical product specification (gps) – surface texture: Areal part 70: Physical measurement standards. Technical report, International Organization of Standardization, 201

[41] ISO 5436 part 1:200, 2000, Geometrical product specification (gps) – surface texture: profile method – measurement standard – material measures. Technical report, International Organization of Standardization

[42] ISO 25178-2: 2012 Geometrical product specification (GPS) – surface texture: Areal – Part 2: Terms, definitions and surface texture parameters. ISO 25178 part 2 (Geneva: International Organization for Standardization)

[43] Quagliotti, D., Tosello, G., Salaga, J., Baruffi, F., Hansen, H.N., Gasparin, S., Metrology of submicron structured polymer surfaces. *3rd Int. Conference Polymer Replication on Nanoscale* (2016)

[44] Calaon, M., Hansen, H.N., Tosello, G., Garnaes, J., Ravn, C., Tang, P.T., Characterization of large area nanostructured surfaces using afm measurements. *Proceedings of the 12th euspen International Conference* (2012), pp. 288–291

[45] Calaon, M., Hansen, H.N., Tosello, G., Garnaes, J., Li, W., Replication fidelity assessment in Nano Moulding. *Proceedings of the 7th International Conference on Multi-Material Micro Manufacture 4M* (2015), p. 4

[46] Domine, D., Haug, F.-J., Battaglia, C., Ballif, C., Modeling of light scattering from micro and nanotextured surfaces. *J. Appl. Phys.* (2010) 107(044504), p. 8

[47] Harvey, J.E., Vernold, C.L., Krywonos, A., Thompson, P.L., Diffracted radiance: a fundamental quantity in nonparaxial scalar diffraction theory. *Appl. Opt.* (2010) 38(31), pp. 6469–6481

[48] Refractive index database. http://refractiveindex.info/. Accessed: 2017-06-06

[49] Calaon, M., Madsen, M.H., Weirich, J., Hansen, H.N., Tosello, G., Hansen, P.E., Garnaes, J., Tang, P.T., Replication fidelity assessment of large area sub-μm structured polymer surfaces using scatterometry. *Surf. Topogr.: Metrol. Prop.* (2015), p. 7

[50] Calaon, M., Hansen, H.N., Tosello, G., Garnaes, J., Nørregaard, J., Li, W., Microfluidic Chip Designs Process Optimization And Dimensional Quality Control. *J. Microsyst. Technol.* (2013) p. 10, DOI: 10.1007/s00542-013-2025-3

[51] Liou, A.C., Chen, R.H., Injection molding of polymer micro- and submicron structures with high-aspect ratios. *Int. J. Adv. Manuf. Technol.* (2006) 28, p. 1097

[52] Theilade, U.A., Hansen, H.N., Surface microstructure replication in injection molding. *Int. J. Adv. Manuf. Technol.* (2007) 33, p. 157

[53] Mönkkönen, K., Hietala, J., Pääkkönen, P., Pääkönen, E.J., Kaikuranta, T., Pakkanen, T.T., Jääskeläinen, T., Replication of Sub-Micron Features Using Amorphous Thermoplastics, *Polym. Eng. Sci.* (2002) vol. 42, p. 1600

[54] Pranov, H., Rasmussen, H.K., Larsen, N.B., Gadegaard, N., On the Injection Molding of Nanostructured Polymer Surfaces. *Polym. Eng. Sci.* (2006) vol. 46, p. 160

[55] Calaon, M., Tosello, G., Elsborg, R., Hansen, H.N., Precision manufacturing of thermoplastic elastomer micro rings. *30th Annual Meeting of the American Society for Precision Engineering* (2015)

[56] Tosello, G., Griffiths, C.A., Hansen, H.N., Dimov, S.S., The COTECH knowledge database: assessment of critical factors in micro moulding. *5th International Conference on Multi-Material Micro Manufacturing 4M* (2012), pp. 229–236

[57] Tosello, G., Marinello, F., Hansen, H. N., Characterisation and analysis of microchannels and submicrometre surface roughness of injection moulded microfluidic systems using optical metrology. *Plastics, Rubber and Composites* (2012) 41(1), pp. 29–39

[58] Sha, B., Dimov, S. S., Griffiths, C. A., Packianather, M. S., Investigation of micro-injection moulding: Factors affecting the replication quality. *J. Mater. Process. Technol.* (2007) 183(2–3), pp. 284–296

[59] Tosello, G., Gava, A., Hansen, H. N., Lucchetta, G., Marinello, F., Characterization and analysis of weld lines on micro injection moulded parts using Atomic Force Microscopy (AFM). *Wear* (2009) 266(5–6), pp. 534–538, DOI: 10.1016/j.wear.2008.04.077

[60] Griffiths, C. A., Dimov, S. S., Scholz, S., Tosello, G., Rees, A., Demoulding force in micro-injection moulding. *Proceedings of the 7th International Conference on Micromanufacturing* (2012) Evanston (IL, USA)

[61] Calaon, M., Tosello, G., Hansen, H. N., Ravn, C., Islam, A., Packing parameters effect on injection molding of polypropylene nanostructured surfaces. *Society of Plastics Engineers – 71st Annual Technical Conference (ANTEC2012)* (2012) Orlando, Florida, USA

[62] Matschuk, M., Larsen, N. B., Injection molding of high aspect ratio sub-100 nm nanostructures. *J. Micromech. Microeng.* (2013) 23, 025003

[63] Tofteberg, T. R., Amedro, H., Grytten, F., Andreassen, E., Effects of injection molding holding pressure on the replication of surface microfeatures. *J. Int. Polym. Process. XXV* (2010) 3, pp. 236–239

[64] Carpenter, B., Patil, S., Hoffman, R., Lilly, B., Castro, J., Effect of machine compliance on mold deflection during injection and packing of thermoplastic parts. *J. Polym. Eng. Sci.* (2006) 46(7), pp. 844–852

[65] Whiteside, B. R., Spares, R., Jose, N., Coates, P. D., Realising Nano-scale Structured Surfaces in Micromoulding. *Proceedings of the 8th International Conference on Multi-Material Micro Manufacture 4M* (2011), Stuttgart (Germany), 8–10 November 2011, Research Publishing, ISBN: 978-981-07-0319-6, DOI: 10.3850/978-981-07-0319-6 257, pp. 306–309

[66] Calaon, M., Tosello, G., Garnaes, J., Hansen, H. N., Injection and injection-compression moulding replication capability for the production of polymer Lab-on-a-Chip with nano structures. *J. Micromech. Microeng.* (2017), https://doi.org/10.1088/1361-6439/aa853f

PART 2

Tooling Technologies for Micro Mold Making

5 Micro Machining Technologies for Micro Injection Mold Making

Giuliano Bissacco

5.1 Introduction

Micro injection molding technology is used for the cost effective mass production of components that find applications in different areas, ranging from sensor technology to medical devices including implantable elements, micro mechanics, communication, and infotainment, etc. Examples are shells for hearing aid devices, pacemakers, drug delivery systems, and polymer optics. Such components are characterized by a wide spectrum of requirements and specifications critical to their functionality with respect to materials, geometry, dimensions of their parts and functional features, feature accuracy, surface finish, and subsurface integrity. As micro injection molding requires the fabrication of molds carrying the negative geometry of the molded component, a wide range of micro manufacturing technologies is necessary to enable the generation of molds with high geometrical complexity, high dimensional and geometrical accuracy, high surface finish, and good subsurface integrity in a wide range of materials.

With few exceptions, micro molding is normally aimed at mass production and, therefore, requires molds capable of withstanding a large number of repeated thermomechanical loading cycles. For this reason, employed mold materials are normally characterized by high hardness and wear resistance in combination with sufficient toughness to avoid fragile rupture of the bulk mold or of its functional micro features. Requirements in terms of thermal properties also influence the selection of the mold material. Typical micro mold materials are, therefore, metals and metallic alloys such as hardened and pre-hardened tool steels, nickel, aluminum alloys, and copper alloys.

The implementation of replication techniques relies on the availability of tooling technologies for manufacturing tools and molds [1]. Micro machining technologies have been developed over the past 20 years to a level that matches, to a large extent, the requirements for manufacturing micro injection molding molds as well as molds with micro features. Such technological processes have primarily arisen from the miniaturization of precision engineering technologies mainly based on material removal. A key condition for the adaptation of such processes to the needs of micro manufacturing is the reduction of the "material removal unit" [2], enabling increases in the geometrical resolution of the process. The concept of a material removal unit was first introduced by Taniguchi [3], referring to the smallest volume of material addressed by a single cycle of removal action. Such a characteristic is common to precision engineering technologies, where the achievable part accuracy is the characterizing aspect. However, micro manufacturing technologies must retain one additional characteristic, namely the capability to access small volumes of materials such as holes, slots, and generic cavities with small access areas, typically below 1 mm^2. This implies the use of small tools.

Micro manufacturing processes can be grouped according to their working principle as shown in Figure 5.1 [4]. Thus, material removal action at the micro scale can be achieved by different working principles such as mechanical force, thermal input, or chemical dissolution. Corresponding micro manufacturing processes finding wide applications in micro mold making are micro cutting, micro electrical discharge machining (EDM), and, to a minor extent, laser ablation and micro electrochemical machining. Each of these processes can be implemented in different configurations, such as, for instance, micro milling and micro turning, exploiting mechanical force, sinking micro EDM and micro EDM milling, using a thermal input.

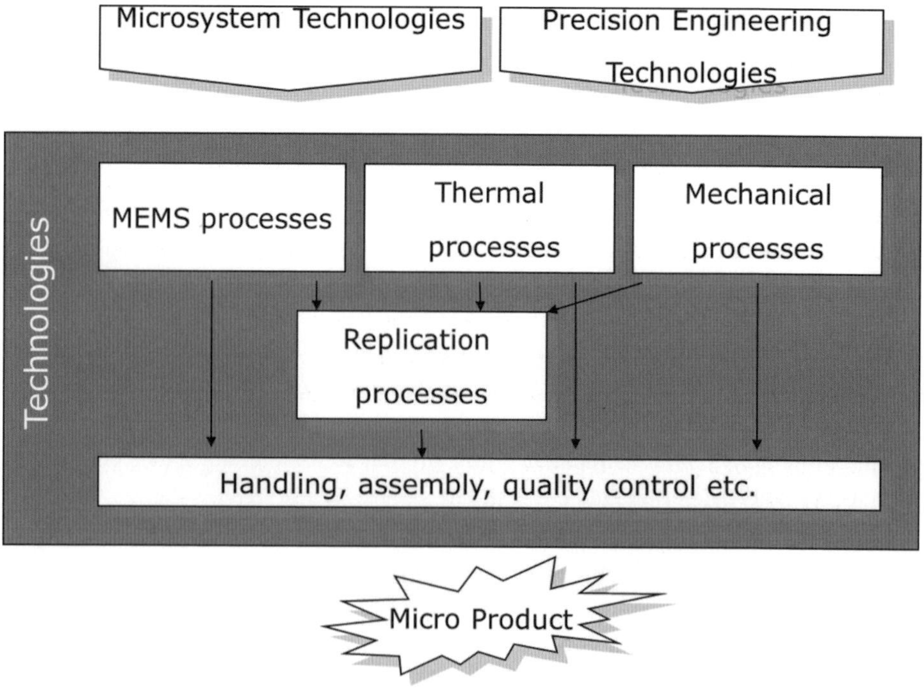

Figure 5.1 Process technologies for machining precision parts and micro structures [4]

Recently, additive manufacturing processes have shown applicability for tooling applications (namely the manufacturing of tools, molds, dies, and fixturing equipment) owing to the increased process resolution related to the reduction and increased stability of the unit of material accretion, opening up new possibilities in micro mold design with benefits in terms of improved replication quality and reduced cycle time. However, in most cases, post processing is still necessary to reach the required accuracy and surface quality.

5.2 Process Chains for Micro Mold Making

A micro injection mold system includes all the necessary elements for providing the required flow path and cavity for the melt polymer, as well as the necessary connections and interfaces to the injection molding machine, sensors, and regulation devices to ensure the desired process conditions. Thus, the mold assembly consists of several parts with different functions. While conventional mold making is used for the complementary (but essential) parts that do not come into contact with the polymer melt, the application of micro manufacturing processes is normally applied to the generation of the cavity, which is often obtained on interchangeable inserts that constitute the active part of the mold. In the following, the focus will be on technologies and process chains used to generate the micro mold cavity, while conventional tool making for the generation of complementary mold elements will not be treated.

"Process chain" is the term used to indicate the sequence of processing steps employed to manufacture a specific component. Thus, in its broadest sense, the term process chain includes not only the operations necessary to obtain the part geometry starting from a substrate or workpiece (in the case of using material removal processes), but also the necessary operations to generate the substrate [5] as well as post processing of the final part. Since each process step influences, in principle, the results of both the previous and the following process steps, the process sequence has to be checked for consistency, and incompatibilities must be identified. In most cases, several alternative process chains are viable for the manufacturing of a given component. As an example, Figure 5.2 shows alternative processing routes for the manufacturing of a polymer micro component.

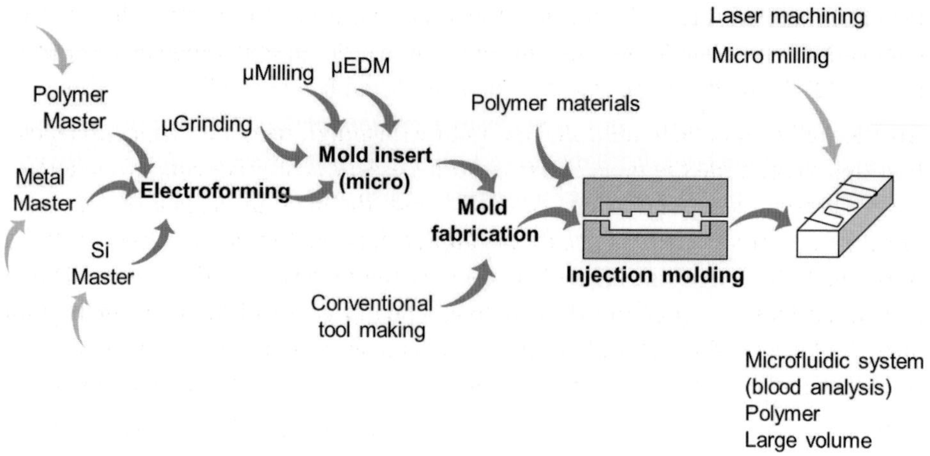

Figure 5.2 Example of process chain alternatives to produce a polymer microfluidic system [5]

As micro injection molding requires a tool to transfer the microstructures to the polymer material, a key issue is the manufacturing of the tool. Depending on the number of replications specified (production volume), accuracy, surface topography, and subsurface integrity requirements, a number of very different processes can be utilized to obtain the tool insert [6].

When the outcome of the process chain is the production of a mold, we refer to a "tooling process chain". The selection of the material of the mold insert is critically affecting the selection of the main process steps of the process chain. For large production volumes, typical materials are hardened tool steels with different alloy elements, with a hardness in the range of 45–58 HRC. For small series production, mold inserts in copper or aluminum alloys can be realized, with a hardness of up to 40 HRC.

A micro mold insert is normally obtained by direct processing of a substrate by means of material removal processes (e.g., micro milling or micro EDM) or by consolidation of metal powder via additive processes (e.g., selected laser melting, electron beam melting, jet binding) and subsequent post processing. As the processing action is obtained directly on the material constituting the insert (workpiece or powder), this approach is called "direct tooling". Due to the limitations of the resolution, accuracy, and surface quality of metal additive processes, micro feature generation is always carried out by means of micro material removal processes. For steel molds, micro milling and micro EDM are the most widely used and commonly available. In fact, such processes are characterized by a great flexibility, allowing the generation of basically any geometry provided that access to the tool can be ensured. The selection of the specific process for the generation of the micro mold geometry depends on a number of factors. Mechanical micro milling ensures the highest degree of flexibility and shorter processing time due to a material removal rate 5–10 times larger than micro EDM at comparable surface quality. However, smaller features by almost one order of magnitude can be generated by EDM (cavities of 10–15 μm can be produced by micro EDM).

An alternative approach, indicated as "indirect tooling", uses a bulk electrodeposition step to generate the mold insert (with necessary post processing) from a master geometry in a suitable material. In this case, the mold insert is made of nickel or copper. A direct consequence of this approach is the limitation in the hardness of the mold insert. When nickel is chosen as the material for the electroforming step, a hardness of up to 450 HV can be reached in virtue of the very small grain size [7]. However, the material for the master can be chosen in a way that optimizes the performance of the micro manufacturing processes used for its manufacture.

Defining as the "positive" geometry the geometry of the molded polymer part, the shape of the mold insert, being the opposite of that of the polymer part, can be defined as the "negative" geometry. This is the geometry to be generated when using a direct tooling approach. In the indirect tooling approach, instead, a pure metal (nickel or copper) is grown over a conductive master, which has essentially the same geometry as the molded polymer part. This is schematically represented in Figure 5.3 [8]. Indirect tooling process chains are inherently longer in time as a consequence of the increased number of processing steps connected with the master generation and subsequent insert electroforming; therefore, whenever possible, a direct tooling process chain is to be preferred. There are, however, situations where an indirect tooling approach is preferable or indeed represents the only possibility to generate the mold insert.

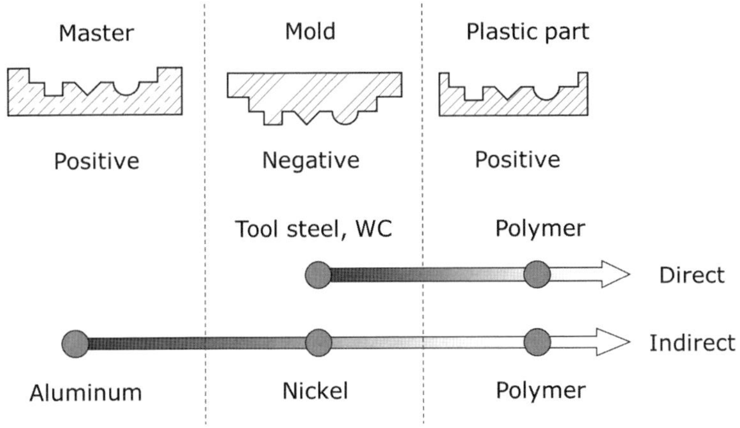

Figure 5.3 Direct and indirect tooling concepts [8]

One additional advantage of the indirect tooling approach is that the master can be generated in a wide variety of materials opening up more options in terms of the selection of the main geometry generating process. As an example, when a silicon master is chosen, the functional features can be generated using MEMS processing routes with the capability to generate smaller and more accurate features than mechanical or thermal micro material removal processes. Masters for electroforming can be realized in soft metals like aluminum or copper, or polymers, or silicon. Masters made in non-conductive materials require an additional preparation step to enable electroforming. The process, in fact, requires a conductive surface, which can be obtained, for instance, by PVD coating on the master with a typical coating thickness of a few tens of nanometers.

As the master and the electroformed mold insert must be separated, care must be taken in ensuring separation without damage to the insert micro features. This can be achieved if the master material is chosen in such a way that it is possible to selectively dissolve it by chemical action, without application of separation forces. As an example, silicon masters can be dissolved in an alkaline solution without damaging the nickel insert.

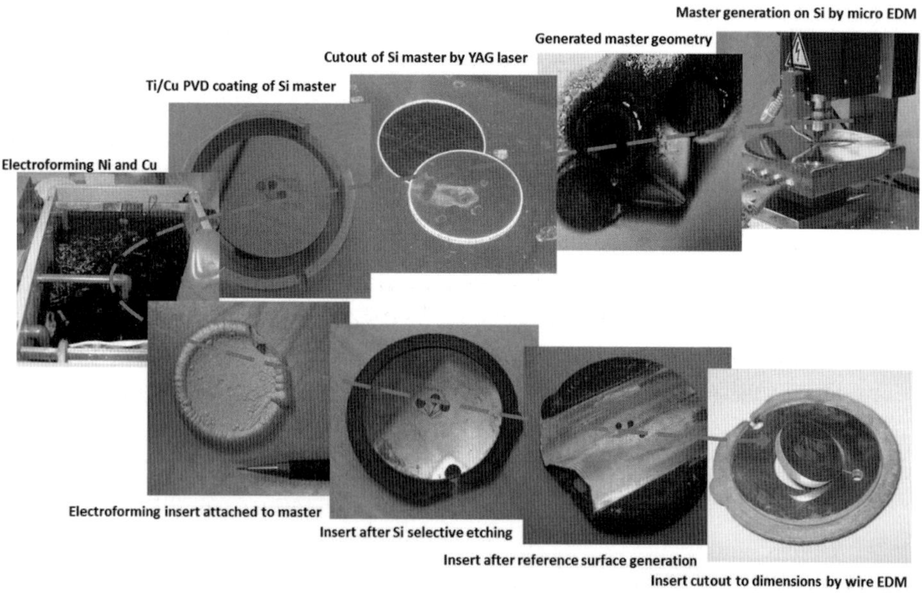

Figure 5.4 Example of indirect tooling process chain [8]

5.3 Micro Mechanical Material Removal

Micro manufacturing processes based on mechanical material removal are the most widely used for manufacturing micro molds in steel and other metallic alloys. Such processes include micro milling, micro drilling, micro turning, micro grinding, and micro polishing, and consist of the downscaling of the corresponding well-known conventional-sized processes. In mechanical material removal processes, the mechanical interaction between a tool and the workpiece exerts a mutual force that generates the material removal action. Mechanical material removal processes can be divided into two groups: material removal with geometrically defined cutting edges, normally referred to as cutting processes, and material removal with geometrically undefined cutting edges, referred to as abrasive processes. The presentation and discussion of cutting processes are treated in this chapter, while abrasive processes are dealt with in Chapter 6.

Micro cutting processes are characterized by a mechanical interaction of a sharp tool with the workpiece material, causing breakage inside the material along defined paths, eventually leading to removal of the useless part of the workpiece in the form of chips [4]. The mechanical action is provided by means of relative motion between the tool and the workpiece, causing a rapid shear deformation of a portion of the workpiece that flows over the tool rake face and becomes the chip. The process is schematically represented in Figure 5.5, depicting a single edge orthogonal cutting process. While sliding over the tool, the chip material must overcome frictional forces at the tool–chip interface. Due to the high speed shear deformation and friction, a high amount of specific energy is dissipated in the form of heat, with an increase in the temperature of the chip as well as the tool. The process, therefore, requires the use of tool materials characterized by a high hardness that must be retained at high temperatures. The combination of friction forces with the high tool temperatures activates multiple wear modes (abrasive, adhesive, diffusive wear) that imply a deterioration of the performance of the tool.

Micro cutting includes several process configurations, such as micro milling, micro turning, and micro drilling, which differ for the specific process kinematics and tool geometries, with the capability to generate workpieces with different geometrical complexities.

As the size of the tool and the material removal unit are downscaled, a number of size effects arise and the required positioning accuracy and rotational speed are largely increased. The whole material removal geometry changes as compared to conventional-size cutting processes and higher tool deflections, high tool wear, and high risk of tool breakage are observed, particularly in connection with hard workpiece materials. Furthermore, the formation of burrs may lead to the rejection of the manufactured item due to the difficulties of burr removal operations for micro components [1].

5.3.1 Size Effects

Size effects are essentially related to the limited scalability of tool geometry, workpiece and tool material microstructure, and tool and workpiece surface topography. The reduction of the material removal unit implies the reduction of the uncut chip thickness. As a result, in micro cutting, the uncut chip thickness is often comparable to, or smaller than, the cutting edge radius and the chip forms in the area of the edge radius, experiencing a highly negative rake angle (Figure 5.5). This is indicated as the cutting edge radius size effect. The unfavorable geometrical configuration occurring at the cutting edge in such conditions strongly affects the cutting mechanics so that the strain pattern inside the removed material does not correspond to that of the macro process with the same nominal geometry. The apparent relative bluntness of the tool increases the specific cutting forces, and above a certain value of the edge radius to chip thickness ratio, a ploughing action dominates [9]. The minimum cutting depth at which chip formation is observed is defined as the minimum uncut chip thickness. Such a limit value depends on the material being cut and on the actual cutting edge radius, so that the minimum uncut chip thickness can be expressed through a ratio between the uncut chip thickness and the cutting edge radius (minimum uncut chip thickness ratio). Specific values of the minimum uncut chip thickness ratio can be identified for different materials, ranging from 0.4 for aluminum to 0.1 for steels [10]. Harder materials typically yield lower minimum uncut chip thickness ratios.

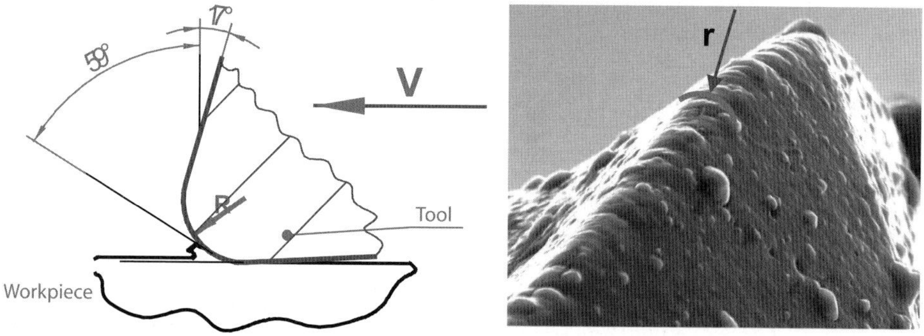

Figure 5.5 Left: cutting edge radius size effect [1]; right: image of the cutting edge of a micro end milling tool [9]

Ploughing occurring at an uncut chip thickness below the minimum value produces an increase of the surface roughness as the plastically deformed material is smeared over the surface [11]. Figure 5.6 shows the accumulation of smearing on specific locations on a micro milled surface. Therefore, the availability of tools with sharp cutting edges is critical to the micro cutting process performance. Commercial tools in micrograin tungsten carbides have typical values of edge radius in the range 1–4 µm.

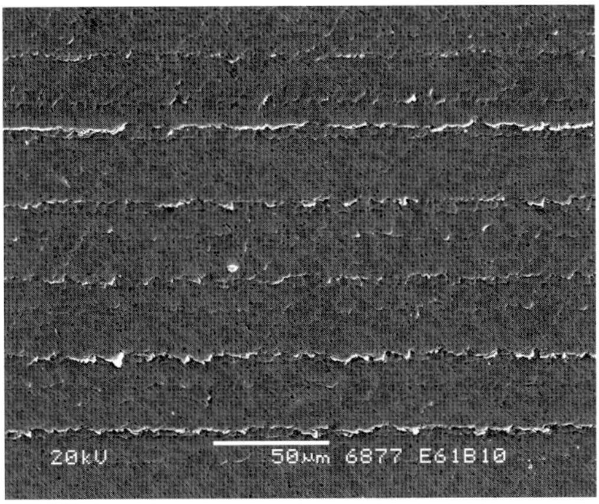

Figure 5.6 Surface in a hardened tool steel 58 HRC finish milled with a 200 µm ball end mill. The accumulations of plastically deformed material to the side of each linear tool path (feed direction) are clearly visible [11]

An additional source of size effect is related to the workpiece material's micro structure. The reduction of the material removal unit and, therefore, of the uncut chip thickness, leads to a reduction of the ratio between the uncut chip thickness and the workpiece material's grain size, often reaching a situation where the chip forms within single grains or phases with variations of the material's response to the cutting action (Figure 5.7). Such a situation can induce differential material spring back, increased surface roughness, and high frequency force variations.

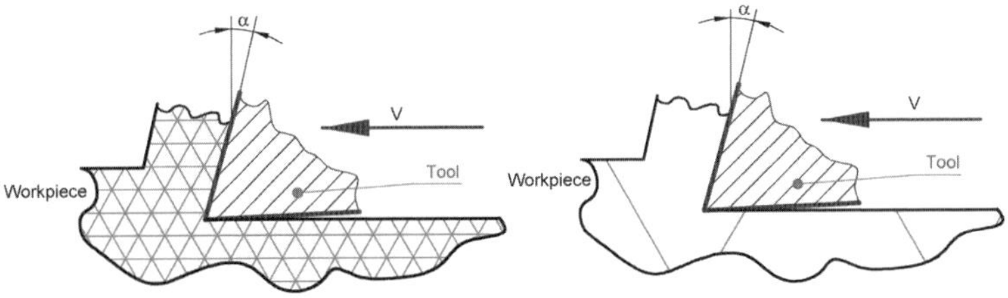

Figure 5.7 Relation between grain size and chip thickness for conventional cutting and micro cutting [1]

5.3.2 Cutting Forces and Tool Deflection

The cutting action requires the exchange of force between the tool and workpiece. As a consequence of the edge radius size effect, which dramatically affects the chip formation, cutting forces are comparatively higher in micro cutting as compared to conventional cutting. More specifically, the large negative rake angle at the tool-chip interface (in Figure 5.5) requires a comparatively larger thrust force component (force orthogonal to the machined surface) to initiate and maintain the cutting action. This force component tends to separate the tool and workpiece. On the other hand, manufacturing of small concave features requires tools with a small cross section and often a high aspect ratio. Thus, slender tools used in micro cutting experience rather high deflections, with a direct impact on the accuracy of the machined features. While this is not a big problem in micro turning and micro drilling, in micro milling, where the cutting forces load the tool in bending mode, tool deflection critically limits the accuracy of machined parts.

5.3.3 Machine Tools

Implementation of micro cutting processes requires suitable machine tools. Machine tool motion accuracy, spindle run out, thermal distortions, static, and dynamic stiffness directly affect the accuracy of the generated features. Sub-micrometer positioning accuracy is required to follow the part contour and maintain control of engagement conditions, which in turn affect cutting forces and tool deflections. Rotating axes and spindles must ensure that the radial deviations at the tool-workpiece contact are less than 0.5 μm in order to maintain form accuracy and consistency of tool-workpiece engagement. Thermal distortions caused by environmental heat sources, and internal heat sources, such as the spindle motor, drive units, and complementary electrical systems, as well as cutting fluids, must be minimized and compensated. The reduced dimensions of tool diameters in milling and drilling require spindles with high rotational speeds in order to maintain sufficient cutting speed. Machine tools for micro machining are essentially machines for ultra-precision machining, with the additional requirement of high rotational speeds. Machines of this type implement design solutions and construction materials for minimization of thermal deformations, real-time active compensation of spindle thermal errors as well as high-precision drive systems.

Modern ultra-precision machine tools applied in micro cutting exhibit linear axes with oil hydrostatic bearings driven by linear motors, motorized air-bearing spindles, high-resolution glass scales for positional feedback, and PC-based controllers [12].

5.4 Micro Milling

Micro milling has become an established process for manufacturing three-dimensional meso and micro components in metals and alloys and is by far the most widely applied process in micro mold making. The large process versatility obtained through the possible tool-workpiece relative motion and tool geometries enables the generation of virtually any geometry and is, therefore, very well suited for the generation of micro molds containing complex and free-form geometries.

Micro milling is defined as the downscaling of the conventional milling process, involving the use of end mill diameters in the sub-millimeter range [1]. The milling tool, provided with cutting edges, rotates at high speeds while moving along defined trajectories and engages the workpiece removing material in the form of chips.

When applied to the manufacture of micro molds, the capabilities of the micro milling process vary largely depending on the machine tool and spindle used, workpiece material, tool geometry, process parameter selection, and tool path strategy.

Figure 5.8 Micro end milling: process of machining a micro mold insert [13]

5.4.1 Cutting Tools

Commercial micro milling tools for machining steels for micro mold making are made of ultra-fine-grain tungsten carbides with a cobalt binder and grain size of less than 0.5 µm. Micro tool geometries include flat end mills, ball nose end mills, and toroidal end mills, normally with two flutes and a 30° helix angle. Flat end mills and toroidal end mills are used to machine flat surfaces and for roughing operations, while ball nose end mills are used for finish machining of curved surfaces and also flat inclined surfaces when using three axis machine tools.

The geometrical configuration of micro end mills is different from that of conventional size end mills (Figure 5.9). The shank diameter can be more than 10 times bigger than the cutting diameter and a soft connection is required in order to avoid notches and thereby dangerous stress concentrations.

Figure 5.9 Conventional size 6 mm flat end mill and 200 µm micro flat end mill [1]

Ball nose and flat micro end mills with diameters down to 30 µm are commercially available. However, most cutting tool manufacturers limit the diameter range to 100 µm. One critical aspect of micro milling tools is the neck length, which defines the maximum depth machinable in a cavity. For end mills with diameters smaller than 200 µm, the neck length is limited to 7–8 times the diameter. A larger neck length induces higher bending moments due to the cutting forces, and the risk for tool breakage is increased. Therefore, with long neck tools, a more conservative selection of the process parameters must be operated so that the uncut chip cross section is reduced. This implies slower material removal and, consequently, a longer machining time. Long neck tools will display larger deflections and the impact on the final part accuracy must be taken into account. Micro end mills are produced by means of a grinding process. The capabilities of the tool manufacturing process limit the minimum cutting edge radius obtainable, which is in the order of 1–4 µm and varies largely among manufacturers.

Micro end mills for machining in tool steel (up to 70 HRC) are coated to increase the surface hardness and reduce the tool wear rate. Coatings are applied to the tool tips by a PVD process and consist of different types of nitrides, such as TiN, TiAlN, TiAlCN, or CrN. The coating thickness can be in the order of 1 µm. One side effect of the application of coatings is an increase in the edge radius.

The small dimensions of the end mills and their geometrical features imply that very small wear volumes create a dramatic change in the cutting edge geometry, so that the tool becomes unable to cut. Tool wear cannot be detected with conventional vision systems for tool inspection or tool presetting systems based on laser interruption or high resolution cameras as they do not provide sufficient accuracy. Therefore, conservative assumptions need to be made on tool life. The consequence of undetected tool wear is mainly an increase in the cutting edge radius leading to ineffective cutting. Tool wear monitoring techniques have been developed in the past decade. Uhlmann et al. [14] applied acoustic emission (AE) sensors to monitor tool wear in the micro milling of steel (0.45% C), and they observed significant peaks in the frequency spectrum correlating with tool wear.

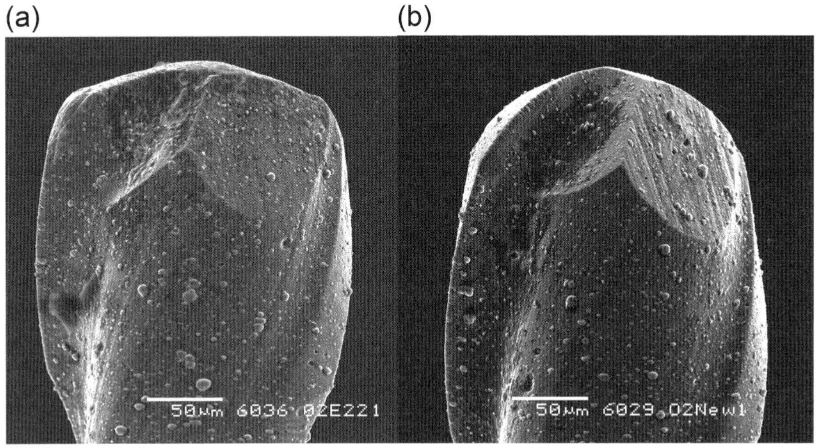

Figure 5.10 Effect of wear volume on the tool geometry of a 200 µm diameter ball nose end mill: (a) new tool; (b) worn tool after machining 58 HRC tool steel [13]

5.5 Micro Turning

Micro turning is used to generate micro scale axialsymmetric components. In this cutting process, the workpiece is given a rotational motion while the tool has translational motions in the axial and radial directions. Normally, the workpiece rotational motion is not synchronized with the axial and radial linear axis motion of the tool. However, in high-precision machines for diamond turning, the synchronization of the axes allows the generation of geometries that are not axialsymmetric.

In micro turning, the tool has generally higher stiffness than the workpiece, which is, therefore, the element subjected to larger bending. Finishing passes are, therefore, necessary to reduce engagement and cutting forces and thereby limit part deflection. The cutting edge radius size effect is, therefore, a limiting factor for part accuracy in micro turning. Achievable aspect ratios can be larger than 30 and achievable feature sizes are down to a few tens of µm.

5.6 Micro Drilling

Micro drilling is used to produce efficiently through micro holes as well as blind micro holes when the bottom surface of the hole does not have particular geometrical requirements. Holes of less than 50 µm in diameter can be machined by micro drilling, using twist drills very similar to the corresponding macro scale tools. Like micro milling, one requirement for micro drilling is the use of high speed spindles. Air turbine spindles are often used for micro drilling. Such spindles are capable of rotational speeds of over 200,000 rpm. Effective lubrication and chip evacuation are the main limiting factors in the generation of high aspect ratio holes by micro drilling.

5.7 Thermal Material Removal Processes

Thermal material removal processes use a sequence of controlled thermal energy inputs to remove discrete material units from the workpiece. During the thermal action, small volumes of material reach molten or vapor states and are ejected from the workpiece. Due to the process-workpiece interaction, workpiece hardness is not an issue and thermal processes are well suited for machining very hard materials. On the other hand, thermal and electrical conductivity, as well as electromagnetic radiation absorption, are characteristics that affect the performance of such processes. Thermal material removal processes most commonly used in micro mold making by a direct tooling approach are limited to micro electrical discharge machining. Laser processing is mostly used in indirect tooling and will not be treated here.

5.8 Micro Electrical Discharge Machining

Electrical discharge machining (EDM) is the process of machining electrically conductive materials by using precisely controlled sparks that occur between a tool electrode and a workpiece in the presence of a dielectric fluid [15]. The tool and workpiece are called electrodes, as they are essentially part of an electrical circuit. The two electrodes are not in contact, but are separated by a distance called the sparking gap. The gap is filled with dielectric fluid, which acts as an electrical insulator between the two electrodes. During the discharge pulses, a high temperature plasma channel is formed in the gap, allowing electrical current to flow between the two electrodes, causing a local temperature increase leading to melting and evaporation of discrete volumes of workpiece and electrode material. As there is no contact between the tool and workpiece, the process-generated forces acting on the tool and workpiece are almost zero, and the process poses no requirements with respect to the relative hardness of the tool and workpiece materials. Therefore, very hard materials, such as hardened tool steel and sintered tungsten carbides (with a cobalt binder), can be machined using electrodes in soft metals like copper or in graphite, which are typical tool electrode materials. Discharges occur at high frequencies (up to 500,000 discharges per second). Each discharge increases the local distance between the tool and workpiece electrodes. As the workpiece must be electrically conductive, the range of machinable materials includes all metals and metallic alloys, as well as silicon and electrically conductive ceramics [16].

Each discharge cycle melts and vaporizes small volumes of material. Not all the material melted by the discharge is vaporized or ejected, and a part of it remains on the surface and solidifies. Therefore, the EDMed surfaces are characterized by craters of melted and re-solidified material as is clearly visible in Figure 5.11 showing the comparison between an EDMed surface and a ground surface. The surface texture is non-reflective and non-directional. The subsurface integrity of the workpiece is affected by the EDM process. Due to the thermal input, the surface material withstands a hardening and tempering cycle with decreasing maximum temperatures and cooling speeds at increasing distance from the surface. As a consequence, residual stresses are built in the surface layer and the hardness profile is affected. The depth of the "heat affected zone" (HAZ) is related to the process parameter settings. High energy discharges will create a higher thermal input and, therefore, higher levels of residual stresses, possibly leading to the generation of surface cracks.

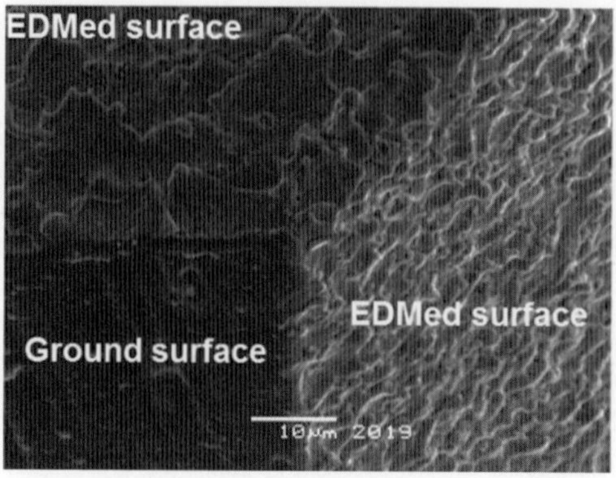

Figure 5.11 Appearance of an EDMed surface compared to a ground surface [17]

Micro EDM can be defined as the downscaling of the conventional EDM process to generate parts and features with critical dimensions below 500 µm. It requires the use of low energy discharge pulses, high accuracy tools and fixtures, and precise machine tool systems.

The application of EDM to micro manufacturing requires the reduction of the material removal unit, namely the amount of material that is removed by a single discharge. As this is largely determined by the discharge energy, the requirement translates into the realization of very short discharge pulses (Figure 5.12). With a discharge energy of less than 1 µJ, material removal units of less than 1 μm^3 can be realized [18]. Discharge energies per pulse as small as approximately 3 nJ, obtained using an RC discharge circuit, have been reported in [19].

Micro EDM enables the manufacturing of complex and free-form micro geometries with high aspect ratios and tight tolerances. Cavities with widths as small as 15 µm can be generated with commercially available EDM machines equipped with fine pulse generators and an accuracy of less than 1 µm can be achieved with the fine tuning of the process. A distinguishing capability of micro EDM when compared to mechanical material removal is the achievable cavity aspect ratio, which can reach a value of 50 or more.

In order to maintain high dimensional accuracy, in micro EDM, the discharge gap is kept small. In fine finishing operations, the gap can be in the order of 2–3 µm, which has to be taken into account when defining the tool dimensions and tool motion.

Current micro EDM configurations include micro EDM sinking, wire micro EDM, micro EDM drilling, micro EDM milling, and wire electrical discharge grinding. The processes are schematically shown in Figure 5.13.

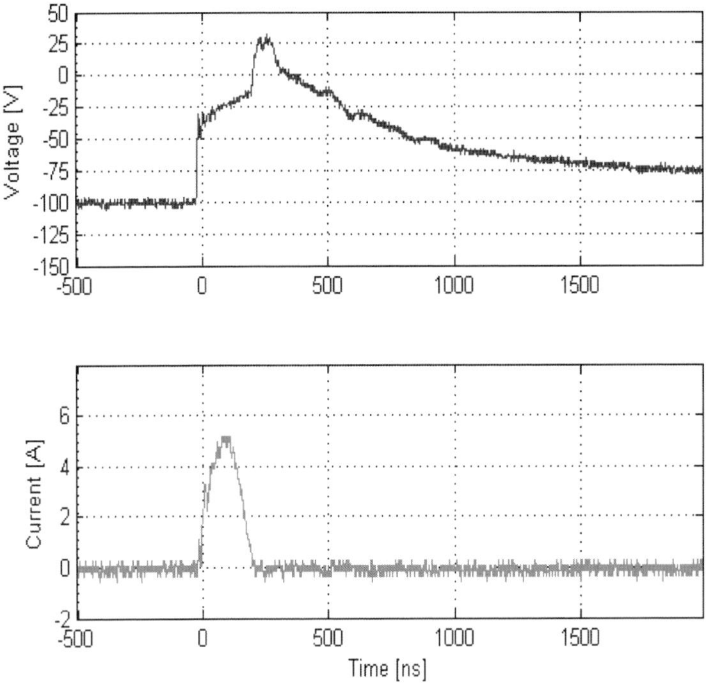

Figure 5.12 Discharge voltage and current profiles of a micro EDM discharge. Pulse duration approximately 200 ns, discharge energy approximately 12 μJ [18]

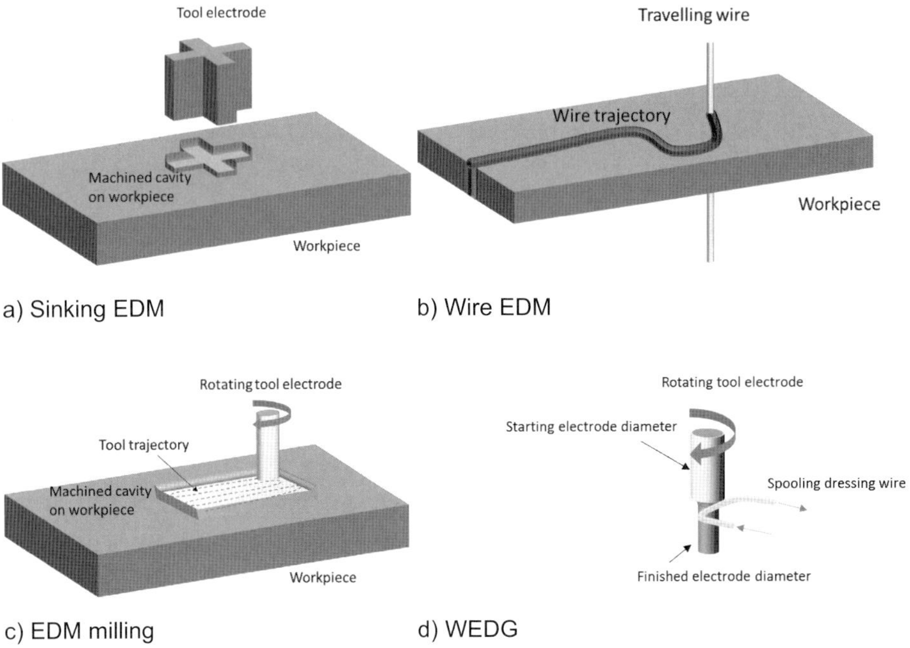

Figure 5.13 Micro EDM process configurations: (a) concept of sinking EDM; (b) concept of wire EDM; (c) concept of EDM milling; (d) concept of WEDG

5.8.1 Micro EDM Sinking

In micro EDM sinking, a shaped tool electrode is used for manufacturing three-dimensional shapes and free forms [20]. The electrode geometry is copied into the workpiece by a unidirectional downward motion of the tool. Machining of blind cavities requires multiple electrodes for high accuracy due to the wear of the tool electrode and corresponding geometrical variation during the process. The tool electrode geometry is the negative of the cavity to be generated. Thus, micro EDM sinking requires the manufacturing of a tool whose geometrical complexity reflects that of the cavity. The electrode must be produced by means of another micro manufacturing technology, typically micro milling. Materials for micro EDM sinking electrodes are mostly copper, tungsten-copper, and graphite. While sinking EDM requires one additional process step when compared to micro milling, advantages are the possibility to generate the electrode on a softer material than the mold tool steel, and the inversion of the geometry so that very small mold cavities (below 100 μm) become protrusions on the insert and the material to be removed becomes accessible by a milling tool.

5.8.2 Micro Wire EDM

Micro wire EDM is the exact match of conventional-size wire EDM where a wire electrode is fed from a spool to another one and cuts the workpiece by relative displacement along a defined contour. A principle sketch of wire EDM is shown in Figure 5.13. The process allows the generation of 2.5D geometries (shapes consisting of a 2D contour drawn in the orthogonal direction). By a relative displacement of upper and lower wire guides, sloped contours can be machined. In micro wire EDM, small diameter electrode wires are used in combination with high accuracy travel motion and wire guiding systems. Wire diameters can be as small as 30 μm. In order to enable proper pretension and avoid frequent breakage, and different from wires used in conventional wire EDM, which are made of copper or brass, wires for micro wire EDM have a core in a stronger material, such as steel or tungsten, with multiple coatings of copper, zinc, and silver. The dielectric fluid is normally deionized water, which is applied to the area of interest through a low pressure flow to avoid wire vibrations while still ensuring proper flushing of the gap. Achievable accuracy is in the order of ±1 μm and aspect ratios can be up to 100 [21].

5.8.3 Micro EDM Drilling

Micro EDM drilling is a special case of micro EDM sinking in which the tool electrode has a constant circular cross section along its length. In this configuration, circular holes with small dimensions and very high aspect ratios can be generated. In machining high aspect ratio holes, the limiting factor is the realization of effective flushing. For hole diameters of 100 µm and above, tubular electrodes are commercially available. Such electrodes enable the supply of dielectric through the electrode and into the hole being machined, ensuring effective debris removal from the sparking gap. In such conditions, aspect ratios of more than 50 can be achieved. The maximum aspect ratio decreases strongly when solid electrodes must be used.

The use of very small electrodes requires careful handling and alignment. It is, therefore, preferable to generate the electrode directly on the machine by means of wire electrical discharge grinding (WEDG).

5.8.4 Wire Electrical Discharge Grinding

In wire electrical discharge grinding (WEDG), a wire tool electrode is used to shape rotating axialsymmetric components, often tools for further micro EDM operations, such as drilling or milling. A sketch of the process is shown in Figure 5.13. The application of WEDG in combination with drilling or milling EDM on the same machine tool removes any electrode mounting error and is, therefore, a preferred approach when generating high accuracy holes or cavities. With this process, electrodes with diameters down to 10 µm can easily be generated. The generation of electrodes with diameters down to 2.3 µm have been reported in [22].

5.8.5 Micro EDM Milling

In micro EDM milling, microelectrodes essentially consisting of cylindrical rods with diameters down to 10 µm are driven along defined paths while rotating, in a way similar to conventional three- and five-axis milling. In this configuration, the material is removed layer by layer, with the layer thickness ranging from a few microns to 0.1 µm, depending on the diameter of the electrode and on the discharge energy. Therefore, micro EDM milling does not require the manufacturing of a tool electrode with a complex geometry [18]. Dedicated CAM software is available for tool path generation in micro EDM milling, enabling easy handling of even very complex geometries and multiple tool electrodes.

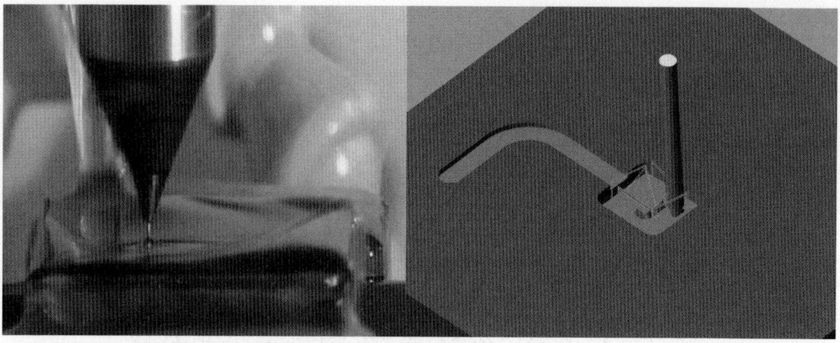

Figure 5.14 Left: micro EDM milling operation; right: visualization of the tool path in a CAM program [23]

Electrode wear must be effectively compensated in order to achieve the required accuracy of the machined features. Since milling EDM is performed in very thin layers compared to the tool diameter, the tool profile stabilizes rapidly and wear correction can be done by a one-dimensional motion parallel to the electrode axis. Electrode wear compensation in conventional milling EDM can be based on estimation of the volumetric wear ratio and continuous compensation proportional to the in-plane displacements (anticipated wear compensation), real-time wear sensing, or a combination of the two [24].

Real-time electrode wear compensation enables a better control of the process as compared to anticipated wear compensation, where assumptions on the process are made prior to the start of the operation. Real-time compensation based on discharge pulse discretization has been investigated by several authors [25, 26]. However, a strong limitation of such an approach is the high acquisition frequency and computational speed necessary when very short discharge pulses are used (less than 100 ns). In such a condition, a discharge population approach is preferable. The process can then be characterized in terms of its population, and effective tool wear compensation can be implemented by counting the discharges and compensating on the basis of the average wear per discharge [27, 28]. As demonstrated by Bissacco and co-authors, electrode wear compensation based on a discharge population approach can be based on the measurement of tool wear per discharge [27] or on the measurement of material removal per discharge [28]. The two approaches have complementary benefits and enable a cavity depth error of less than 1 μm.

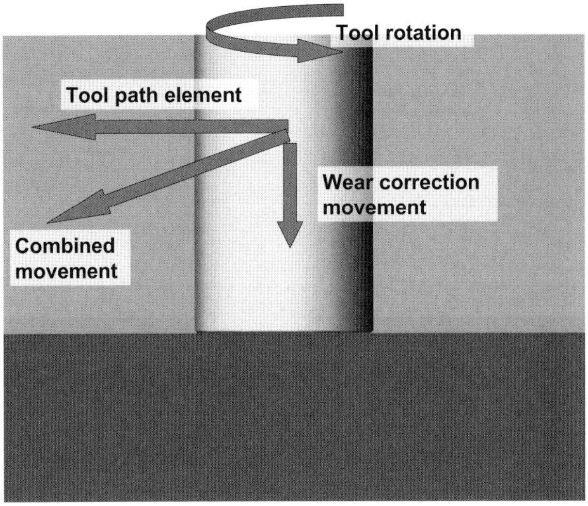

Figure 5.15 Electrode displacements during machining showing an electrode wear correction movement [18]

5.9 Application Examples of Machining Technologies for Micro Mold Making

In this section, two examples of applications of the technologies and process chains described above will be shown. The first example uses a direct tooling approach, where micro milling is used as the main process for the generation of the micro mold geometry. In the second example, indirect tooling is used to generate a nickel-copper micro mold for a polymer microfluidic system.

5.9.1 Micro Mold Produced by Direct Tooling

This example, described in [29], refers to the fabrication of a mold insert for a polymer microfluidic system by means of a direct tooling approach. The design of the polymer micro chip incorporates four main chambers as well as a series of channels connecting the chambers with each other and with the inlets. The channels' width and depth is 200 µm (Figure 5.16). The depth of the chambers is up to 1.5 mm. The shape of the mold insert in tool steel (35 HRC) corresponds to the negative of the polymer chip geometry, thus the features corresponding to the chambers and the channels protrude from the base of the insert. Details of the insert design require the use of long neck micro end mills of 200 µm in diameter and 1.5 mm in neck length, yielding an aspect ratio of 7.5. The use of high aspect ratio micro end mills, in combination with the workpiece material hardness, is a challenging aspect with respect to the achievable machining accuracy, due to tool deflection, leading to errors in the lateral dimensions of features in the order of more than 10 µm. One relevant aspect when machining tool steel mold inserts is the progress of tool wear during the machining operation, leading to a progressive increase of burr formation, as can be observed in Figure 5.16.

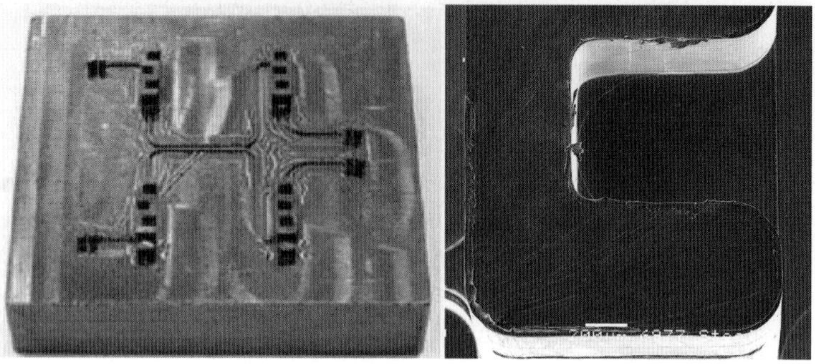

Figure 5.16 Left: tool steel mold insert for a microfluidic system produced by micro milling; right: SEM picture of a detail showing formation of burrs at the top edges [29]

5.9.2 Micro Mold Produced by Indirect Tooling

This example refers to the fabrication of a mold insert for polymer microfluidic systems using an indirect tooling approach. Microfluidic systems consist of chambers and channels on larger area substrates. The mold insert, therefore, consists of protrusions, with different aspect ratios, on a larger substrate. If the insert is to be realized by a direct tooling approach, the protrusions corresponding to the channels are particularly vulnerable when realized by mechanical micromachining, due to the risk of bending. By contrast, in the case of indirect tooling, the features on the master consist of cavities (corresponding to the chambers) and long grooves (corresponding to the channels), thus reducing the risk of spoiling the features during the manufacturing process.

In the specific case of this example, described in [30], the chosen material for the master is a silicon wafer and the geometry-generating process is micro EDM milling. The chosen substrate for the mold master was a 500-µm-thick silicon wafer provided with a thin gold coating. The chambers and channels representing the microfluidic system were generated on the master by means of micro EDM milling. For the generation of 20-, 50-, 75-µm-wide and 90-µm-deep channels, electrodes with diameters of 15, 45, and 68 µm were generated by means of on-the-machine dressing through the WEDG process. Following the generation of the master geometry, the machined silicon surface is cleaned and the residuals of the gold coatings are removed. A new coating of titanium and copper is applied by PVD to make the surface electrically conductive and ensure good adhesion to the silicon surface in preparation for the subsequent electroforming step. The next process step consists of deposition by electroforming a nickel layer approximately 50 µm in thickness, followed by copper electroforming of the bulk of the insert to a total thickness of approximately 3 mm. The insert is released from the silicon master by selective etching. Reference surfaces for mounting the insert on the mold assembly are obtained by milling and wire EDM. Finally, the insert is cleaned and the residual copper layer is removed by selective etching. Figure 5.17 shows the schematic representation of the process chain, while Figure 5.18 shows details of the master and mold insert.

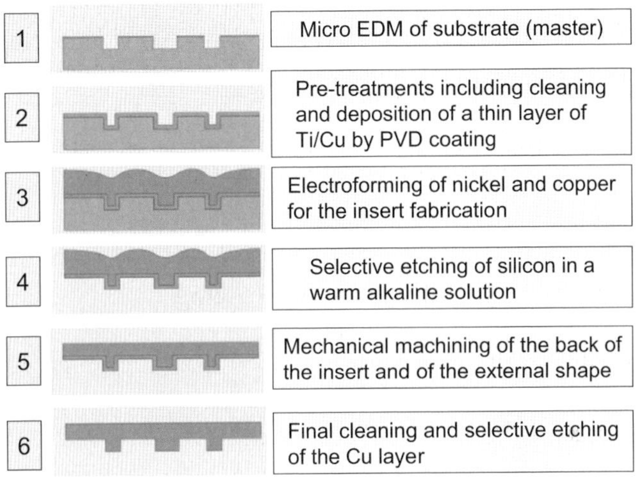

Figure 5.17 Schematic representation of the process chain [30]

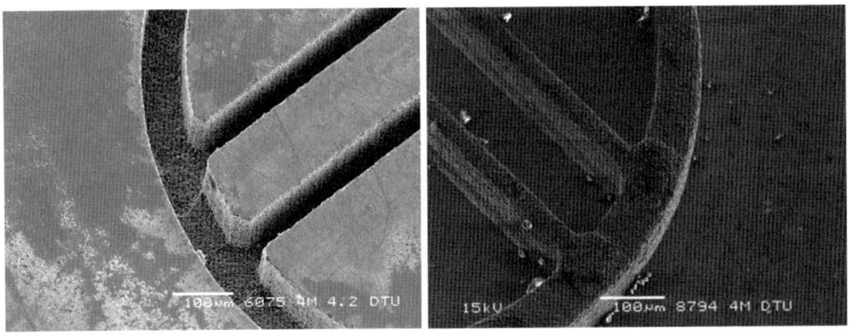

Figure 5.18 Left: details of channels on the silicon master; right: corresponding features on the electroformed mold insert [30]

References

[1] Bissacco, G., Hansen, H.N., De Chiffre, L., Micromilling of hardened tool steel for micro mould making applications. *Int. J. Mater. Process. Technol.* (2005) 167(2–3), pp. 201–207

[2] Masuzawa, T., State of the Art of Micromachining. *CIRP Ann. – Manufacturing Technology* (2000) 49(2), pp. 473–488

[3] Taniguchi, N., Current Status in, and Future Trends of, Ultraprecision Machining and Ultrafine Materials Processing. *CIRP Ann.* (1983) 32(2), pp. 573–582

[4] Alting, L., Kimura, F., Hansen, H.N., Bissacco, G., Micro Engineering. *CIRP Ann.* (2003) 52(2), pp. 635–658

[5] Arentoft, M., Eriksen, R.S., Hansen, H.N., Tooling Process Chains and Concepts, in *Micro Manufacturing Engineering and Technology*, Qin, Y. (Ed.) (2010), Elsevier, pp. 287–297

[6] Bissacco, G., Hansen, H.N., Tang, P.T., Fugl, J., Precision manufacturing methods of inserts for injection moulding of microfluidic systems. *Proc. of the ASPE Spring Topical Meeting* (2005) Columbus, Ohio, USA, pp. 57–63

[7] Bissacco, G., Hansen, H.N., Tang, P.T., Holme, C., Validation of alternative tooling process chains for nickel based glass moulding moulds. *Proc. of the 11th Int. Conf. of the European Soc. of Precision Engineering and Nanotechnology, EUSPEN* (2011) 21–27 May, Como, Italy

[8] Bissacco, G., lecture notes, PhD Summer School on Micro Mechanical Systems Design and Manufacture, lecture on tooling process chains (2007) Lyngby, Denmark

[9] Bissacco, G., Hansen, H.N., Slunsky, J., Modelling the Cutting Edge Radius Size Effect for Force Prediction in Micro Milling. *CIRP Ann.* (2008) 57(1), pp. 113–116

[10] De Oliveira, F.B., Rodrigues, A.R., Coelho, R.T., De Souza, A.F., Size effect and minimum chip thickness in micromilling. *Int. J. Mach. Tools Manuf.* (2015) 89, pp. 39–54

[11] Bissacco, G., Hansen, H.N., De Chiffre, L., Size Effects on Surface Generation in Micro Milling Hardened Tool Steel. *CIRP Ann.* (2006) 55(1), pp. 593–596

[12] Uhlmann, E., Mullany, B., Biermann, D., Rajurkar, K.P., Hausotte, T., Brinksmeier, E., Process chains for high-precision components with micro-scale features. *CIRP Ann. – Manufacturing Technology* (2016) 65, pp. 549–572

[13] Bissacco, G., *Surface Generation and Optimization in Micromilling* (2004), Technical University of Denmark, IPL246.04, PhD Thesis

[14] Uhlmann, E., Mahr, F., Lowenstein, A., Raue, N., Oberschmidt, D., Performance Characteristics of Coated Micro Milling Tools. *Proceedings of 9th International Conference on Coatings in Manufacturing Engineering* (2011), pp. 173–182

[15] Jameson, E.C., *Electrical Discharge Machining*, Society of Manufacturing Engineers SME (2001) USA

[16] Ferraris, E., Reynaerts, D., Lauwers, B., Micro-EDM Process Investigation and Comparison Performance of Al_3O_2 and ZrO_2 Based Ceramic Composites. *CIRP Ann.* (2011) 60(1), pp. 235–238

[17] Bissacco, G., lecture notes, PhD Summer School on Micro Mechanical Systems Design and Manufacture, lecture on modelling of micro material removal processes (2010) Lyngby, Denmark

[18] Bissacco, G., Valentincic, J., Hansen, H.N., Wiwe, B.D., Towards the effective tool wear control in micro EDM milling. *Int. J. Adv. Manuf. Technol.* (2010) 47(1), p. 3, DOI: 10.1007/s00170-009-2057-0

[19] Egashira, K., Matsugasako, A., Tsuchiya, H., Miyazaki, M., Electrical discharge machining with ultralow discharge energy. *Precis. Eng.* (2006) 30(4), pp. 414–420, DOI: 10.1016/j.precisioneng.2006.01.004

[20] Uhlmann, E. et al, Micro EDM, in *Micro Manufacturing Engineering and Technology*, Qin, Y. (Ed.) (2010), Elsevier, pp. 39–58

[21] Uhlmann, E., Piltz, S., Doll, U., Machining of micro/miniature die and moulds by electrical discharge machining - recent development. *J. Mater. Process. Technol.* (2005) 167(2), pp. 488–493

[22] Kawakami, T., Kunieda, M., Study on Factors Determining Limits of Minimum Machinable Size in Micro EDM. *CIRP Ann.* (2005) 54(1), pp. 167–170

[23] Bissacco, G., Presentation at CIRP GA (2011) Budapest, Hungary

[24] Bleys, P., Kruth, J. P., Lauwers, B., Sensing and Compensation of Tool Wear in Milling EDM. *J. Mater. Process. Technol.* (2004) 149, pp. 139–146

[25] Jung, J. W., Jeong, Y. H., Min, B. K., Lee, S. J., Model-based Pulse Frequency Control for Micro-EDM Milling using Real-time Discharge Pulse Monitoring. *J. Manuf. Sci. Eng.* (2008) 130(3), pp. 1–11

[26] Aligiri, E., Yeo, S. H., Tan, P. C., A New Tool Wear Compensation Method Based on Real-time Estimation of Material Removal Volume in Micro-EDM. *J. Mater. Process. Technol.* (2010) 210, pp. 2292–2303

[27] Bissacco, G., Hansen, H. N., Tristo, G., Valentincic, J., Feasibility of wear compensation in micro EDM milling based on discharge counting and discharge population characterization. *CIRP Ann.* (2011) 60(1), pp. 231–234

[28] Bissacco, G., Tristo, G., Hansen, H. N., Valentincic, J., Reliability of electrode wear compensation based on material removal per discharge in micro EDM milling. *CIRP Ann.* (2013) 62(1), pp. 179–182

[29] Bissacco, G., Hansen, H. N., Fugl, J., Tang, P. T., De Chiffre, L., Manufacturing of an injection moulding mould insert with three-dimensional micro features by micromilling. *Proc. of the 5th Euspen International Conference* (2005) Montpellier, France, pp. 505–508

[30] Tosello, G., Bissacco, G., Tang, P. T., Hansen, H. N., Nielsen, P. C., Micro tools manufacturing for polymer replication with high aspect ratio structures using µEDM of silicon, selective etching and electroforming. *Int. J. Microsyst. Technol.* (2007) 14(9–11), pp. 1757–1764

6 Ultra-Precision Machining Technologies for Micro Injection Mold Making

Oltmann Riemer, Carla Flosky

Ultra-precision machining technologies have been developed into flexible, fast, and reliable techniques for generating complex high-quality surfaces that either cannot be produced in any other way or are more expensive to manufacture using alternative techniques. Products with optical features that require ultra-precision manufacturing are a part of our daily life, for example, mobile phones, barcode scanners, reflective tapes, and contact lenses. All these goods are mass produced by replication processes like injection or compression molding and depend on the quality of ultra-precise molds [1, 2].

Taniguchi defined ultra-precision machining as technologies that achieve the highest possible dimensional accuracy [3]. The achievable form accuracy is in the sub-micrometer range (0.1 to 1 μm, depending on the size and shape of the workpiece) and the feasible surface roughness reaches down to a few nanometers, often in the range between 1 and 10 nm, reflecting the requirements for applications for visible light. In order to achieve this accuracy, highly precise diamond tools and ultra-precise machine tools are combined in the manufacturing process [2, 4].

A single crystal diamond is the most commonly used tool material for ultra-precision machining as it allows for sharp, accurate cutting edges with radii down to a few tens of nanometers and its very low wear when machining materials such as non-ferrous metals. In addition, it has many properties that suit ultra-precision processes, such as the highest hardness, high thermal conductivity, tool edges with nanometric shape accuracy, high wear resistance, and low friction [5]. The anisotropy of the abrasive wear of diamond as a crystalline material must be considered in the tool design, since the tool's life depends on the hardness of its rake and clearance faces [6].

Machine tool performance is one of the main sources of form deviation and insufficient surface roughness of machined workpieces. In ultra-precision machining, natural granite, polymer concrete, and other materials with high stiffness and damping properties are adopted in order to minimize or eliminate vibration and deformation effects during surface generation [7–9]. Additionally, high internal damping [8], aero-/hydro-static slides with low friction [9], high thermal stability

[10], and nanometric tool positioning [11] have been applied to fulfill and improve the machining requirements of ultra-precision technologies. Nowadays, ultra-precise machine tools with nanometric resolution are commercially available [2, 4].

Typical workpiece materials for ultra-precision machining with defined cutting edges are aluminum alloys, copper alloys, electroless nickel-phosphor plating, and polymers [2, 12, 13]. Nowadays, due to the ever-growing demands for specialized products and the implementation of very low cutting depths and specialized machining techniques, the spectrum of workpiece materials has expanded to hard and brittle materials such as silicon [14] as well as wear-inducing metals such as steel [15]. However, rapid diamond tool wear is still a decisive drawback.

In order to address different material properties and accuracy requirements, ultra-precision machining encompasses a large range of manufacturing technologies, which can be classified according to workpiece material characteristic as well as material removal, see Figure 6.1. Diamond machining technologies, like ultra-precision turning and milling, play a dominant role when ductile and certain hard materials have to be machined in optical quality. Materials exhibiting a significant amount of brittle fracture, like optical glasses, ceramics, carbides, semi-conductors, and certain crystals, require abrasive machining, i.e., ultra-precision grinding and polishing.

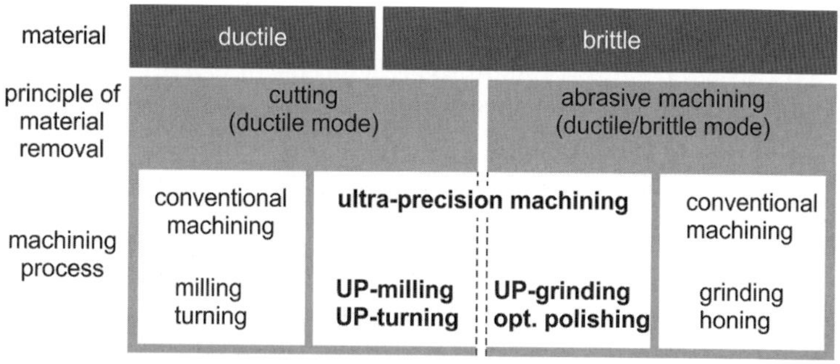

Figure 6.1 Classification of ultra-precision machining depending on material, material removal mechanism, and manufacturing process, adapted from [16]

6.1 General Aspects of Ultra-Precision Machining

Ultra-precision machining comprises several manufacturing technologies to generate functional surfaces and structures, usually in optical quality. Therefore, premachining has to guarantee the main material removal that provides the part shape beforehand. Usually, workpiece preparation involves the following steps: generating the overall three-dimensional geometry by conventional machining processes; machining the back side of the workpiece so that it acts as a clamping face as well as any further surfaces for clamping and handling; and machining of alignment reference features. In addition to workpiece preparation, the vast majority of ultra-precise machining processes require the functional geometry to be made up by CAD or specialized optical design suites into the machining code by dedicated CAM tools. Also to be considered for ultra-precision manufacturing technologies is the careful handling and cleaning of parts throughout the manufacturing chain to prevent damaging the high-quality surfaces machined.

For ultra-precision machining technologies, accurate workpiece clamping is vital for a high surface quality and shape conformity. Clamping includes workpiece centering and alignment as well as precision balancing; the latter is of highest importance when large, heavy workpieces or off-axis parts are machined in order to avoid process-induced vibrations.

Tools, including diamond turning or milling tools as well as grinding wheels or pins, must also be precisely aligned with respect to the workpieces' and machine tools' coordinate systems. Deviations can lead to significant aberrations in the machined parts' geometry. In some cases, the machining of witness samples might be required to determine the tool radius and the exact tool position within the machining system. Additionally, diamond machining does not require high cutting speeds, e. g., in turning on-axis parts: an identical nanometric surface finish can be achieved even in the center of the part, where cutting speed v_c reaches zero. This is a major difference to conventional cutting processes.

The cooling and lubricating technologies used in ultra-precision machining vary strongly according to the characteristics of the manufacturing process: diamond turning and milling processes require a spray mist for lubrication and chip removal from the cutting zone to avoid any scratching on the final surface and must be appropriately adjusted to achieve these purposes; grinding usually involves higher amounts of cooling fluid using mostly conventional cooling methods for oil- or water-based coolants; and polishing processes use water- or oil-based fluids with added abrasive grains for material removal, which are then named polishing suspension or slurry. The amount of lubricant or slurry, chemical composition, way of delivery, pressure, etc., determine the effectiveness of these abrasive processes.

6.2 Diamond Machining

6.2.1 Diamond Turning

In diamond turning processes, the cutting motion is provided by the rotating workpiece, while the tool is moved in relation to the surface, thereby copying the geometry of the diamond tool into the surface to determine the shape of the desired part. In the most basic set-up, only two controlled axes (for generating the feed and infeed motion) and a spindle are necessary for the generation of plane and rotationally symmetric surfaces or structures, such as Fresnel lenses. Diamond turning can also be used for on-axis, symmetrical part geometries, such as spherical shapes, but also for off-axis geometries with an asymmetrical design, i. e., aspheric mirrors like parabolic or elliptical surfaces.

The structure's shape can either be determined by the geometry of the diamond tool (profile-turning) or by modulation of the infeed depth (form-turning). Profile-turning can generate circular or spiral-type structures by plunging the diamond tool into the surface according to the radial position of the tool on the surface (Figure 6.2a) or by superposition of the tool geometry and the applied feed (Figure 6.2b). More complex shapes can be generated when contouring the desired shape by controlling both linear axes (Figure 6.2c). Applying a dynamic modulation of the cutting depth (Figure 6.2d), with so-called tool servo technologies, further enhances the geometrical, and particularly the structuring, capabilities in ultra-precision turning (see next section). Examples of diamond-turned functional surfaces include Fresnel lenses [17], and, in the case of fast tool servo turning, facetted mirrors and micro lens arrays [18].

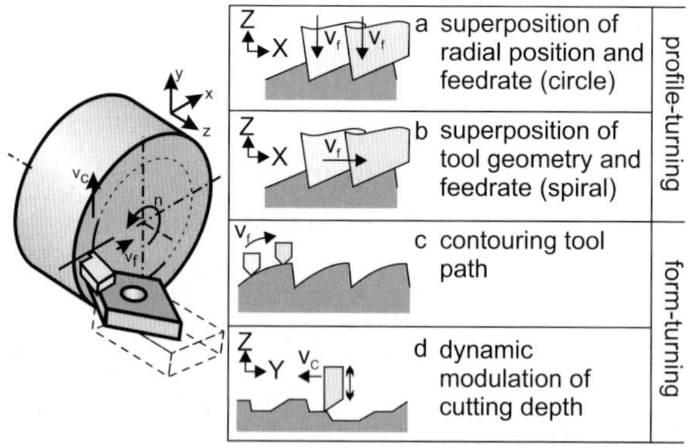

Figure 6.2 Principle of diamond turning and kinematic variations for structure generation

The inherent advantage of turning operations is the high machining speed, since the process time is scaling only linearly depending on the diameter of the workpiece. The constraints of diamond turning are mostly the kinematic roughness (i.e., influence of feed and tool nose radius), the dynamic machine performance, as well as the intrinsic material response.

Structures without rotational symmetry can be produced by turning operations when the cutting depth is dynamically modulated according to the radial and angular position on the surface (Figure 6.2d). Depending on the frequency of the modulation and the applied device, these processes are called slow slide, when using the machine tool's own slides, or fast tool servo turning, when specific accessory devices are used. In both cases–slow slide and fast tool servo–the devices are operated in a master-slave configuration, usually with the main spindle ruling the machining process.

The fast tool servo (FTS) concept was originally conceived for increasing the accuracy of an ultra-precision lathe [18] and was applied later for the non-circular turning of aspheres with a small deviation from rotational symmetry [2, 19]. FTS machining has been widely applied to manufacture micro components and complex micro structured surfaces such as micro prisms, lens arrays, torics, and off-axis aspheric surfaces with high accuracy and cost effectiveness [20]. The depth of cut is modulated dynamically according to the radial and angular position of a cutting tool on the workpiece surface (see Figure 6.2d). The high shape accuracy of cutting tools as well as the high resolution of synchronous multiaxial motion in ultraprecise machine tools are key advances in achieving the high surface and structuring qualities required. According to the drive mechanism of actuators, FTSs can be classified into hydraulic FTSs, piezoelectric FTSs, magnetostrictive FTSs, Lorentz force FTSs, and electromagnetically-driven FTSs. These systems have strokes from several micrometers to a few millimeters with frequencies from 20 Hz to 23 kHz [21].

Figure 6.3 Machine set-up for diamond turning with a nano fast tool servo

FTS systems have been further developed for dedicated operations and structuring applications. Brinksmeier et al. [22] developed the nano fast tool servo (nFTS), featuring a stroke of up to 500 nm at a bandwidth of 5 kHz and higher (see Figure 6.3). The designed nFTS can generate a variation of the undeformed chip thickness within the nanometer range. This enables the processing of diffractive micro structures by combining a classical diamond turning process with a nano fast tool servo unit with a positioning accuracy of 4 nm. For the machining of blaze (saw tooth-shaped) structures, diamond tools with a knife geometry are used (see Figure 6.4). Due to this tool geometry and tool servo technology, it is possible to generate 2000 structure elements per revolution, each at an individual height level. Thus, it is possible to machine diffractive optical elements (DOEs) like holograms, in a single step process with sub-micron precision and optical grade surface roughness ($Sa < 10$ nm) [23]. These parts can be used directly as reflective metal optics [24] or applied as mold inserts for injection molding [25], see Section 6.4.3.

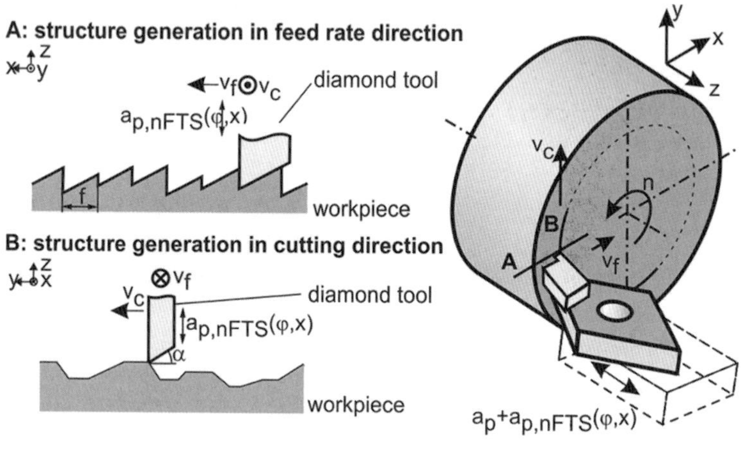

Figure 6.4 Working principle of the nano fast tool servo

End-fly-cutting-servo (EFCS) systems feature four-axis servo motions that combine the concepts of a fast/slow tool servo and end fly-cutting (see Section 6.2.2). A variety of nano structures can be generated with high accuracy (as a secondary structure on free-form surfaces or micro aspheric arrays), which is conventionally difficult to achieve. The method is useful to generate hierarchical micro nano structures [26].

The virtual spindle-based tool servo (VSTS) diamond turning method is capable of generating discontinuously structured micro optics as arrays by a fast or slow tool servo, greatly increasing the feasibility for machining discontinuous arrays of micro optics on both planar and free-form surfaces [27]. A virtual spindle axis is defined at a specific position by combining the translational and rotational servo motions of the machine tool. Thereby, it is possible to generate a micro structure cell array with a shape that can be machined in a fast or slow tool servo by controlling the virtual spindle axis to sequentially pass through the center of each cell. The aspect ratio of the structures is up to 1. All these dimensional parameters are dependent on the complexity of the structure to be machined.

6.2.2 Diamond Milling

Ultra-precision milling or diamond milling is one of the most flexible and efficient processes in ultra-precision machining. As in conventional milling, the tool rotates while the workpiece performs a comparably slow translational or rotational motion defined by a numerically controlled path. By using three numerically controlled axes, a wide range of shapes can be manufactured in optical quality by ultra-precision milling. Due to the various milling processes available, there is a large number of possibilities for manufacturing complex shapes and structured surfaces. These milling processes are classified as face and peripheral milling, as shown in Figure 6.5 [28]. In face milling, the tool rotates perpendicularly to the workpiece surface, while in peripheral milling, the tool's rotational axis is parallel to the surface to be machined.

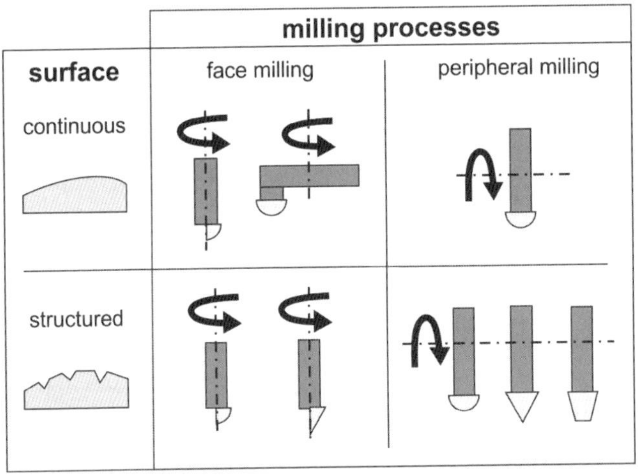

Figure 6.5 Classification of ultra-precision milling processes

Usually, milling tools used for the manufacture of plane, spherical, aspherical, and free-form surfaces have ball-end or half-arc geometry. The achievable kinematic surface roughness in face milling is directly constrained by the cutting edge radius as well as the overall fly-cutting radius, while peripheral milling allows a faster machining of the same surface compared to face milling. Ball-end milling, however, is more advantageous when manufacturing free forms, since it offers a smaller cutting edge radius and, therefore, smaller radii of curvature as well as steeper surface slopes are feasible.

Furthermore, peripheral milling is differentiated into linear and punctiform tool contact. Linear contact occurs when the peripheral milling radius of the tool is the same as the radius of curvature of the workpiece surface. The tool engagement is located in a discrete line on the surface to be machined and the contour of the tool or the tool path is directly copied on the workpiece. In contrast, in a point contact, the material removal takes place in a very limited and toric area. The size of this area depends on the ratio between infeed, rotational speed, and distance between parallel tool paths, but is always several orders of magnitude smaller than the surface to be machined [28].

The achievable quality of peripheral milling depends strongly on the choice of machining parameters and milling strategy, which is divided into raster and spiral milling. Raster milling can be carried out with alternating milling direction (bi-directional) or in a single direction (uni-directional) [28]. An example of a free-form surface generated by raster milling on a state-of-the-art three-axis ultra-precision machine is shown in Figure 6.6. A shape deviation of less than 0.3 µm (peak-to-valley) and a surface roughness, *S*a, of less than 4 nm is the current standard. The machining of free-form surfaces in such an optical quality is only possible due to fast computer numerical control of machine tools, since the rate at which slide positions have to be updated during machining is of the order of a hundred commands per second [2].

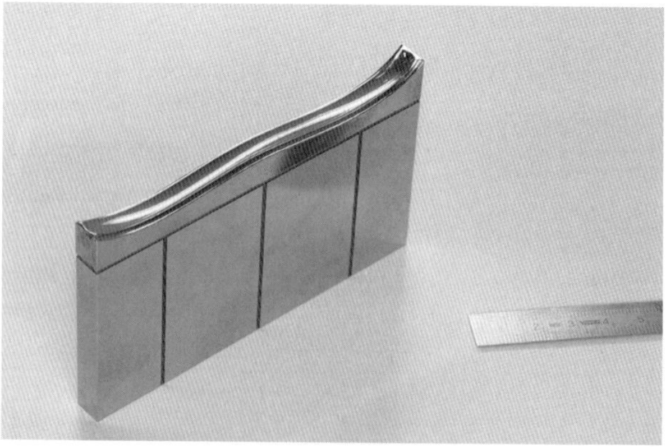

Figure 6.6 Mold insert for an F-theta lens, machined by diamond raster milling on a three-axis ultra-precision machine

Several technological applications require ultra-precise gratings and prismatic structures, for example in the field of display technologies and for retroreflective components. In order to manufacture surfaces with grooves, arrays, or intersecting structures, tools with a cutting edge with a trapezoidal or V shape are used. Parallel grooves can be machined by peripheral milling, while prismatic structures are generated by superposition of several intersecting groove systems. The geometry of the diamond tool defines the cross-sectional profile of the grooves [28].

The orientation of the milling tool, the workpiece, and tool alignment are crucial in ultra-precision milling. For example: to machine a prismatic structure with triangular corner cubes (retroreflectors), it is necessary to align the axis of the round table precisely to the milling plane to ensure that the single grooves will intersect precisely in one point and generate a high quality prismatic structure. The geometrical tolerance is usually only a few micrometers or below [28].

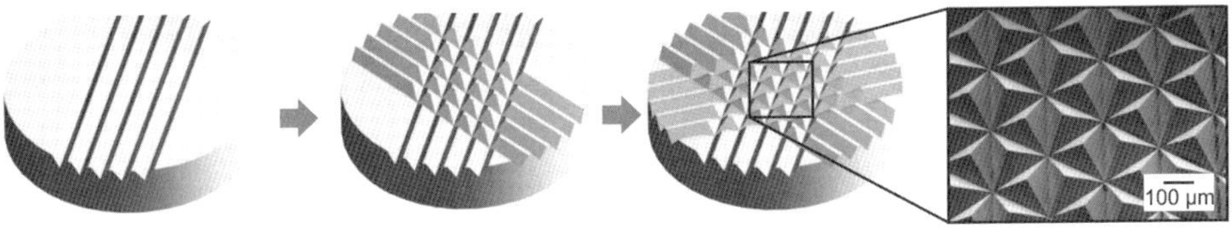

Figure 6.7 Principle for the generation of a retroreflector array and SEM image of a retroreflective prismatic structure with triangular corner cubes

Ultra-precise face milling, also called ball-end milling, is the most flexible process regarding the spectrum of machinable geometries. The achievable radii of curvature are only limited by the tool radius itself. However, the machining time is in an inverse proportion to the tool radius. Additionally, the shape accuracy of face milling depends on the alignment between the milling tool and tool spindle: the rotational axis of the cutting edge must be perfectly aligned to the axis of the tool spindle. Otherwise, distorted structure geometries or artifacts will occur.

Ultra-precision fly-cutting is essentially a further development of ultra-precision diamond milling, where the cutting velocity is constant and the material removal is discontinuous. Due to these characteristics, the cutting efficiency of fly-cutting processes is lower than that of diamond turning. Due to the discontinuous cutting mechanism, the surface generation in a fly-cutting process is governed by its intermittent tool–workpiece contact, the reproduction of tool geometry on the machined surface, material removal involving pile-up, burr, as well as material deformation [4].

A machine tool suitable for fly-cutting must have high stiffness and an effective damping system due to the highly dynamic characteristics of the process. Cutting strategies, tool path, and kinematic motion errors are vital to the achievement of high form accuracy. Additionally, cutting conditions, tool geometry, cutting strategies, and tool wear affect surface generation [29]. Face milling was originally used to machine large, flat surfaces in ultra-precise quality [30–32]. In the most recent research, this fly-cutting method was used with tool servo technologies to manufacture free-form surfaces with structures ranging from micro to nanometers [26, 33, 34]. Face milling processes are able to generate large, flat [35], and free-form [29] surfaces as well as grooves and micro structures [36, 37] with continuous, high surface quality [4].

6.2.3 Shaping

Shaping processes, also called planing of chiseling processes, are cutting technologies that apply no rotational motion to generate the desired surface, which can be either continuous or structured. The cutting motion is carried out at a comparably low speed and is usually carried out by the machine axes, i. e., usually linear axes, but also rotary axes might be applied. A main advantage of shaping processes is that machine dynamics are not crucial for the achievement of high quality surfaces. Due to the relatively low speeds and small accelerations, discontinuous tool paths can be processed to process steep and unsteady structures or contours. But, therefore, the major disadvantage of shaping processes is their long machining times.

Diamond micro chiseling (DMC) is a shaping process that enables the generation of prismatic micro structures, such as cube corner retroreflectors in sizes between 50 and 500 µm, by combining V-shaped diamond tools with dedicated process kinematics. The general concept is to machine each facet of a structure individually, by performing a triangular or trapezoid cutting motion (see Figure 6.8). The workpiece is rotated to the required angular orientation by the C-axis and the tool is repositioned between every cut in a different direction [38].

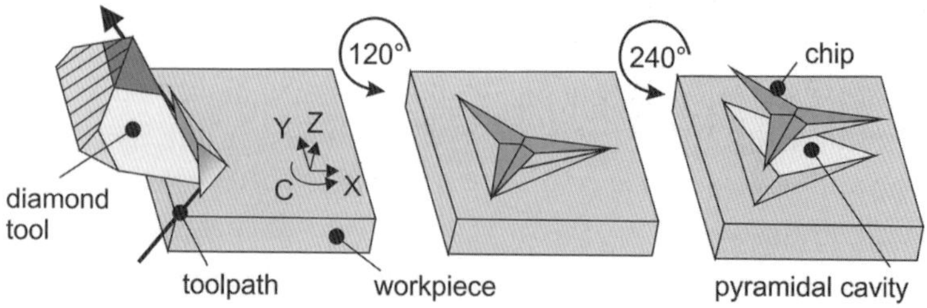

Figure 6.8 Process kinematics of diamond micro chiseling (DMC) for pyramidal structures. Adapted from [39]

Due to the kinematics of diamond micro chiseling, the primary motion of the tool is sidewise, which allows the rake face of a diamond tool to switch with the clearance face of conventional tools. The rake and clearance angles are directly reproduced in the workpiece, which requires a complex diamond shape. Tools suitable for diamond micro chiseling have a nose radius for wear resistance and cutting edges that are either sharp or slightly rounded with $r_\beta \approx 200$ nm [38].

A great advantage of DMC machined micro structures as functional optical surfaces is the surface finish that can be generated on the facets of the cavities. In previous investigations, the influence of the undeformed chip thickness h and the cutting speed v_c indicates that an optical surface finish can be achieved by low values of $h < 3$ µm and $v_c < 6$ mm min^{-1} [40].

A typically sized cavity with a structure size of 100 µm, usually used for machining cubical retroreflectors, is shown in Figure 6.9 for comparison. For machining functional surfaces, a large amount of uniform structures is usually necessary. Thus, large arrays of triple sided cavities were machined by DMC to generate hexagonal cube corner retroreflectors of a miniature size in contrast to triangular cube corner reflectors, which can be machined by fly-cutting (see Figure 6.7). The largest structure machined with this technique is a pyramidal cavity with a structure size of 1 mm, which is shown in Section 6.4.2 [38].

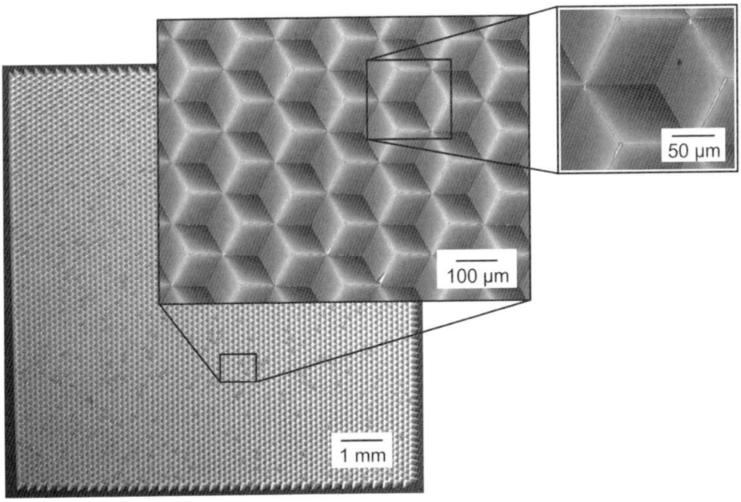

Figure 6.9 Typical retroreflective mold surface machined by diamond micro chiseling (DMC)

6.3 Abrasive Machining

As it is usual in conventional machining, hard and brittle materials like glass, carbides, and ceramics have to be machined by abrasive processes (see the classification in Figure 6.1). Because of their non-deterministic aspect and the necessary additional machining devices (spindles, pumps, dressers, etc.), the manufacturing of high precision parts by grinding or polishing processes is more complex than by processes with geometrical defined cutting edges, such as turning and milling. In this sense, ultra-precision grinding is primarily deployed in the machining of high quality, functional parts from hard and brittle materials, with the primary goal of manufacturing parts with a high surface finish, high form accuracy, and high surface integrity.

6.3.1 Process Chains

In typical process chains for generating optical surfaces in hard and brittle materials, like certain mold materials, consecutive grinding and polishing steps have to be applied (see Figure 6.10). Even though many efforts have been made in the development of grinding processes to achieve ductile mode machining, a significant amount of sub-surface damage usually remains. Thus, ultra-precision grinding as a path-controlled machining process with stiff and precise machine tools provides the required shape accuracy; subsequent polishing, generally a force-controlled process, removes the top, most damaged, material layer and ensures the required nanometric surface roughness.

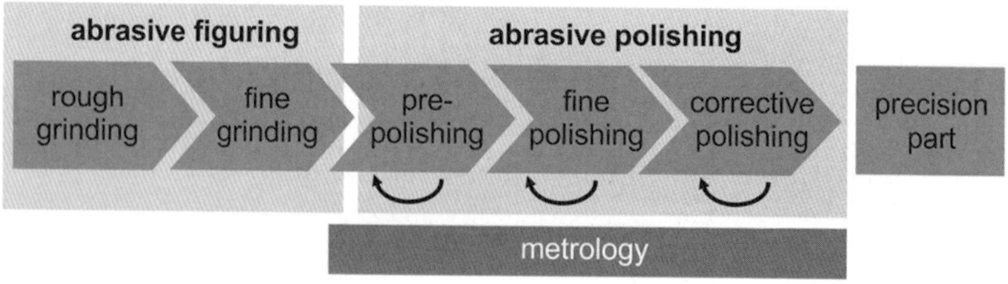

Figure 6.10 Principal process chain for the ultra-precision machining of hard and brittle materials

In this sense, polishing is first carried out to keep the part's shape generated by grinding, remove sub-surface damage, and smooth the surface. This is done by applying polishing tools larger than the part's aperture with an averaging characteristic. Nevertheless, polishing with sub-aperture tools is used not only for smoothing, but also for removing material in a controlled way and, thus, it changes the shape of the workpiece intentionally; this process employs particular dwell times of the polishing pad in relation to the position of the part and is called correction polishing. Stable material removal functions, also called "footprints", iterative measurement steps, and a comprehensive understanding of the polishing process including material removal mechanisms is essential.

6.3.2 Ultra-Precision Grinding

Although there is no unanimous definition for ultra-precision grinding, Brinksmeier et al. [16] attempted to summarize most of the available definitions by stating: *ultra-precision grinding is a material removal process with fixed abrasives that are in intermittent contact with the workpiece surface and where the abrasive tool is characterized by at least one statistical distributed parameter.* Additionally, an ultra-precision grinding process should be able to deliver a surface finish suitable for optical applications, such as shape accuracy in the micrometer range, nanometric surface roughness, and avoidance of sub-surface damage. Generally, grinding of hard and brittle materials leads to micro cracks and lower surface quality compared to ultra-precision machining of ductile materials. This is due to the brittle nature of the material removal process. Therefore, the transition from brittle to ductile material removal is considered to be essential in ultra-precision grinding, and machining in ductile regime is intended. Figure 6.11 summarizes the determining requirements and decisive criteria for the characterization of ultra-precision grinding according to Brinksmeier et al. [16].

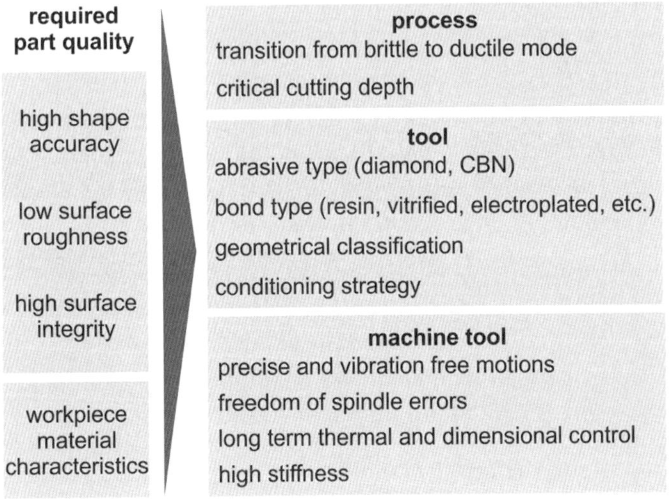

Figure 6.11 Requirements and criteria for the characterization of ultra-precision grinding, adapted from [16]

Damage-free and, thus, ductile grinding calls for a maximum chip thickness $h_{cu,max}$ below the critical and material specific chip thickness $h_{id,crit}$ in order to avoid crack initiation. According to Marshall et al. [41], the critical material specific chip thickness $h_{id,crit}$ depends on the Young's modulus E, the Knoop hardness HK, and the critical fracture toughness K_c:

$$h_{id,crit} \sim \left(\frac{E}{HK}\right)\left(\frac{K_c}{HK}\right)^2 \tag{6.1}$$

The critical chip thickness can be estimated from Eq. 6.1. In order to define the ductile–brittle transition, Bifano et al. [42, 43] determined the proportionality factor by indentation tests and the material specific chip thickness $h_{id,crit}$ based on grinding processes. In various grinding experiments of different brittle materials, this factor was found to be approximately 0.15 [16].

In ultra-precision grinding, diamond and cubic boron nitride (CBN) are the two most frequently applied types of abrasives [44, 45]. Both natural and synthetic diamond can be used, having high wear resistance, heat conductivity, hardness, and a low coefficient of friction [46, 47]. A significant disadvantage of diamonds is their chemical affinity to a number of metals, like iron-based materials such as steel, which leads to the diffusion of carbon into the metal and, thus, to high tool wear when grinding at high temperatures [44, 48]. This effect means that diamond tools are restricted to non-ferrous metals for diamond turning and milling. For this reason, diamond abrasives are mainly used for the grinding of non-ferrous materials, such as carbides, silicon, glass, and ceramics. Cubic boron nitride (CBN) has a higher thermo-chemical stability compared to diamond [44, 45] and is the best alternative when grinding metals that react with diamond.

Besides abrasive grains, the bonding of abrasive tools is critical for achieving high surface quality and superior grinding performance. The bonding system holds the abrasive grains of the grinding tool as long as they are sharp and, ideally, releases the grains when they are blunt [49]. The major bond systems are metal-, resin-, and vitrified bonding [46, 49]. Further information on different bonding types can be found in [16].

Most grinding wheels applied in ultra-precision grinding are made with diverse design concepts, e.g., with an undefined or defined grain setting, grooved wheels, or cup wheels. Typically, wheel diameters range from 50 to 400 mm and abrasive grain sizes from fine (~0.025 mm) to coarse (~200 mm). Generally, grinding tools have an undefined grain distribution [16]. Besides wheels, grinding pins are widely used to generate micro structures and small cavities with complex, three-dimensional features. These grinding tools enable the structuring of wear resistant and chemically inert materials that can be of great advantage in molding applications [50]. Grinding pins can be manufactured in various ways, e.g., cylindrical blanks with diamond grains in an electroplated bonding system with diameters down to 5 µm [51]; pins with a sintered bronze layer as bonding with diameters of about 200 µm [52]; and CVD-coated grinding pins, where a cylindrical blank is coated with poly-crystalline diamond grinding layers that are bonding and abrasive grains at the same time (minimal tool diameter is 50 µm) [53]. Figure 6.12 shows grinding tools with electroplated bonding and CVD-coated grinding pins.

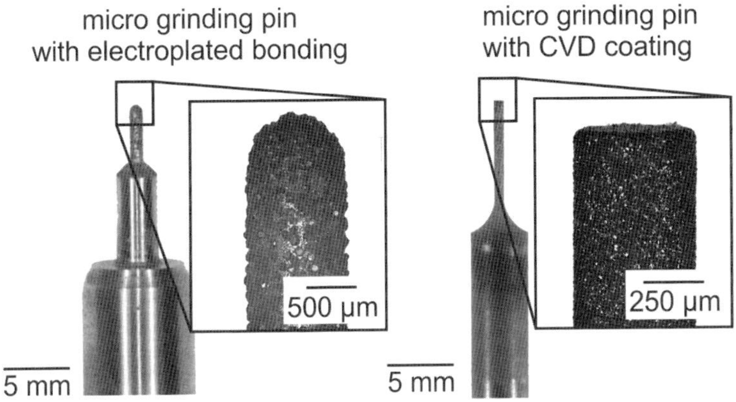

Figure 6.12 Micro grinding pins with electroplated bonding (left) and CVD diamond coating (right)

In ultra-precision grinding, the shape and sharpness of the grinding tool before and during the process has a considerable influence on grinding performance. Therefore, a successful grinding process requires a stable tool condition throughout the grinding process. Generally, there are two tasks to be accomplished by a tool conditioning process: profiling of the tool to achieve the desired shape (macro geometry); and tool sharpening to ensure a suitable grain protrusion (micro geometry). The techniques for tool conditioning are various and range from mechanical processes, like block sharpening, diamond dressing with a single diamond dresser or multiple grain dresser, and crushing, to electrochemical and photonic technologies such as EDM/ECM technologies (i.e., ELID grinding), and laser trueing [16, 49]. However, the dressing of micro grinding tools is often impractical or even impossible.

The demand for parts with features and structures at the micro- and nano-scale level is growing continuously [54]. At the same time, parts that must be made of hard and brittle materials, such as silicon, glass, and ceramics, e.g., as mold materials, can only be achieved by grinding. For this reason, micro grinding processes have been developed with the goals of decreasing grinding tool dimensions and increasing the achievable aspect ratio of surface structures. For the manufacturing of micro structures, two grinding methods are mostly used: peripheral grinding wheels for machining continuous structures (e.g., pyramid arrays) and grinding pins for discontinuous structures (see Figure 6.13) [55].

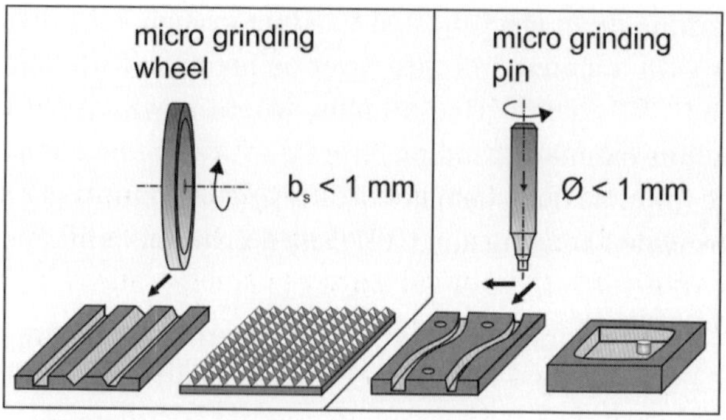

Figure 6.13 Micro grinding methods to manufacture continuous structures using grinding wheels (left) and discontinuous structures using grinding pins (right), adapted from [55]

In order to machine cavities in the micrometer range, Aurich et al. [51, 56] have developed a method for manufacturing grinding pins with a diameter of down to 13 µm, with diamond grains of about 1 to 3 µm electroplated onto a carbide blank in the desired grinding pin shape. Other works show that a feasible range for tool diameters lies at about 50 to 100 µm for micro grinding pins [57]. Dressing strategies can also be further developed to manufacture micro structures. Denkena et al. [58] used diamond profile rollers to generate micro profiles on vitrified grinding wheels, enabling the machining of micro grooves with a width of 40 µm and depth of 20 µm on the surface of jet turbine compressor blades.

Grinding pins of millimeter size also have dedicated applications, where it is necessary to machine cavities (usually for mold inserts) with an aperture in the millimeter range and with the same high surface quality and low shape deviation that can be expected from ultra-precision grinding processes. Figure 6.14 shows an ultra-precision grinding process using a grinding pin (diameter: 6 mm) to machine a spherical cavity in a mold insert made of tungsten carbide for application in glass pressing [59].

Figure 6.14 Ultra-precision grinding of a mold insert made of tungsten carbide with a grinding pin

6.3.3 Polishing

One of the characteristics of polishing processes is the ability to achieve low surface roughness, usually at comparably low material removal rates, which makes them ideal as a final manufacturing step for finishing ultra-precision parts. In addition, the low material removal with polishing processes allows the reduction, and even elimination, of surface or sub-surface damage caused by earlier machining processes [60–62] and therefore ensures the surface integrity of molds made from hard and brittle materials. Application fields for polishing processes are the finishing of mold inserts for parts with paramount shape accuracy and roughness, e. g., polymer optics, and parts with well-determined surface integrity like components for medical parts.

There are several ways to realize a polishing process, which can be based on different material removal mechanisms (see Figure 6.15) that, according to Nogawa et al. and Komanduri et al., can be divided into chemical and physical working principles, as well as a combination of both [63–65]. Physical working principles can be further divided into mechanical and non-mechanical principles. This diversity of applicable mechanisms justifies the plethora of polishing processes established in science and industry. Anyhow, mechanical abrasive polishing is still the most common and widespread variant.

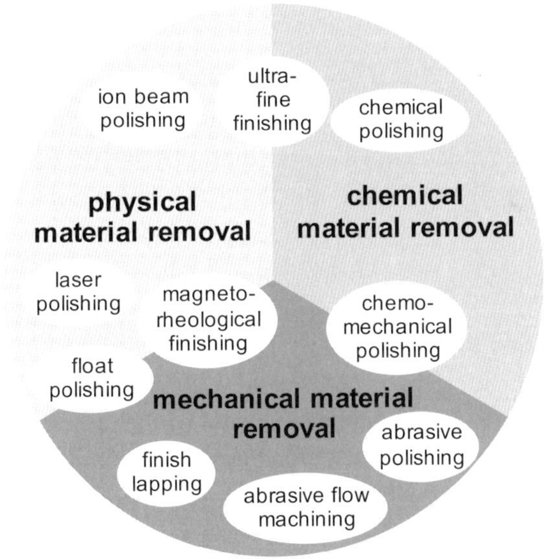

Figure 6.15 Classification of polishing processes according to their material removal mechanisms

The material removal of a mechanical-abrasive polishing process occurs mostly through abrasive grains that are added to the process in the form of a suspension, the so-called slurry. The motion of the polishing pad generates a motion relative to the surface to be machined that induces the material removal [66]. Therefore, the actual polishing tool is only established during the polishing process and originates from the interaction of the polishing pad, suspension, and abrasive grains. The polishing pad is pressed against the workpiece surface with a pre-defined force, which causes the grains to either roll between the pad and the workpiece or they are incorporated into the polishing pad to be carried by the pad over the workpiece's surface. As basic mechanisms, two- and three-body abrasion are distinguished [67]. By rolling, the abrasive grains can induce surface hardening and consequently material fatigue. The grains embedded into the polishing pad cause material chipping and ploughing that resemble a grinding process [68].

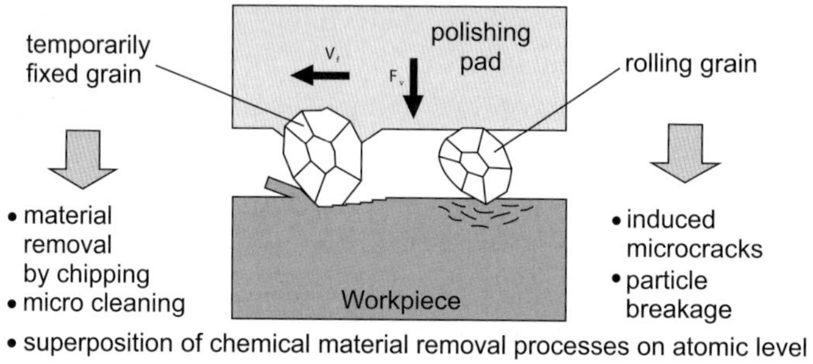

Figure 6.16 Material removal mechanism during abrasive polishing, adapted from [62]

The properties of the polishing pad are partially critical for the machining of structured surfaces. These need to have a stable form to allow a uniform contact to the workpiece surface, and be soft enough to accommodate abrasive grains. Common pad materials are pitch, polymers, and felt. Even hard wood is suitable for polishing structured surfaces. The abrasive grains, usually diamond with a grain size ranging from 0.5 µm up to a few micrometers, are added to oil- or water-based suspensions or pastes [67].

The Preston hypothesis [69] is one of the most commonly used equations to perform a quantitative analysis of a polishing process and describe the material removal:

$$\frac{dz}{dt} = K_\mathrm{p} \cdot p \cdot v_\mathrm{r} \tag{6.2}$$

where ${dz}/{dt}$ is the material removal per time unit, K_p is a proportionality factor (also called the Preston coefficient), p is the surface pressure, and v_r is the relative velocity between the workpiece and the polishing pad. Although this equation was originally developed for the machining of glass, it principally applies to all materials. The material properties as well as further relevant process parameters, such as suspension, polishing grains, properties of the polishing pad, etc., are summarized in the Preston coefficient. This coefficient must be determined through experiments for every different combination of polishing pad, suspension, polishing grains, workpiece material, etc., every so-called polishing system. The majority of characterization models for material removal by polishing are based on the Preston hypothesis [67].

The polishing process is affected by several factors that restrain the development of a deterministic process, such as suspension properties, grain size and shape, material of the polishing pad, process kinematics, workpiece material properties, machine stiffness, etc. Although polishing is one of the oldest manufacturing processes, the mechanisms of material removal are still not completely determined [70].

For finishing small and narrow cavities, as is usually the case with mold inserts, the use of rotating polishing pads is restricted. In most cases, the rotational motion is substituted by a linear motion (implemented through vibration) to achieve the required relative velocity between the pad and workpiece. Thus, the shape of the polishing pad can be adapted to the desired cavity's shape and it is possible to polish corners of concave features. The additional, typically sinusoidal vibration with micrometer amplitude range is superposed to the feed motion and even enables an increase in the relative velocity.

Robot-assisted polishing has become a common machining process for surface finishing with applications focused on rotational polishing processes for molds and dies without inner corner features [71–73]. The robot-assisted polishing of structured mold surfaces remains a novel research field that is currently being developed [74, 75]. In order to meet this challenge, metrology and machining technology are being closely linked to process measuring data to execute the polishing tasks.

6.4 Applications of Ultra-Precision Machining

In this chapter, four different examples of ultra-precision technologies and the manufacturing of mold inserts will be shown: Fresnel lenses manufactured by diamond turning, a micro beam splitter made by diamond micro chiseling, holograms machined with a nano fast tool servo, and retroreflectors that use a polishing process for surface finishing. All the different ultra-precision technologies can be used for machine inserts for micro injection molding applications in various fields of technology.

6.4.1 Fresnel Lens

Fresnel lenses are compact optical components with rotational symmetric structured surfaces and, therefore, diamond turning is the machining process of choice. Due to their space saving properties and vast application fields, i. e., imaging, photography, illumination, projection, etc., Fresnel molds may have sizes varying from several centimeters down to a few millimeters to fit in a wide range of products.

Fresnel lenses are thin optical elements mimicking the refracting function of a thick lens surface with the particular curvature separated into rings. Depending on the radial pitch and structure height, both can be the determining structure parameters, and whether the curved surface is transferred entirely into the structures or approximated by facets, a large scale of different dimensions is covered. In the example shown in Figure 6.17, a Fresnel structure with varying pitch, but constant structure height of 15 µm, was diamond turned with a 1 µm radius tool into a brass mold; the insert will be used for the replication of microscope lenses by injection molding.

Figure 6.17 Mold insert with diamond-turned Fresnel structure for the replication of plastic lenses

6.4.2 Micro Beam Splitter

Diamond micro chiseling can be used to manufacture micro structures with several different prismatic shapes. However, two micro optical components have been identified as promising: large arrays of miniaturized cube corner retroreflectors (i.e., a high amount of small structures, see Figure 6.9) and prismatic beam splitters (i.e., large-scale singular structures, see Figure 6.18).

The mold for the prismatic beam splitter consists of a pyramidal cavity, with dimensions of $2 \times 2 \times 2\ mm^3$, as can be seen in Figure 6.18, left. Such a mold cannot be manufactured by diamond milling since the tool radius does not allow for the sharp corners required by this optical component. Thus, the mold is machined in two separate parts: the cube structure along with the gate and runner for the injection molding process is machined by diamond milling, and the pyramidal part of the mold by diamond micro chiseling.

The pyramidal structure was divided into 176 layers in order to be machined by DMC with satisfactory results. The maximum chip width in the final layer was 1414 µm and, so far, it is the largest singular structure machined using diamond micro chiseling. Despite the high chip width, no tool wear was identified. An average surface roughness of 8.57 nm and a waviness of 5.48 nm were measured on the structure facets [76].

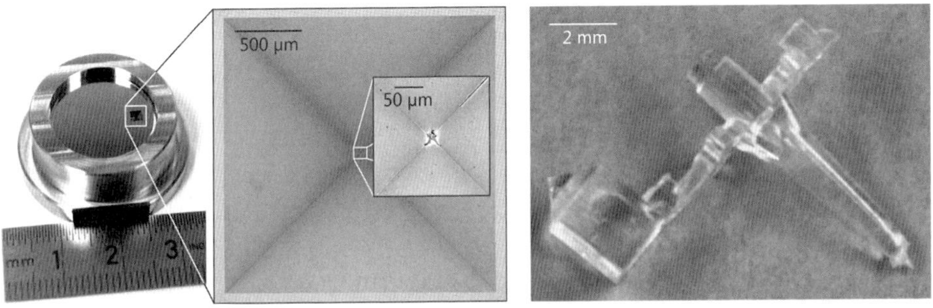

Figure 6.18 Mold insert with a prismatic cavity for the beam splitter (left) and molded part (right)

6.4.3 Diffractive Optical Elements

Diffractive optical elements (DOEs) are used to produce complex light patterns with precise dimensions on a specific plane. The dimension of the structures that is necessary to generate the optical effect is either within the range of the light's wavelength or is smaller, thus, usually in the sub-micron range for applications of visible light. The applications for DOEs are various, from three-dimensional metrology to laser technology and holograms [77]. The machining of diamond-turned holograms (DTH) can be achieved by a face-turning process combined with a piezo-driven nano fast tool servo (nFTS) for cutting depth modulation (see Figure 6.2d). Due to local height levels, diffractive holograms are capable of modulating the phase of laser radiation to generate a defined light intensity distribution. The maximum stroke of the nFTS measures 350 nm and is able to modulate the cutting depth with 4 nm positioning accuracy, which is necessary in order to achieve the high shape accuracy of the hologram. In combination with a spindle speed of 100 min^{-1}, the nFTS is able to machine 2000 segments per revolution, each with a different height level. The wedge-shape diamond tool combined with the process kinematics generates a spiral-shaped blaze structure on the surface [78].

There are several requirements regarding the mold material in order to achieve diffractive optics with an excellent surface finish. The functionality of holographic structures depends highly on the contouring accuracy and a minimal burr formation. Nickel silver shows a high hardness and does not cause any significant tool wear, so nickel silver is advantageous when machining sharp edges and guarantees fewer surface defects, a high contouring accuracy, and lower roughness values (Sa = 5 nm on top of the optical effective surfaces of the blaze structure) [22]. The optically effective surface, shown in Figure 6.19, provides the required structure angle and height level to reflect the incident light by shifting the phase of the reflected wavefront (see Figure 6.4, Section 6.2.1) [25].

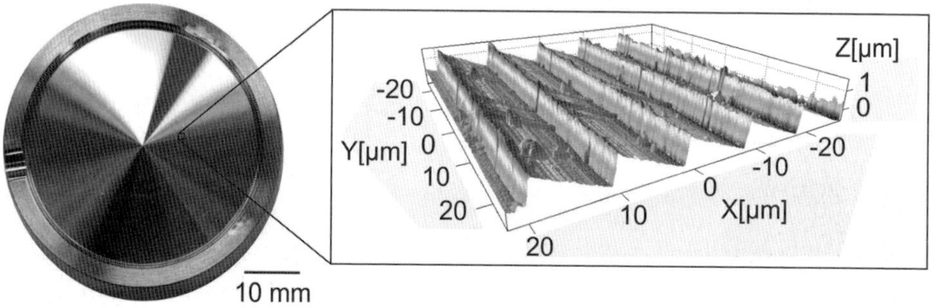

Figure 6.19 Diamond-turned mold insert and surface topography measured with a white light interferometer. The figure is presented in the color supplement

6.4.4 Retroreflectors

Retroreflectors are optical elements that redirect light beams antiparallel to their source. Important fields of application are traffic and lighting systems, e. g., traffic signs, safety clothing, and car lighting, as well as industrial measurement systems. The structure of a retroreflector is based on three perpendicular mirror surfaces in a cube corner arrangement. This structure can be miniaturized by assembling a large number of micro retroreflectors in an array either in a triangular or hexagonal fashion. The triangular shape can be machined by diamond fly-cutting with specific V-shaped tools (see Figure 6.7). However, these structures exhibit a limited range of intensity. On the other hand, retroreflectors with a hexagonal shape exhibit superior optical performance.

Hexagonal retroreflectors with a wrench size below approximately 1 mm require dedicated manufacturing processes like DMC (see Section 6.2.3); traditionally, molds for retroreflectors are assembled from millimeter-sized hexagonal pins in a mechanical frame, which is then replicated in a galvanic process and the galvanic replicated parts are then shaped as an injection molding insert. These galvanic retroreflector molds suffer from fabrication errors and considerable wear due to the manufacturing process and their subsequent use in injection molding. To remove degradations like flaws and wear marks from the mold's optical surfaces and avoid manual labour, a partially automated process chain using a six-axis robot is being developed, which combines surface metrology and machining technology. The geometrical constraints of the retroreflectors exclude the application of rotational polishing motions. Therefore, vibration polishing is used to eliminate surface flaws and improve surface roughness to values lower than 50 nm. The experimental set-up of the polishing system and tool design were developed as shown in Figure 6.20. The polishing of galvanized NiCo facets resulted in surfaces with near optical quality, therefore complying with the targeted objective: a surface roughness, *S*a, of less than 30 nm was achieved [74].

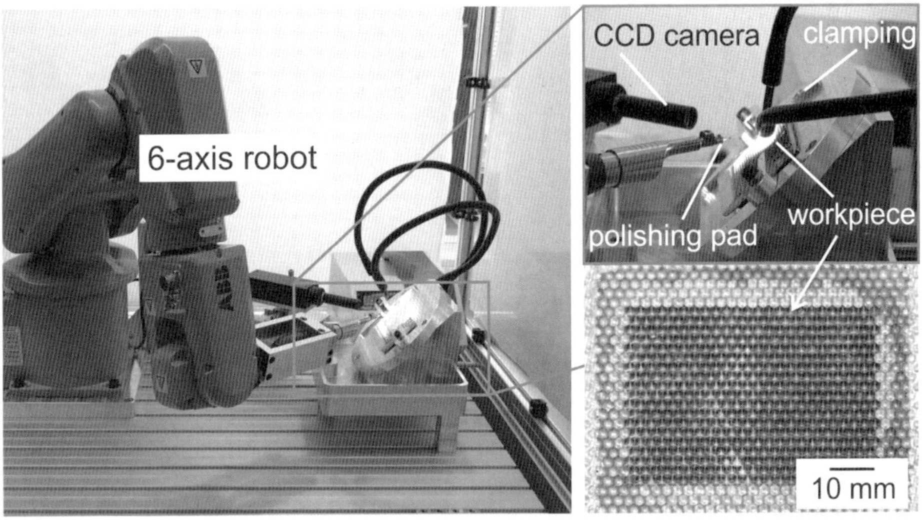

Figure 6.20 Robot system set-up with integrated vibration polishing device (left and upper right) as well as polished retroreflectors (lower right), adapted from [74]

References

[1] Bäumer, S., *Handbook of plastic optics* (2010) Wiley-VCH Publishing Co., Weinheim, Germany

[2] Brinksmeier, E., Preuss, W., Micro-machining. *Philos. Trans. R. Soc., A* (2012) 370, pp. 3973–3992

[3] Taniguchi, N., Current Status in, and Future Trends of Ultraprecision Machining and Ultrafine Materials Processing. *CIRP Ann. Manuf. Technol.* (1983) 32(2), pp. 573–582

[4] Zhang, S.J., To, S., Zhu, Z.W., Zhang, G.Q., A review of fly cutting applied to surface generation in ultra-precision machining. *Int. J. Mach. Tools Manuf.* (2016) 103, pp. 13–27

[5] Erdemir, A., Donnet, C., *Tribology of diamond, diamond-like carbon, and related films*, Bhushan, B. (Ed.) (2001) Modern Tribology Handbook, CRCPress, United States, pp. 871–908

[6] Grillo, S.E., Field, J.E., van Bouwelen, F.M., Diamond polishing: the dependency of friction and wear on load and crystal orientation. *J. Phys. D: Appl. Phys.* (2000) 33, pp. 985–990

[7] Moore, W.R., *Foundations of mechanical accuracy* (1970) Moore Special Tool Company, Bridgeport, USA

[8] McKeown, P.A., Wills-Moren, W.J., Read, R.F.J., Modjarrad, H., The design and development of a large ultra-precision CNC diamond turning machine. *Mater. Manuf. Processes* (1986) 1, pp. 133–157

[9] Furukawa, Y., Moronuki, N., Kitagawa, K., Development of ultra-precision machine tool made of ceramics. *CIRP Ann. Manuf. Technol.* (1986) 35, pp. 279–282

[10] Sriyotha, P., Nakamoto, K., Sugai, M., Yamazaki, K., Development of 5-axis linear motor-driven super-precision machine. *CIRP Ann. Manuf. Technol.* (2006) 55, pp. 381–384

[11] Mehta, M., Mital, A., *Recent Developments in Production Research* (1988) Elsevier, Amsterdam, pp. 52–58

[12] Gerchman, M.C., Specifications and manufacturing considerations of diamond machined optical components. *Proceedings of the Society of Photo-Optical Instrumentation Engineer's O-E Laser Conference (SPIE 607)* (1986), pp. 36–45

[13] Paul, E., Evans, C.J., Mangamelli, A., McGlauflin, M.L., Polvani, R.S., Chemical aspects of tool wear in single point diamond turning. *Precis. Eng.* (1996) 18, pp. 4–19

[14] Puttick, K.E., Whitmore, L.C., Zhdan, P., Gee, A.E., Chao, C.L., Energy scaling transitions in machining of silicon by diamond. *Tribol. Int.* (1995) 28, pp. 349–355

[15] Brinksmeier, E., Gläbe, R., Osmer, J., Ultra-precision diamond cutting of steel molds. *CIRP Ann. Manuf. Technol.* (2006) 55, pp. 551–554

[16] Brinksmeier, E., Mutlugünes, Y., Klocke, F., Aurich, J.C., Shore, P., Ohmori, H., Ultra-precision Grinding. *CIRP Ann. Manuf. Technol.* (2010) 59, pp. 652–671

[17] Wills-Moren, W.J., Modjarrad, H., Read, R.F.J., McKeown, P.A., Some aspects of the design and development of a large high precision CNC diamond turning machine. *CIRP Ann. Manuf. Technol.* (1982) 31, pp. 409–414

[18] Patterson, S.R., Magrab, E.B., Design and testing of a fast tool servo for diamond turning. *Precis. Eng.* (1985) 7, pp. 123–128

[19] Dow, T.A., Miller, M.H., Falter, P.J., Application of a fast tool servo for diamond turning of non-rotationally symmetric surfaces. *Precis. Eng.* (1991) 13, pp. 243–250

[20] Fang, F., Zhang, X., Weckenmann, A., Zhang, G., Evans, C., Manufacturing and measurement of freeform optics. *CIRP Ann. Manuf. Technol.* (2013) 62, pp. 823–846

[21] Lu, X.D., Trumper, D.L., Ultrafast Tool Servos for Diamond Turning. *CIRP Ann. Manuf. Technol.* (2005) 54(1), pp. 383–388

[22] Brinksmeier, E., Riemer, O., Gläbe, R., Lünemann, B., Kopylow, C., Dankwart, C., Meier, A., Submicron functional surfaces generated by diamond machining. *CIRP Ann. Manuf. Technol.* (2010) 59, pp. 535–538

[23] Meier, A., Diamond turning of diffractive micro structures. *Precis. Eng.* (2015) 42, pp. 253–260.

[24] Meier, A., *Diamantdrehen komplexer diffraktiver Mikrostrukturen* (2016), University of Bremen, Germany, PhD Thesis

[25] Holthusen, A.-K., Riemer, O., Schmütz, J., Meier, A., Mold machining and injection molding of diffractive microstructures. *J. Manuf. Process.* (2017) 26, pp. 290–294

[26] To, S., Zhu, Z.W., Zeng, W.H., Novel end-fly-cutting-servo system for deterministic generation of hierarchical micro-nanostructures. *CIRP Ann. Manuf. Technol.* (2015) 64(1), pp. 133–136

[27] To, S., Zhu, Z.W., Wang, H.T., Virtual spindle based tool servo diamond turning of discontinuously structured microoptics arrays. *CIRP Ann. Manuf. Technol.* (2016) 65(1), pp. 475–478

[28] Gläbe, R., Schönemann, L., Riemer, O., Herstellung funktionaler Oberflächenstrukturen mit Diamantwerkzeugen. *Jahrbuch 2010 Optik und Feinmechanik*, Prenzel, W.-D. (Ed.), Schiele&Schön (2010) 56, pp. 109–129

[29] Kong, L.B., Cheung, C.F., To, S., Lee, W.B., An investigation into surface generation in ultra-precision raster milling. *J. Mater. Process. Technol.* (2009) 209(8), pp. 4178–4185

[30] Montesanti, R.C., Locke, S.F., Thompson, S.L., Cass, C.L., Wulff, T.A., Vertical-axis diamond fly-cutting machine for producing flat half-meter-scaleoptics. *Optical Fabrication and Testing,* OSA Technical Digest (2000) Optical Society of America

[31] Liang, Y., Chen, W., Sun, Y., Luo, X., Lu, L., Liu, H., A mechanical structure-based design method and its implementation on a fly-cutting machine tool design. *Int. J. Adv. Manuf. Technol.* (2014) 70, pp. 1915–1921

[32] Li, M., Chen, M., Cheng, J., Xiao, Y., Jiang, W., Two important mechanisms damaging KH_2PO_4 crystal processed by ultraprecision fly cutting and their relationships with cutting parameters. *Appl. Opt.* (2013) 52, pp. 3451–3460

[33] Zhu, Z.W., To, S., Zhang, S.J., Theoretical and experimental investigation on the novel end-fly-cutting-servo diamond machining of hierarchical micro-nanostructures. *Int. J. Mach. Tools Manuf.* (2015) 94, pp. 15–25

[34] Zhu, Z.W., To, S., Zhang, S.J., Large-scale fabrication of micro-lens array by novel end-fly-cutting-servo diamond machining. *Opt. Express* (2015) 23, pp. 20593–20604

[35] Brinksmeier, E., Gläbe, R., Osmer, J., Ultra-precision diamond cutting of steel molds. *CIRP Ann. Manuf. Technol.* (2006) 55, pp. 551–554

[36] Fang, F.Z., Liu, Y.C., On minimum exit-burr in microcutting. *J. Micromech. Microeng.* (2004) 14, pp. 984–988

[37] Wu, Y., Peng, W., Liu, Y., A novel fabrication method for micro optical wave guide mold based on fly-cutting technology. *Optik* (2013) 124, pp. 867–869

[38] Brinksmeier, E., Schönemann, L., Generation of discontinuous microstructures by Diamond Micro Chiseling. *CIRP Ann. Manuf. Technol.* (2014) 63(1), pp. 49–52

[39] Schönemann, L., *Potenziale und Grenzen des Mikroschneidens zur Herstellung optischer Funktionsflächen* (2014) University of Bremen, Germany, PhD Thesis

[40] Brinksmeier, E., Gläbe, R., Schönemann, L., Diamond Micro Chiseling of Large-scale Retroreflective Arrays. *Precis. Eng.* (2012) 36, pp. 650–657

[41] Marshall, D.B., Lawn, B.R., Cook, R.F., Microstructural Effects on Grinding of Alumina and Glass-ceramics. *J. Am. Ceram. Soc.* (1987) 70(6), pp. 139–140

[42] Bifano, T.G., *Ductile-Regime Grinding of Brittle Materials* (1988) North Carolina State University, US, PhD Thesis

[43] Bifano, T.G., Dow, T.A., Scattergood, V., Ductile-Regime Grinding – A new Technology for Machining Brittle Materials. *Trans. ASME* (1991) 113(2), pp. 184–189

[44] Chen, X., Rowe, W.B., Cai, R., Precision Grinding Using cBN Wheels. *Int. J. Mach. Tools Manuf.* (2001) 42(5), pp. 585–593

[45] Ichida, Y., Mechanical Properties and Grinding Performance of Ultrafine-crystalline cBN Abrasive Grains. *Diamond Relat. Mater.* (2008) 17(7–10), pp. 1791–1795

[46] Anderson, J., Hollman, P., Jakobson, S., Abrasive Capacity of Thin Film Diamond Structures. *Key Eng. Mater.* (2001) 196, pp. 141–148

[47] Liu, Y.K., Tso, P.L., The Optimal Diamond Wheels for Grinding Diamond Tools. *Int. J. Adv. Manuf. Technol.* (2003) 22(5–6), pp. 396–400

[48] Tanaka, T., Esaki, S., Nishida, K., Nakajima, T., Ueno, K., Development and Application of Porous Vitrified Bonded Wheel with Ultrafine Diamond Abrasives. *Key Eng. Mater.* (2004) 257, pp. 251–256

[49] Klocke, F., *Manufacturing Processes 2* (2009) Springer-Verlag, Berlin Heidelberg, Germany

[50] Engmann, J., Schüler, G.M., Haberland, R., Walk, M., Aurich, J.C., Efficient Technique for 3rd Micro Structuring of Carbide and Brittle Materials with Diamond Micro-shaft Grinding Tools. *Proceedings of the 9th euspen International Conference* (2009) vol. 2, San Sebastian, Spain, pp. 97–100

[51] Aurich, J.C., Engmann, J., Schueler, G., Haberland, R., Micro Grinding Tool for Manufacture of Complex Structures in Brittle Materials. *CIRP Ann. Manuf. Technol.* (2009) 58(1), pp. 311–314

[52] Haefeli, R., Anwendungsmöglichkeiten von Schleifstiften. *Ind. Diamond Rev.* (1997) 2(97), pp. 164–168

[53] Gäbler, J., Schäfer, L., Menze, B., Hoffmeister, H.W., Micro Abrasive Pencils with CVD Diamond Coating. *Diamond Relat. Mater.* (2003) 12, pp. 707–710

[54] Dornfeld, D., Min, S., Takeuchi, Y., Recent Advances in Mechanical Micromachining. *CIRP Ann. Manuf. Technol.* (2006) 55(2), pp. 745–768

[55] Wenda, A., *Jahrbuch Schleifen Honen Läppen und Polieren*, Hoffmeister, H.-W., Tönshoff, H.-K. (Ed.), Vulkan Verlag, Essen (2000) 59, pp. 273–284

[56] Aurich, J.C., Schüler, G.M., Engmann, J., Komplexe Mikrostrukturierung von Hartmetall mit 20 mm Mikroschleifstiften. *Diamond Business* (2009), pp. 22–27

[57] Onikura, H., Inoue, R., Okuno, K., Ohnishi, O., Fabrication of Electroplated Micro Grinding Wheels and Manufacturing of Microstructures with Ultrasonic Vibration. *Key Eng. Mater.* (2003) 238–239, pp. 9–14

[58] Denkena, B., de Leon, L., Wang, B., Grinding of Microstructured Functional Surfaces, A Novel Strategy for Dressing of Microprofiles. *Prod. Eng.* (2009) 3, pp. 41–48

[59] Brinksmeier, E., Riemer, O., Rickens, K., Meiners, K., Kinematics in ultra-precision grinding of WC moulds. *Int. J. Nanomanuf.* (2011) 7(3–4), pp. 199–213

[60] Hambücker, S., *Technologie der Politur sphärischer Optiken mit Hilfe der Synchrospeed-Kinematik* (2001) RWTH Aachen, Germany, PhD Thesis

[61] Leitel, A., *Ausbildung der Feingestalt optischer Funktionsflächen im Poliervorgang* (1983) Friedrich Schiller University, Jena, Germany, PhD Thesis

[62] Wagemann, A., *Wirkzusammenhänge beim Planparallelpolieren von Hochleistungskeramik* (1994) RWTH Aachen, Germany, PhD Thesis

[63] Nogawa, H., *Ceramic Processing - State of the Art of R & D in Japan* (1988), ASM International, Metals Park, OH, US

[64] Komanduri, R., Lucca, D.A., Tani, Y., Technological Advances in Fine Abrasive Processes. *CIRP Ann. Manuf. Technol.* (1997) 46(2), pp. 545–596

[65] Kasai, T., Horio, K., Karaki-Doy, T., Kobayashi, A., Improvement of Conventional Polishing Conditions for Obtaining Super Smooth Surfaces of Glass and Metal Works. *CIRP Ann. Manuf. Technol.* (1990) 39(1), pp. 321–324

[66] Dambon, O., Demmer, A., Surface Interactions in Steel Polishing for the Precision Tool Making. *CIRP Ann. Manuf. Technol.* (2006) 55(1), pp. 609–612

[67] Gessenharter, A., *Polieren strukturierter Oberflächen optischer Bauteile* (2006) University of Bremen, Germany, PhD Thesis

[68] Klocke, F., Dambon, O., Behrens, B., Analysis of defect mechanisms in polishing of tool steels. *Prod. Eng.* (2011) 5(5), pp. 475–483

[69] Preston, F.W., The Theory and Design of Plate Glass Polishing Machines. *J. Soc. Glass Technol.* (1927) 11, pp. 214–256

[70] Evans, C.J., Paul, E., Dornfeld, D., Lucca, D.A., Byrne, G., Tricard, M., Klocke, F., Dambon, O., Mullany, B.A., Material Removal Mechanisms in Lapping and Polishing. *CIRP Ann. Manuf. Technol.* (2003) 52(2), pp. 1–24

[71] Feng-yun, L., Tian-sheng, L., Development of a robot system for complex surfaces polishing based on CL data. *Int. J. Adv. Manuf. Technol.* (2005) 26(9–10), pp. 1132–1137

[72] Tsai, M.J., Huang, J.F., Kao, W.L., Robotic polishing of precision molds with uniform material removal control. *Int. J. Mach. Tools Manuf.* (2009) 49(11), pp. 855–859

[73] Ryuh, B.S., Park, S.M., Pennock, G.R., An automatic tool changer and integrated software for a robotic die polishing station. *Mech. Mach. Theory* (2006) 41, pp. 415–432

[74] Elsner-Dörge, F., Riemer, O., Brinksmeier, E., Robot-assisted vibration polishing of NiCo retroreflector moulds. *Adv. Mater. Res.* (2014) 1018, pp. 175–181

[75] Riemer, O., Elsner-Dörge, F., Investigation of material removal in vibration polishing of NiCo alloys with millimetre-sized tools. *Int. J. Adv. Manuf. Technol.* (2017), 92(9–12), pp. 3117–3122

[76] Brinksmeier, E., Riemer, O., Schönemann, L., Ultra-precision mold manufacture for micro injection molded micro optics. *Proceedings of the 70th Annual Technical Conference of the Society of Plastics Engineers 2012 (ANTEC 2012)* (2012), Society of Plastics Engineers (SPE), Orlando, USA

[77] Hermerschmidt, A., Krüger, S., DIFFRACTIVE OPTICS: Diffractive elements generate complex light patterns (2006) http://www.laserfocusworld.com/articles/print/volume-42/issue-12/features/diffractive-optics-diffractive-elements-generate-complex-light-patterns.html, 15.03.2017

[78] Brinksmeier, E., Riemer, O., Gläbe, R., Lünemann, B., von Kopylow, C., Dankwart, C., Meier, A., Submicron functional surfaces generated by diamond machining. *CIRP Ann. Manuf. Technol.* (2010) 59, pp. 535–538

7 Surface Treatment of Mold Tools in Micro Injection Molding

Christian A. Griffiths

7.1 Introduction

Micro injection molding has been applied successfully to manufacture a wide range of products in the micro scale and is also an appropriate technology for manufacturing meso parts with micro and nano features. The replication accuracy of molded parts made from polymers has been successful by determining the optimum process conditions for mass replication. Improving the part quality is key and it is also important to note that, as new high resolution micro/nano surface structures and advanced products that incorporate specific micro quality requirements have to be satisfied, the need for process control becomes ever more critical. Mass manufacture requires robust replication technologies to meet these demands and micro injection molding is a key technology for achieving micro parts in large scale production. The high performance micro molds used in the molding process are critical for delivering high quality parts for the duration of the product life cycle. Tool coatings are commonly used to improve the life of large scale molds, and recently diamond-like carbon (DLC) coatings have been shown to enhance the performance of micro molds. This chapter reports on novel surface treatments used in micro injection molding to improve the factors that affect parts' quality and to increase part functionality. The chapter advises on the various surface treatments and their influence regarding the behavior of the micro injection molding process.

The chapter is organized as follows: The next section highlights research conducted on DLC surface treatment and its use in micro injection molding; this is then followed by the molding process effects on DLC-coated micro molds; and, finally, the use of novel surface treatments for texturing micro mold tools. For each section, the current research findings are discussed and conclusions are made.

7.2 Investigation of DLC Surface Treatment Effects in Micro Injection Molding

7.2.1 Surface Treatment for Improved Demolding

An important stage in micro injection molding which can affect the accuracy and mechanical properties of the produced components is part demolding. This section of the chapter looks at countering the negative aspects of demolding by the application of different surface treatments. In particular, the demolding of a representative micro part will be demonstrated as a function of two different tool coatings, diamond-like carbon and silicon carbide, employing the design of experiment approach. In addition, the results obtained using different combinations of process parameters are shown to identify the best processing conditions in regards to demolding behavior when utilizing these two coatings. This section will investigate the micro mold surface treatments and highlight proven experiments for improved part quality.

The part-mold forces that develop at the demolding stage in micro injection molding have to be overcome for part removal from the mold. These forces are due to an interaction between the polymer and cavity surfaces, and they are a combination of the pressure P from material shrinkage and the coefficient of friction of the polymer and mold material [1]. An ejector system with ejector pins is used to apply a force to overcome the friction at the surfaces. Each ejector pin has to overcome the part and mold forces without causing damage to the part. Ejector pins are well-known for introducing high local stresses; these areas of increased stress can result in part defects such as stress marks, deformation, and fracture of the part features [2]. Micro parts have a high surface to volume ratio (SV_R) and this can lead to reduced mechanical strength. During demolding, this reduction makes the parts and features susceptible to damage. To improve the producibility of micro parts, it is important to consider surface treatment as a means to reduce the ejection forces. Two types of ejection forces (F_E) have been identified [1, 3]. A tool with simple geometrical features where the total friction between the tool and the polymer interface was investigated by Menges and Mohren [1]. Their work was based on part surface area (A_C), the coefficient of friction of the polymer (μ) [ISO 8295], the identification of the contact pressure (P_A) [ISO 294-4] during molding. Coefficients of static (μ_s) and kinetic (μ_k) friction are also important in determining part-mold forces. μ_s is used to describe ejection breakaway motion [4]; subsequent motion to remove the part is described by μ_k.

Toolmaking processes available to produce mold cavities have a range of different surface finishes, from polished to visible machine cutter lines. The F_E can vary depending on the toolmaking method and the resulting surface finish of the tool.

Also, as the tool gets used more and more, the tool surface will change over time; this degradation will also result in F_E changes during the tool life. One counter measure for improving tool wear resistance is the application of surface treatment. Traditional methods such as heat treatment and nitriding are well-known but, recently, techniques such as physical vapor deposition (PVD) and chemical vapor deposition (CVD) of titanium nitride (TiN) and chromium nitride (CrN) have proven to significantly improve wear resistance [5–7]. Nano composite surface coatings such as polytetrafluoroethylene (PTFE) have also been used for part and mold force reduction. However, it has been found that these surfaces in a mass production environment have a poor wear resistance [8]. The surface treatment of tools using novel materials such as diamond-like carbon (DLC) coatings can produce tools with hard surfaces of up to 70 GPa. In addition to this high hardness, these surfaces can have friction coefficients below those of ceramics and in the range of 0.05–0.2 μ [9].

7.2.2 An Experimental Case Study for Improving Part Demolding Using Surface Treatment

In the following case study the influence of two surface treatments, DLC and SiOC, is shown. The test design used is a 15 mm × 20 mm × 1 mm microfluidics part (Figure 7.1a) with micro channel cross sections of 200 × 100 μm (Table 7.1). Two identical tools were manufactured, one for each surface treatment.

The DLC surface was deposited using a low frequency (LF) plasma enhanced chemical vapor deposition (PECVD) reactor. To aid adhesion, a Si-C:H intermediate layer (0.5 μm) was deposited on the tool surface using a plasma of tetramethylsilane (TMS) and argon. After this, a 2 μm DLC layer was deposited onto the Si-C:H (Figure 7.2). The second tool treatment to be investigated is an organosilicon coating (SiOC) deposited using a PECVD process that was designed for industrial scale high quality hydrophobic applications.

For the injection molding trials, acrylonitrile butadiene styrene (ABS) and polycarbonate (PC) materials were chosen and the machine used was a Battenfeld Microsystem 50. The demolding force of the molding cycle was measured using a Dynisco PCI 4011 piezoelectric force transducer. The sensor response signals were collected using a National Instruments cDAQ-9172 USB data acquisition unit and processed using National Instruments Labview 8 software. The mold and mold inserts were specially machined to accommodate the force sensor such that when the ejector moves forward and the part is demolded the transducer receives a mechanical load that generates an electric potential. The charge is converted using a Kistler charge amplifier (type 5039A222) into a manageable voltage that is then monitored using a National Instruments NI 9205 16-bit module. For the planned experiments, the melt temperature (T_b), mold temperature (T_m), cooling time after part filling (t_c), and ejection time delay (t_e) for cooling control are used. A Taguchi L9 orthogonal array (OA) for four factors at three levels was selected for each tool and polymer. The behavior of the surface treatments to each experiment was analyzed by measuring FE during demolding. Six L9 OAs were used and a total of 540 trials were performed.

Table 7.1 Part Design Properties

Design properties	Design with micro features	Design without features
Volume	3.10 mm^3	3.37 mm^3
Surface area	8.33 mm^2	7.62 mm^2
SV_R	2.68	2.26

Figure 7.1 The test design: (a) microfluidic part and (b) FEA mesh of the microfluidic part; (c) micro mold

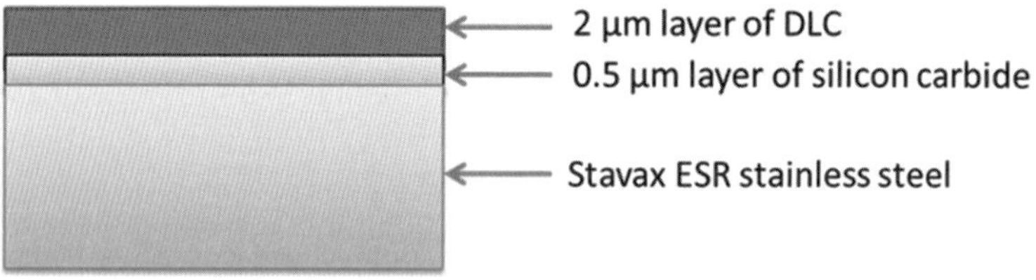

Figure 7.2 DLC coating deposition configuration

7.2.3 Validation, Verification, and Results

In the case study an experimental method to investigate part demolding in micro injection molding, with a focus on the surface treatment influences on the demolding force [10], is shown. The demolding forces were studied as a function of mold surface treatment and four process parameters. The influence of the surface treatment on F_E was calculated using a wide range of process parameter combinations;

these were then analyzed to identify optimum processing conditions for the demolding of micro parts. The main findings are:

- The average demolding forces for PC and ABS materials show that surface treatments significantly reduce F_E when compared to untreated tools. The DLC coating showed a 40% reduction of F_E when processing ABS, and PC was also high at 16%. The experiments also showed that the overall quality of the parts improved with the surface-treated tools.
- From all of the experiment there was not a unique selection of parameters levels that can be considered optimum for demolding with the selected surface treatments or polymers used in this study. In this way it is not possible to make conclusions about the optimum set of process parameters for these polymers, but it remains a possibility for other polymers. Therefore, systematic experimental studies should be considered when new combinations of tool treatments and polymers are considered.
- By performing an ANOVA analysis, the individual process parameters' contribution to optimum performance is shown. The lowest possible demolding forces for each tool treatment and polymer combination showed that DLC and SiOC coatings significantly reduce the demolding forces. Furthermore, by performing a Taguchi analysis engineers can determine the optimum process parameters for reduced demolding forces, using the investigated surface treatment and polymer combinations.

7.3 Temperature Effects on DLC-Coated Micro Molds

7.3.1 DLC Coatings Used on Micro Molds

Novel coatings deposited on tool surfaces can be expected to experience wear and delamination during service. Additionally, the high speed filling and fast cycle times associated with the micro injection molding process mean that friction and heat transfer at the surface increase the likelihood of surface treatment delamination and cracking. DLC films provide surface wear resistance and a reduction of demolding forces. However, it is difficult to ascertain the life of DLC coatings during the high speed thermal cycles in micro products. In this section of the chapter, a micro injection molding process using three different polymers is studied and a FEA simulation model will be used to identify the process temperature factors that lead to tool thermal expansion and dimensional changes that impact the life of the coating. Analysis of the bulk temperature distributions is used to calculate the theoretical heat transfer influence on expansion and identify areas for optimization of the molding process that can be used to reduce the expansion, in order to reduce delamination and cracking. This section will discuss the challenges encountered with novel surface treatments and highlight proven experiments and FEA methods for improved reliability.

Well-established and novel coating technologies for structured/engineered surfaces have been used to provide an increase in wear resistance and also to functionalize the surfaces [10–13]. The previous section demonstrated that DLC coatings improve the replication performance of the micro-structured masters. This is achieved using a combination of superior tribological and mechanical properties that include low friction and wear and high hardness when compared to standard tool materials [14–16]. The ability to identify and predict the delamination of coatings is technically significant when judging the lifetime of the surface in service conditions. In addition to mechanical interactions from the process including pressure, force, and friction between the tool surface and polymer melt, there are several heating and cooling stages with a wide range of cyclical temperature gradients. Coating delamination can occur with temperature changes and the associated thermal expansion experienced by the tool and the coating. The change in length experienced when a solid body is subjected to a temperature depends on the original length of the body and the temperature range during heating and cooling. The relationship between these factors is the average coefficient of linear thermal expansion $\propto_{t_2}^{t_1}$, defined as:

$$\propto_{t_2}^{t_1} = \frac{L_2 - L_1}{L_R\left(t_2 - t_1\right)} \quad (7.1)$$

where t_1 and t_2 are initial and final temperatures, L_1 and L_2 are initial and final lengths, and L_R is the length at a reference temperature.

The instantaneous coefficient of linear expansion at any temperature t can be defined by the Equation (7.2) as:

$$\propto_t = \lim_{t_1 \to t_2}\left(\frac{L_2 - L_1}{L_R\left(t_2 - t_1\right)}\right) = \frac{dL}{L_R\left(dt\right)} \quad (7.2)$$

where α_t is the instantaneous coefficient of linear expansion at temperature t [17].

The fractional change in length ΔL of a solid subjected to linear thermal expansion is:

$$\frac{\Delta L}{L_0} = \alpha \Delta T \quad (7.3)$$

where α is the coefficient of thermal expansion, L_0 is the initial length, and ΔT is the change of temperature the material is subjected to.

Due to thermal expansion of the tooling during processing, the features will be subjected to superficial thermal expansion. Therefore, the calculation is expanded to allow for the superficial change in length L, defined by a new equation:

$$L = L_0 + L_0 \alpha \Delta T = L_0 (1 + \alpha \Delta T) \quad (7.4)$$

A rectangular micro rib feature with an area A_0 can be calculated from the height H_0 and width W_0. The influence of the temperature in both H_0 and W_0 results in an increased area, defined by the following equation:

$$A = HW = H_0 (1 + \alpha \Delta T) W_0 (1 + \alpha \Delta T) = H_0 W_0 (1 + \alpha \Delta T)(1 + \alpha \Delta T)$$

$$A = A_0 \left[1 + 2\alpha \Delta T + (\alpha \Delta T)^2\right] \quad (7.5)$$

Since the value of α is typically small, for a small temperature change $\alpha\Delta T$ is significantly less than 1. Therefore, the term $(\alpha\Delta T)^2$ in Equation (7.5) is negligible compared to the $2\alpha\Delta T$ term. Hence, the following equation defines the new area of an object subjected to thermal expansion, which is given by:

$$A = A_0 (1 + 2\alpha \Delta T) \quad (7.6)$$

7.3.2 An Experimental Case Study for Identifying the Temperature Effects on DLC-Coated Micro Molds

During the micro injection molding process, wide variations in temperature through cyclic heating and cooling and localized heating from polymer shear heating are experienced. The following case study shows the effect that the thermal and processing variables can have on a DLC coating. In particular, the investigation analyzes the difference in thermal expansion between the polymer melt, mold, and the coating has on the potential for surface delamination. First, a model for the thermal expansion of the tool and surface treatment is shown. This is done using the microfluidic part design (Figure 7.1a) from the previous case study. The mold material used in this case is Stavax ESR stainless steel and the surface coating is a double layer coating of silicon carbide (Si-C:H) (0.5 µm thickness), and then a 2 µm layer of DLC. The mechanical properties of the DLC coating include a micro hardness of 22 ± 2 GPa and a Young's Modulus of 160 ± 10 GPa. The second section of the case study uses Autodesk Simulation Moldflow Insight/Synergy 2014 finite element analysis to simulate the process. After optimizing the FEA mesh of the microfluidic part (Figure 7.1b), the focus of the simulation will be on results for bulk temperature at end of fill, part and cavity surface temperature, and frozen layer fraction at end of fill. ABS and polypropylene (PP) and polyether ether ketone (PEEK) polymers are the materials being simulated using high speed manufacturing processing settings.

7.3.3 Validation, Verification, and Results

7.3.3.1 Linear and Superficial Thermal Expansion During Processing

Using the upper limit of the known polymer melt temperatures, the fractional change in length ΔL can be calculated from Equation (7.3) for the mold design (Figure 7.3). It can be seen that the Stavax ESR mold, SiC layer, and DLC coating experience different dimensional changes during thermal cycles. In addition to the different fractional length changes, each material changes dimensions at different rates due to the different thermal expansion coefficients. The expansion and contraction mismatch can result in micro structural problems between the tool and coatings. The model has identified the possibility of the formation of residual stresses in the coating that lead to micro cracking, loss of adhesion, and subsequent delamination of the coating. Figure 7.3 also shows the T_b ranges for the three polymer materials.

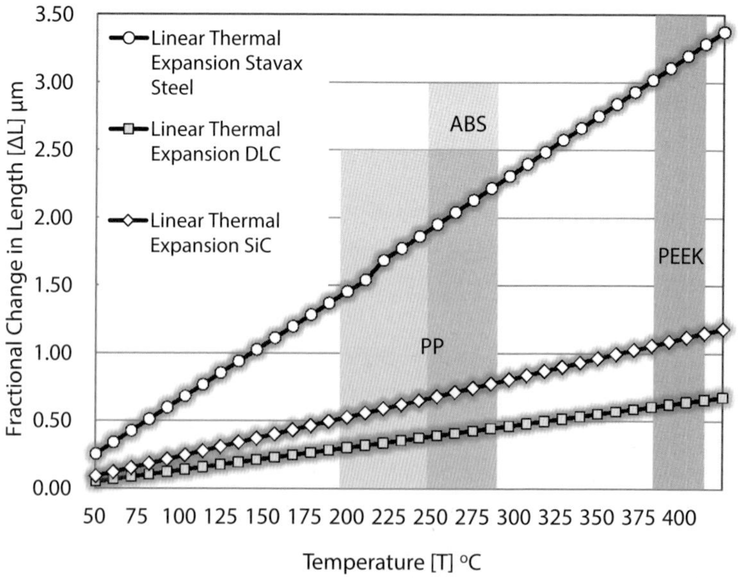

Figure 7.3 Linear thermal expansion for the DLC coating, SiC intermediate layer, and steel substrate

The result shows that, when using PEEK, the tool surface expansion is between 3.11–3.37 μm and the DLC will have a ΔL of 0.62–0.68 μm. The difference between the changes in length is around 2.5 μm. For the material processing temperature ranges there is a difference in the change in length. PP has the widest range, and in this range the steel and DLC can experience a change in ΔL of 0.76 μm and 0.14 μm, respectively. For the superficial thermal expansion, Figure 7.4 shows the influence of temperature on the micro mold. It can be seen that the tool and coating have different superficial growth gradients, where a slope of 1.64 μm^2 is seen for the steel, compared to that of 0.35 μm^2 for the DLC coating. The dissimilar expansion from temperature cycles can lead to bending and/or compression on the coating; this can also lead to coating damage. It is also a possibility that expansion effects do not adversely affect the coating due to the adhesion strength of the DLC to the substrate.

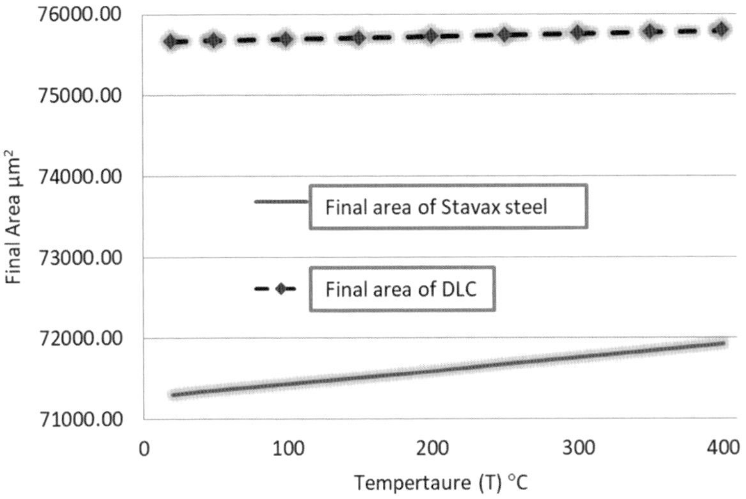

Figure 7.4 Superficial thermal expansion of DLC coating and steel substrate

7.3.3.2 Finite Element Analysis Results

The process setting temperature influence on bulk temperature shows that the maximum temperature was 271.3 °C for PP, 262.3 °C for ABS, and 406.4 °C for PEEK (Figure 7.5). These values are all higher than the set starting temperatures of 270 °C for PP, 260 °C for ABS, and 400 °C for PEEK. This is due to shear heating of the polymers during filling. To show the bulk temperature variation, a distribution across the part is shown (Figure 7.6). For the ABS example shown, red indicates the highest temperature and green and yellow the lowest. For part quality, a uniform distribution is preferable, and in this example this is not the case. The result can be used to show areas of increased thermal expansion and can be critical for inspecting possible DLC delamination (Figure 7.7). To highlight the difference between the bulk temperature of the part and the surface temperature of the mold, the simulation investigates part and cavity surface temperature variations.

The result of this investigation shows that the surface temperatures are much lower than the bulk temperatures, meaning that the thermal expansion that the DLC experiences is not extreme. The assumptions used in the simulation are based on a macro scale, where a large layer of frozen polymer at the mold part interface is preventing bulk heat transfer from part to mold surface. This assumption is based on the development of a large frozen layer; however, this is not the case for micro injection molding. Micro parts have a higher surface to volume ratio, meaning that a high heat loss is experienced by the parts as heat transfers to the mold surface. A cooling stage heat distribution can be seen for the mold (Figure 7.8b) and part (Figure 7.8a); the results show that there is a similar distribution of heat and, also, it can be seen that the part micro features are cooled and the micro features of the tool have an increased temperature. Both are high surface to volume ratio area features and sensitive to heat change. This result confirms that a frozen layer on the part surface is not fully developed as an insulator. This then demonstrates that the bulk temperature variations seen in Figure 7.5 are being directly transferred to the DLC surface, causing increased thermal expansion.

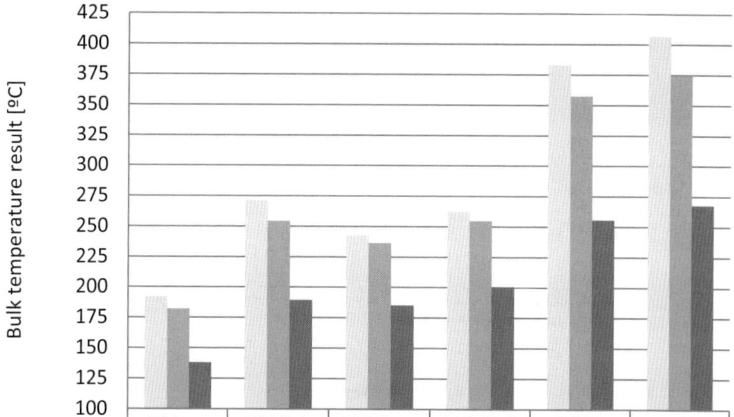

	PP Low Tb	PP High Tb	ABS Low Tb	ABS High Tb	PEEK Low Tb	PEEK High Tb
Maximum	191.9	271.3	242.8	262.3	382.9	406.4
Average	182	254.5	236.5	254.8	357.5	374.6
Minimum	138.1	189.4	185.2	200.8	255.7	267.8

Figure 7.5 Bulk temperature results for PP, ABS, and PEEK

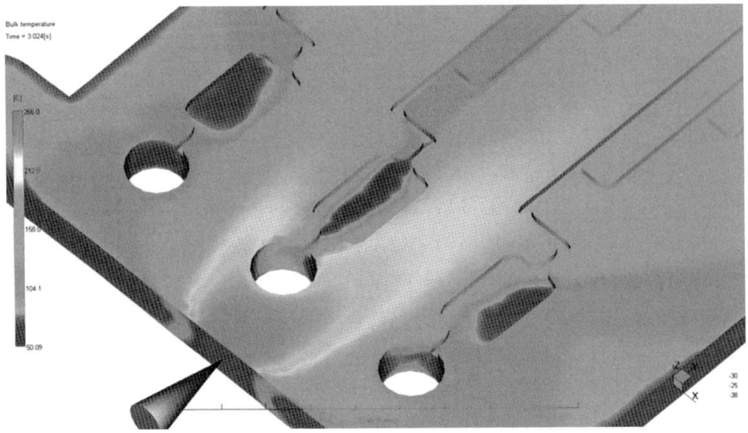

Figure 7.6 Bulk temperature distribution at the end of the fill for ABS. The figure is presented in the color supplement

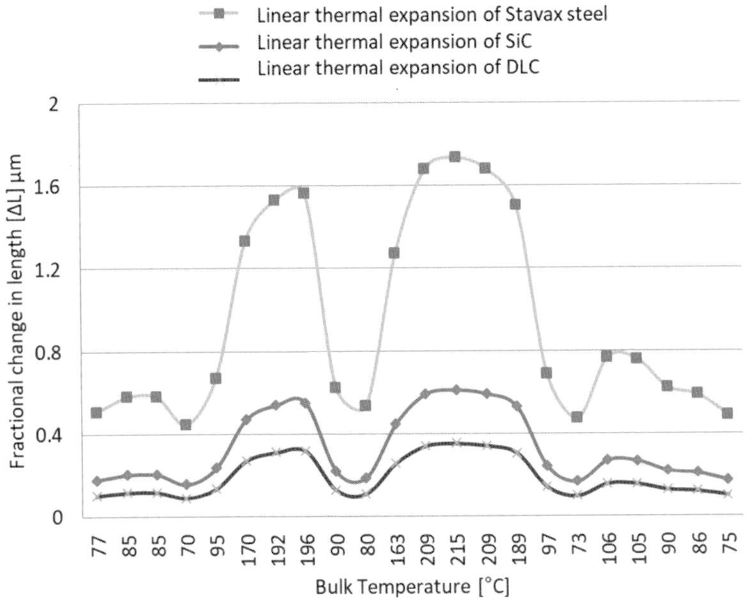

Figure 7.7 Thermal expansion distribution across the microfluidic channels

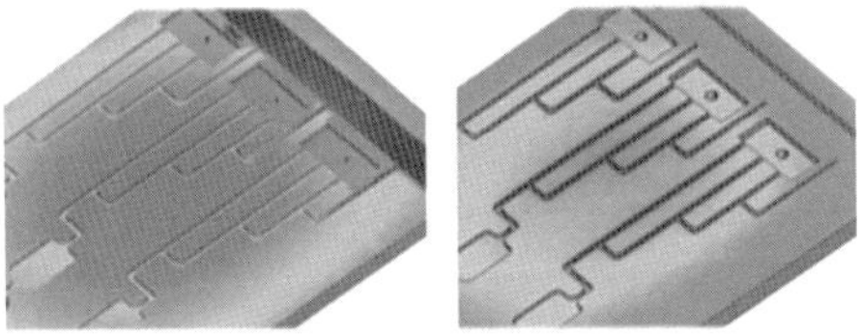

(a) (b)

Figure 7.8 Simulation of part (a) and mold (b) surface temperature distribution after filling. The figure is presented in the color supplement

7.3.4 Main Findings

The case study looks at the simulation of the high speed micro injection molding process used to produce a micro part in three different polymer materials [18]. A simulation model is also used to identify the process temperature factors that lead to tool thermal expansion and dimensional changes that impact the life of the coating.

- FEA analysis is used to identify the heat transfer that can result in thermal expansion and to highlight areas on the mold surface where expansion could be critical. The study shows that the mold material and coatings experience very different thermal expansions. Also for processing the different polymers, an expansion of up to 3.4 μm could be experienced by the steel, compared to 1 μm for the coatings. This variation could lead to highly stressed regions in the coating, resulting in coating failure due to cracking and delamination. Adhesion of DLC to the mold requires that the materials be thermally aligned. This is not possible with dissimilar materials, but there are opportunities in the part design and processing window for optimization to reduce the areas of increased expansion.
- The results show that at the micro scale a heat transfer can be seen and this rapid variation in heat could negatively influence the materials' thermal expansion rates. The DLC-coated micro rib features are susceptible to high heat transfer; this makes them areas of interest for inspection of surface cracking and delamination.
- The results show that part and mold designers should consider process and part optimization to reduce areas of sustained and sudden heat variation across the part surface. With the target of reducing thermal expansion at the mold surface, process optimization of the injection speeds could be used to control part filling with reduced polymer shear rates and shear stresses.

7.4 A Novel Surface Treatment Texturing of Micro Injection Molding Tools

7.4.1 Mold Tool Texturing and Demolding Force

This section of the chapter investigates the effect that a novel texturing of replication tools—by applying amorphous hydrogenated carbon (aC:H) coating—has on the processing conditions in micro injection molding. Texturing usually increases the surface area of tools, leading to higher demolding forces acting on the component during the ejection stage of the process. As a consequence of this, the ejection forces can cause stress marks, deformation, fracture, and stretching of the polymer micro features. Therefore, this research studies the effect that a textured aC:H coating with nano scale structures has on the resulting demolding forces in comparison to an untreated tooling surface. The obtained results demonstrate the beneficial effect of the aC:H surface coating on demolding forces in replicating nano scale surface structures. The carried out proof of concept study showed that the applied surface texturing method can be considered as an alternative to existing techniques for surface structuring and, at the same time, to reduce demolding forces. This section will discuss the current state of the art for novel surface treatment texturing and highlight proven experiments used in practical applications.

The structuring of part surfaces with micro and nano scale features can lead to functional behavior, such as anti-reflectivity, surface adhesion, friction control, self-cleaning, hydrophobic and hydrophilic fluid control, and the enhancement of cell attachment and alignment [19–22]. Structured surfaces are considered to be "surfaces with a deterministic pattern of high aspect ratio geometric features designed to give a specific function" [23]. Engineered surfaces, on the other hand, are surfaces where the manufacturing process can generate variations in geometry and/or near surface material properties to obtain a specific function [23]. A number of manufacturing technologies can be used to structure and engineer surfaces, including sand blasting, surface grinding, focus ion beam machining, nano imprint lithography, chemical texturing, and laser machining [24–28]. Employing novel methods for surface modifications using self-assembly, it is possible to create homogeneous and non-homogenous patterns, controlled topography, and depressions on planar surfaces. DLC coatings, as shown in the previous sections, have superior tribological properties including low friction and mechanical properties for low wear, such as high hardness [29–31]. The engineering benefits from the use of DLC coatings in micro replication processes such as micro injection molding are proven [32]. DLC coating technology can be applied innovatively in mold master making. One novel application of DLC is for it to be used to modify the tool surface topography for increased part surface functionality. In this section of the chapter, the deposition of a structured aC:H coating on a mold surface to have surface-engineered nano beads and high aspect ratio nano pillars is demonstrated. The wear resistance of this aC:H coating and the replication of the nano surfaces is demonstrated.

7.4.2 An Experimental Case Study for Tool Texturing

The part design used in this study is the same as the one in Figure 7.1, and the tool insert was manufactured using micro milling on Stavax stainless steel. For the experimental trials, a 3 mm-diameter pin was positioned at the center of the part and a Kistler 9211B miniature force sensor positioned behind the ejector pin. F_e, during the micro injection molding demolding stage of the process, is assessed using a novel condition monitoring system. The system uses National Instruments hardware and Labview software similar to the one in Section 7.2.3 to gather the data. In this case study, the surface treatment was carried out by depositing amorphous hydrogenated carbon (aC:H) onto the tooling surface. The coating was performed in a low frequency PECVD reactor and the parameters were optimized based on research conducted on DLC deposition [48]. Before coating, the tools were ultrasonically cleaned in acetone, rinsed with ethanol, and dried before being placed into the PECVD chamber. To improve adhesion, a hydrogenated silicon carbide bonding layer (0.4 μm thickness) was deposited to ensure the mechanical cohesion to the mold surface. The mechanical properties of the aC:H surface were studied previously [15, 49, 50] and are shown in Table 7.2. For the surface texturing this was performed using a sacrificial hard mask of silica beads followed by anisotropic etching of the aC:H coating. The plasma exposure has a higher etching rate (× 20) for the aC:H coating compared to the SiO_2 beads, so, when the beads are ultrasonically removed, the unetched coating under the bead remains (Figure 7.9). As shown in Figure 7.10–Figure 7.12, the nano beads are deposited at different population densities and it is possible to produce single high aspect ratio pillars.

Table 7.2 aC:H Mechanical Properties

Properties	aC:H
Coating thickness	1 μm
Hardness	21 ± 2 GPa
Young's modulus	160 ± 9 GPa
Residual stress	1.2 ± 0.1 GPa
Friction coefficient	0.15 μ
Wear rate	3×10^{-8} $mm^3 \cdot N^{-1} \cdot m^{-1}$

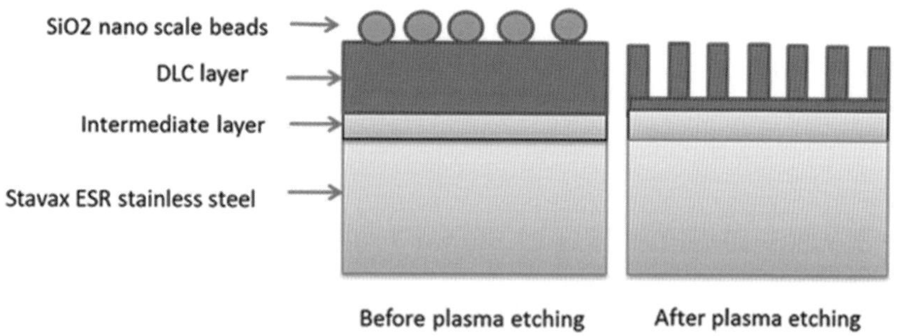

Figure 7.9 Schematic representation for texturing DLC coatings

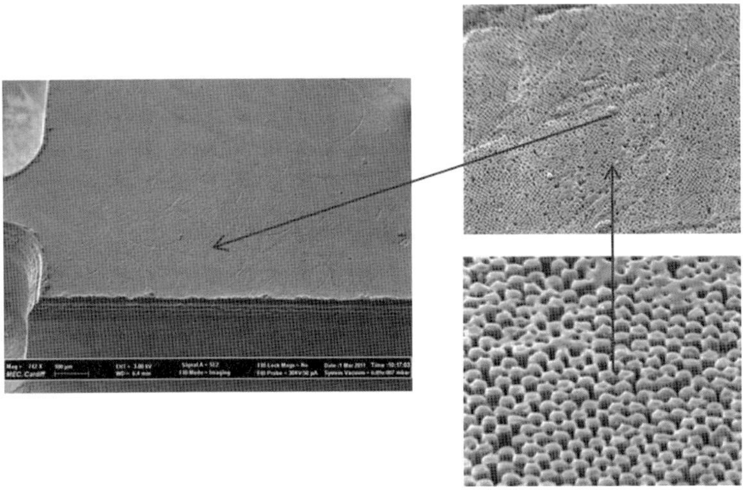

Figure 7.10 Surface treatment of aC:H with a high population of nano texture

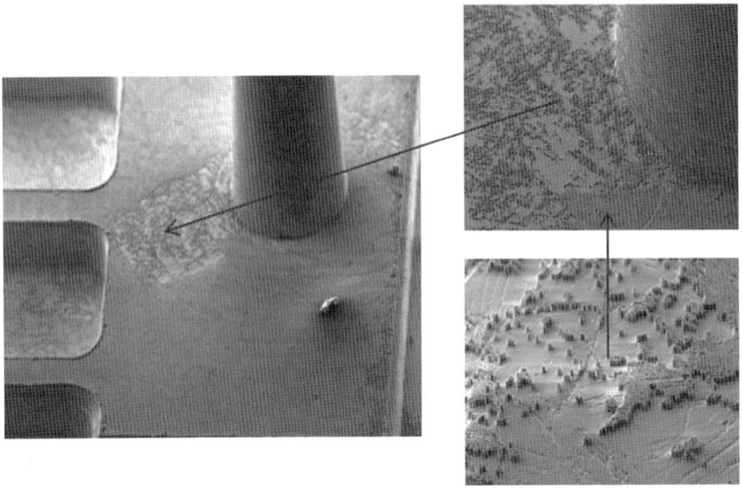

Figure 7.11 Surface treatment of aC:H with a medium population of nano texture

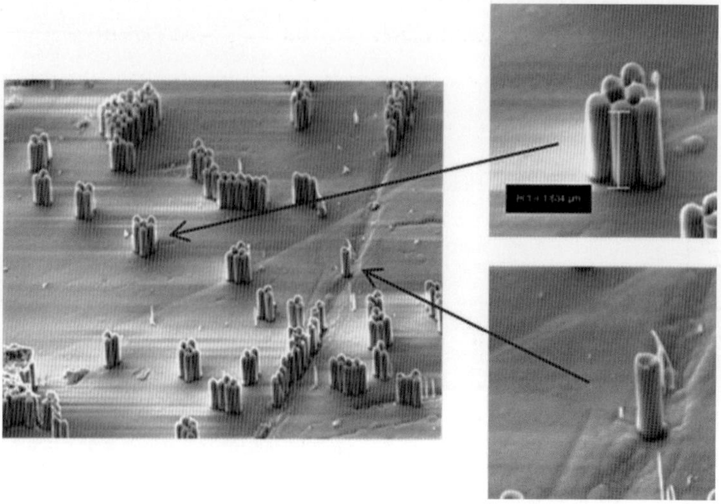

Figure 7.12 Surface treatment of aC:H with a low population of nano texture including a single high aspect ratio pillar of 1634 µm height

7.4.3 Validation, Verification, and Results

7.4.3.1 Nano-Structured Surfaces Replication and Demolding Forces

In this case study, known processing parameters were chosen so that low demolding forces were applied to the textured surfaces. To investigate the F_e influence of the textured coating, 10 parts were made and the F_e was monitored. The tool coating was then stripped using oxygen plasma etching and the experiments repeated. For each trial, F_e^{max} was analyzed and the mean values were calculated for both tools (Figure 7.13). The research showed that the F_e curves for the two tools and the tool with the nano-structured coating has a lower F_e^{max} (15.8%) than the untreated one (Table 7.3). This finding shows that even with a textured surface that should increase pat and tool adhesion, the coating still produces less friction. The SEM inspection of the tool coated surface showed that the aC:H texturing produced densely populated areas of pillars with 0.5 µm diameter and 1.6 µm height. It was also observed that the pillars deposition is irregular in some areas. To find out if the coating could withstand the micro injection molding process, 100 parts were produced using PP material. After the production run, there were no signs that the surface was deteriorating. The bead-like textured coating of the tool was replicated onto the part (Figure 7.14); it was also shown that groups and single pillars were replicated, resulting in hole-like features on the part surfaces. In some cases, the holes are not straight and this is due to an unbalanced demolding of the part, due to the use of a single ejector pin. This is an additional stress on the nano pillars but, due to their strength, they did not fracture and only one isolated pillar was shown to have failed.

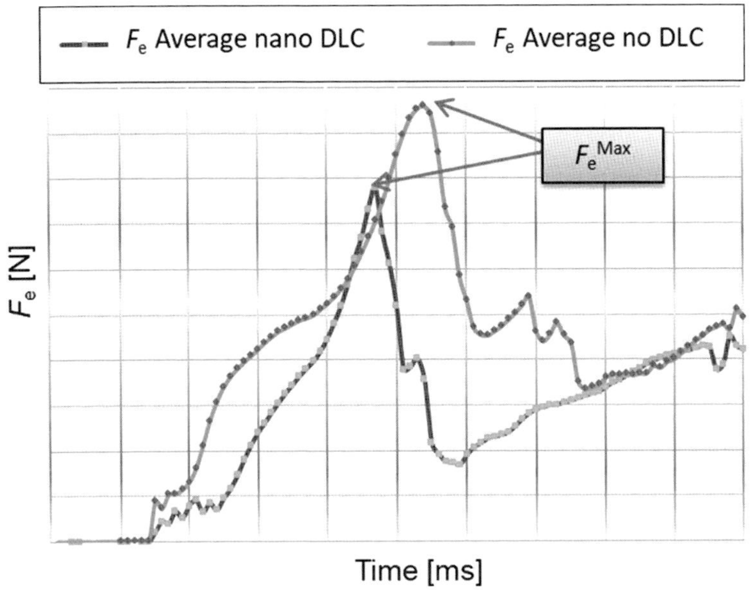

Figure 7.13 Typical demolding curves for treated and untreated tools

Table 7.3 F_e^{max} Results

F_e [N]	Max	Min	Average
aC:H surface textured tool	208	178.1	196.6
Untreated tool	244.9	230.8	233.5
F_e reduction [%]	15.07	22.8	15.8

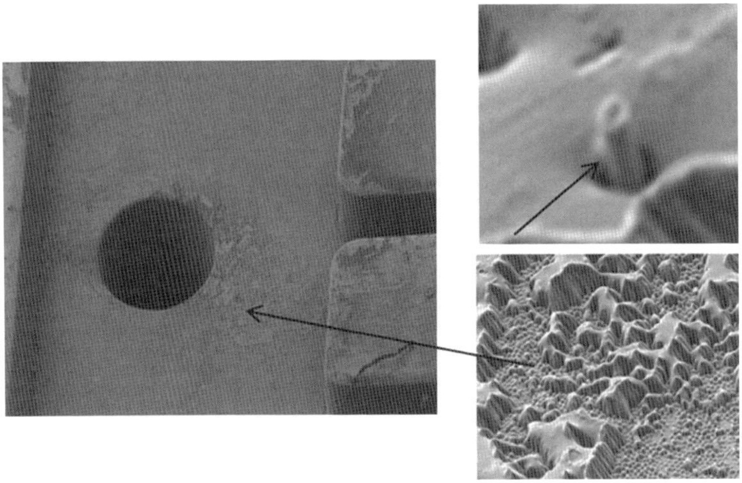

Figure 7.14 A PP part with a replication of the irregular aC:H surface texture and a single nano pillar trapped in the molding

7.4.3.2 Main Findings

The research has shown how a textured coating influences the demolding forces, and a novel condition monitoring system is employed to compare the untreated and treated tool surfaces [33]. In addition to forces, the capability of the molding method to replicate the surface features and nano pillars is investigated. The following conclusions are made:

- When compared to an untreated tool, the textured aC:H coating resulted in an F_E reduction of 15.8%. This is impressive because the nano texturing could have increased the part and tool force surfaces. The significant reductions are due to DLC type surfaces having low friction coefficients and this is a dominant factor in determining the tool performance.
- The aC:H coating had a strong adhesion to the mold surface and no delamination was observed. Importantly, single pillars with high aspect ratio survived over 100 process cycles. One pillar fractured and this was due to the simplicity of the demolding method. Therefore, when considering this method for large scale manufacture, the ejector system must be designed to ensure that the ejection is perfectly balanced for improved part demolding.
- The polymer parts made showed that the surface features and pillars were replicated. In this study, the micro injection molding settings were not tuned for improved surface replication performance; further research could be considered to find the optimum processing parameters together with reduced demolding forces.

7.5 Conclusions

The introduction of miniaturized parts and products to the marketplace requires that current and newly developed manufacturing systems are capable of delivering quality components at high production rates. The aims of this chapter are to highlight the use of surface treatments on mold masters used in micro injection molding as a means to improve the process and to make it more flexible. To demonstrate the application of surface-treated molds, three experimental examples are shown. The main conclusions from the case studies are as follows:

- In the case study focusing on the influence of surface treatments on the demolding force, the average forces for both materials showed that the treatments significantly reduce the demolding force when compared to untreated tools. The experiments also showed that the overall quality of the parts improved with the surface-treated tools. The research approach adopted used a condition monitoring system together with a Taguchi design of experiments for the polymer processing combinations. This method can help engineers to determine the optimum process parameters for reduced demolding forces, and to produce better parts.

- With an increasing recognition that surface treatments can improve the micro injection molding process, it is important that the life cycle of the deposited surface be considered. To investigate this, a case study to identify the process factors that lead to tool thermal expansion and dimensional changes that impact the life of the coating is demonstrated. The FEA analysis identified variations in heat transfer that can result in thermal expansion. Importantly, it is shown that the mold material and coatings experience very different thermal expansions and this could result in coating failure due to cracking and delamination. Based on the findings, it is important that part and mold designers should consider process and part optimization to reduce areas of sustained and sudden heat variation across the part surface.
- The case study investigating the performance of molds with a nano textured aC:H coating against untreated tools showed how a textured coating influences the demolding forces. When compared to an untreated tool, the textured aC:H coating resulted in a significant reduction of force due to the surface having a low coefficient of friction. For the replication of the surface texture, the polymer parts produced showed that the surface features and pillars were replicated. After more than 100 cycles, the coating was shown to have a strong adhesion to the mold surface and no delamination was observed.

In order for engineers to invent new production methods and determine the best production routes for existing manufacturing processes, they must consider new advances in technology. Innovative surface treatments of mold masters are shown to be important for improved productivity and part quality. It is essential that practitioners continue to investigate new surface technologies and to find innovative ways of applying them to micro manufacture. When considering surface treatments for improved functionality and productivity, it is also important to apply a systematic experimental approach to gathering empirical data on the process. This can be achieved by the use of a specially designed system for condition monitoring together with a FEA model. The experimental approach is invaluable to engineers particularly when they need to assess new and complex process behavior.

References

[1] Menges, G. and Mohren, P., eds., *How to Make Injection Molds* 2nd ed. (1993), Hanser Gardner

[2] Heyderman, L. J., Schifta, H., David, C., Gobrecht, J., Schweizer, T., Flow behaviour of thin polymer films used for hot embossing lithography, *Microelectronic Engineering* (2000), 54(3–4), pp. 229–245

[3] Kwak, S., Kim, T., Park, S., Lee, K., Layout and sizing of ejector pins for injection mold design using the wavelet transform, *Proceedings of the Institution of Mechanical Engineers, Part B: Journal of Engineering Manufacture* (2003), 217(4), pp. 463–473

[4] Pouzada, A. S., Ferreira, E. C., Pontes, A. J., Friction properties of molding thermoplastics, *Polymer Testing* (2006), 25(8), pp. 1017–1023

[5] Mitterer, C. et al., Industrial applications of PACVD hard coatings, *Surface and Coatings Technology* (2003), 163–164, pp. 716–722

[6] Heinze, M., Wear resistance of hard coatings in plastics processing, *Surface and Coatings Technology* (1998), 105(1–2), pp. 38–44

[7] Cunha, L. et al., Performance of chromium nitride and titanium nitride coatings during plastic injection molding, *Surface and Coatings Technology* (2002), 153(2–3), pp. 160–165

[8] Sawyer, W. G. et al., A study on the friction and wear behavior of PTFE filled with alumina nanoparticles, *Wear* (2003), 254(5–6), pp. 573–580

[9] Voevodin, A. A., Donley, M. S., Zabinski, J. S., Pulsed laser deposition of diamond-like carbon wear protective coatings: a review, *Surface and Coatings Technology* (1997), 92(1–2), pp. 42–49

[10] Griffiths, C. A., Dimov, S. S., Brousseau, E. B., Chouquet, C., Gavillet, J., Investigation of surface treatment effects in micro injection molding, *International Journal of Advanced Manufacturing Technology* (2010), Volume 47, Issues (1–4), pp. 99–110, DOI 10.1007/s00170-009-2000-4

[11] Martinez, E., Engel, E., Planell, J. A., Samitier, J., Effects of artificial micro-and nano-structured surfaces on cell behaviour, *Ann. Anat.* (2009), 191 (1), pp. 126–135, http://dx.doi.org/10.1016/j.aanat.2008.05.006

[12] Kovalchenko, A., Ajayi, O., Erdemir, A., Fenske, G., Etsion, I., The effect of laser surface texturing on transitions in lubrication regimes during unidirectional sliding contact, *Tribol. Int.* (2005), 38 (3), pp. 219–225

[13] Dimitrov, A. S., Nagayama, K., Continuous convective assembling of fine particles into two-dimensional arrays on solid surfaces, *Langmuir* (1996), 12 (5), pp. 1303–1311, http://dx.doi.org/10.1021/la9502251

[14] Tillmann, W., Vogli, E., Hoffmann, F., Wear-resistant and low-friction diamond-like-carbon (DL-C)-layers for industrial tribological applications under humid conditions, *Surf. Coat. Technol.* (2009), 204 (6–7), pp. 1040–1045, http://dx.doi.org/10.1016/j.surfcoat.2009.06.005

[15] Voevodin, A. A., Donley, M. S., Zabinski, J. S., Comparative Tribology Studies of Hard Ceramic and Composite Metal-DLC Coatings in Sliding Friction Conditions, *Surf. Coat. Technol.* (1997), 92 (1–2) 42–49

[16] Saha, B., Liu, E., Tor, S. B., Hardt, D. E., Chun, J. H., Khun, N. W., Improvement in lifetime and replication quality of Si micromold using N:DLC:Ni coatings for microfluidic devices, *Sensors Actuators B Chem.* (2010), 150 (1), pp. 174–182, http://dx.doi.org/10.1016/j.snb.2010.07.019

[17] Hidnert, P. and Souder, W., Thermal Expansion of Solids, National Bureau of Standards Circular 486, United States Department of Commerce (1950)

[18] Griffiths, C. A., Rees, A., Kerton, R. M., Fonseca, O. V., Temperature effects on DLC coated micro molds, *Surface and Coatings Technology* (2016), Volume 307, Part A, pp. 28–37

[19] Scholz, S. G., Griffiths, C. A., Dimov, S. S., Brousseau, E., Lalev, G., Petkov, P., Manufacturing routes for replicating micro and nano surface structures with bio-mimetic applications, *CIRP Journal of Manufacturing Science and Technology* (2011), 4(4), pp. 347–356

[20] Kim, D. H., Kim, Y., Hwang, S. H., Bang, Y. S., Cho, C. R., Kim, Y. K., Kim, J. M., Experimental and theoretical evaluation of wettability on micro/nano hierarchically engineered surfaces based on robust micro-post-arrayed- and highly ordered nano-rippled-structures, *Applied Surface Science* (2011), Volume 257(21), pp. 8985–8992

[21] Song, Y., Premachandran, N. R., Zou, M., Wang, Y. A., Adhesion and friction properties of micro nano-engineered superhydrophobic hydrophobic surfaces, *Thin Solid Films* (2010), 518(14), pp. 3801–3807

[22] Yeong, W,Y., Yu, H., Lim, K.P., Ng, K.L., Boey, Y.C., Subbu, V.S., Tan, L.P., Multiscale Topological Guidance for Cell Alignment via Direct Laser Writing on Biodegradable Polymer, *Tissue Engineering: Part C* (2010), 16(5), pp. 1011–1021

[23] Evans, C.J., Bryan, J., "Structured", "Textured" or "Engineered" Surfaces, *CIRP Annals – Manufacturing Technology*, (1999) 48(2), pp. 541–556

[24] Martinez, E., Engel, E., Planell, J.A., Samitier, J., Effects of artificial micro- and nano-structured surfaces on cell behaviour, *Annals of Anatomy* (2009), Volume 191(1), pp. 126–135

[25] Kovalchenko, A., Ajayi, O., Erdemir, A., Fenske, G., Etsion, I., The effect of laser surface texturing on transitions in lubrication regimes during unidirectional sliding contact, *Tribology International* (2005), 38(3), pp. 219–225

[26] Bruzzone, A.A.G., Costa, H.L., Lonardo, P.M., Lucca, D.A., Advances in engineered surfaces for functional performance, *CIRP Annals – Manufacturing Technology* (2008), 57(2), pp. 750–769

[27] Chou, S.Y., Krauss, P.R., Renstrom, P.J., Imprint of sub 25nm vias and trenches in polymers, *Applied Physics Letters* (1995), 67(21), pp. 3114–3116

[28] Dimitrov, A.S., Nagayama, K., Continuous Convective Assembling of Fine Particles into Two-Dimensional Arrays on Solid Surfaces, *Langmuir* (1996), Volume 12(5), pp. 1303–1311

[29] Tillmann, W., Vogli, E., Hoffmann, F., Wear-resistant and low-friction diamond-like carbon (DLC)-layers for industrial tribological applications under humid conditions, *Surface and Coatings Technology* (2009), 204(6–7), pp. 1040–1045

[30] Voevodin, A.A., Rebholz, C., Matthews, A., Comparative tribology studies of hard ceramic and composite metal-DLC coatings in sliding friction conditions, *Tribology Transactions* (1995), 38(4), pp. 829–836

[31] Saha, B., Liu, E., Tor, S.B., Hardt, D.E., Chun, J.H., Khun, N.W., Improvement in lifetime and replication quality of Si micromold using N:DLC:Ni coatings for microfluidic devices, *Sensors and Actuators B: Chemical* (2010), 150(1), pp. 174–182

[32] Saha, B., Liu, E., Tor, S.B., Khun, N.W., Hardt, D.E., Chun, J.H. , Replication performance of Si–N–DLC-coated Si micro-molds in micro-hot-embossing, *Journal of Micromechanics Microengineering* (2010), 20(4), pp. 1–8

[33] Griffiths, C.A., Dimov, S.S., Rees, A., Dellea, O., Gavillet, J., Lacan, F., Hirshy, H., A novel texturing of micro injection molding tools by applying an amorphous hydrogenated carbon coating, *Surface and Coatings Technology* (2013), ISSN 0257-8972, http://dx.doi.org/10.1016/j.surfcoat.2013.07.006

PART 3

Micro Molding Key-Enabling Technologies

8 Vacuum-Assisted Micro Injection Molding

Giovanni Lucchetta, Marco Sorgato, Davide Masato

8.1 Introduction

8.1.1 Air Evacuation in Injection Molding

In injection molding, the evacuation of air and non-aqueous volatiles is one of the main factors affecting the quality of the molded parts, the integrity of the tool, and the controllability of the process [1]. During the filling phase, the advancing melt flow front compresses and pushes them toward the end of the cavity where they accumulate. Besides this, the air may become so hot from compression that it burns the surrounding material. The molding compounds may decompose, outgas, or form a corrosive residue on cavity walls. This effect can occasionally be noticed in poorly vented molds at knit lines or in corners or flanges opposite the gate. Burns usually appear as dark discolorations in a molded part and render it useless. Moreover, inadequate venting can affect the quality of the molded parts by affecting the consistency of micro feature replication [2].

In standard applications, air and gases in the cavity are displaced by the pressurized melt flow front (i.e., passive venting), which pushes them out through the parting plane, the ejector pins, or some specifically machined slots [1]. This is particularly true if a certain roughness is provided at the parting surface, such as planing with a coarse-grained grinding wheel. Another solution consists in the use of ejector pins. If they are located in an area where air may be trapped, they can usually be used for venting, which can be facilitated by enlarging the ejector pin hole. Although most molds do not need special design features for venting due to the fact that air has sufficient chances to escape along ejector pins or at the parting plane, there are some applications where these solutions are not feasible and active venting systems are needed. This is the case with thermoset and elastomer injection molding, for which their low viscosity can lead to flashing at the mold parting plane [3]. Active venting is also applied when injection molding at the micro scale, where using conventional passive venting systems is not possible due to the reduced dimensions of micro features [4].

8.1.2 Vacuum-Assisted Micro Injection Molding

The application of vacuum venting for the evacuation of air from the cavity has been introduced in micro injection molding with the intent of improving the replication degree of micro features (i.e., filling length, geometrical accuracy, and feature definition) by reducing the counteracting pressure generated by the entrapped air [5]. Indeed, the reduced dimensions of typical micro injection molding cavities do not allow the venting of air from the mold by means of conventional slots, which are usually machined on the parting plane between the two mold halves [1]. Moreover, these slots might not work properly when adopting a high injection speed, which is usually needed in micro injection molding to completely fill in the micro features [6].

Evacuation of air from the cavity is a crucial factor for the manufacturing of parts characterized by micro- or nano-structured surfaces [7]. Micro or nano features in the mold inserts represent 'blind-holes' in which air tends to accumulate and be compressed during the high-speed filling of the cavity [8]. The entrapped and compressed air pockets (Figure 8.1) can increase the melt flow resistance, thus hindering the achievement of a complete replication and compromising the final functionality of the parts.

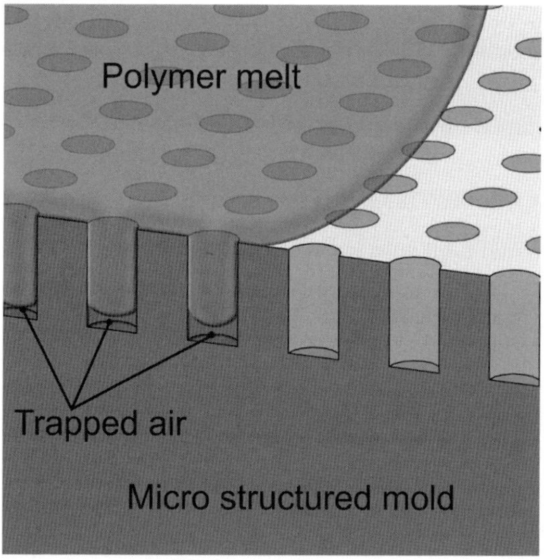

Figure 8.1 Entrapment of air pockets in micro features during the filling phase

8.1.3 Cavity Air Flow in Micro Injection Molding

The gaseous substances that need to be actively evacuated before the injection of the polymer melt are both air and non-aqueous volatiles, which are produced when the polymer is plasticated. These residual substances are generated by the gasification of the injected polymer due to an excessive increase of its temperature. This is a consequence of the severe shearing conditions and the adoption of a high mold temperature that characterizes the micro injection molding process. Moreover, at high shear rates and temperatures, some low-molecular weight additives (e.g., lubricants, plasticizers, flame retardants, fillers, antioxidants, UV stabilizers, anti-microbial additives, and coupling agents) can separate from the base polymer and accumulate on the melt flow front.

If venting is inadequate, the advancing polymer melt confines the air against the cavity walls and between convergent flow fronts. In such locations, the air is subjected to a rapid heating (up to 800 °C) that can lead to ignition and burning of the polymer [9], which ultimately produces a by-product that could degrade the surface of the mold cavity [10].

Short shots and incomplete filling, especially for micro and nano features, can be caused by inappropriate venting of air, when it is pushed into blind features and counteracts the filling. In addition, air trapped within the micro features affects the thermal boundary conditions at the interface between the polymer and the mold. In fact, air gaps can locally insulate the polymer melt, thus leading to differential cooling and consequent warpage [11].

8.2 Advantages and Limitations

Adoption of the vacuum-assisted micro injection molding process over the conventional micro injection molding process can lead to several advantages. Inadequate cavity air evacuation in the mold can result in air pockets trapped against the cavity wall and between converging flow fronts. This can cause problems such as burn marks (i.e., diesel effect) and incomplete replication of the cavity (i.e., short shots), affecting the quality of the molded parts and the mold surface integrity. Considering the effect of cavity air evacuation on the quality of the molded parts and on the process, several studies have been reported in the literature.

Su et al. investigated the polymer flow behavior in the micro mold cavity, focusing on the replication of polymeric microstructures [12]. The results show that the use of cavity air evacuation in micro injection molding helps to avoid possible air traps, improving the replication degree of the microstructures. Yokoi et al. analyzed the transcription ratio (TR), using ultrahigh-speed injection molding, finding a correlation between TR and injection velocity, flow patterns, and vent conditions [13]. Moreover, employing vacuum-assisted micro injection molding, the average TR is higher compared to that obtained with the conventional micro injection molding process.

Ineffective cavity air evacuation could have detrimental effects on part quality where weld lines are formed. Debondue et al. identified a direct correlation between the molecular diffusion and the fracture mechanism of weld lines, which is not only influenced by polymer selection and process conditions, but also by the effectiveness of cavity air evacuation [14].

Griffiths at al. analyzed the effects of varying micro injection molding process parameters on the cavity air flow behavior during cavity filling [2]. The experimental results indicate that, by applying vacuum venting, the cavity backpressure is reduced and, consequently, the filling length is improved due to a lower injection pressure.

In general, it has been reported that, in vacuum-assisted micro injection molding, the benefits are not limited to the quality of the replication of micro- or nano-structured surfaces, but they extend also to process control. For instance, a lower packing pressure and shorter filling and cooling times were observed [15].

Additionally, evacuating air from the cavity before the pressurized injection of the polymer melt can also benefit the mold, which is usually prone to wear and damage especially for micro molding applications. In fact, inadequate mold venting could cause abrasion (leaching) and corrosion of the mold surface through combustion residues in the combustion gas [16]. The adiabatic compression of air during the high-speed filling phase can cause the triggering of localized combustion (i.e., diesel effect), which can damage the surface of the mold. Moreover, the localized heating and combustion can cause the accumulation of gaseous volatiles in the cavities and in venting slots. These issues often lead to the necessity of more frequent tool cleaning and maintenance, if not to permanent damage of the micro or nano features.

Despite the advantages associated with the evacuation of air from the cavity, there are some limitations and issues that should be considered. With the use of vacuum-assisted micro injection molding, an active venting system needs to be implemented, which means higher investments and an increased complexity for process control and maintenance of the tool. The application of vacuum venting for evacuation of air from the cavity also increases the cycle time due to the fact that once the mold is closed, the vacuum system needs some time to depressurize the cavity. Moreover, the application of vacuum venting results in a significant reduction of the melt flow temperature [17]. Therefore, when applying vacuum venting, the consequent reduction of the melt and mold temperature should be considered, as it directly affects the thermal gradient between the melt and the mold.

Considering its effect on the quality of the molded parts and on the process, the main advantages of vacuum-assisted micro injection molding can be summarized as follows:

- reduction of cavity backpressure due to air entrapment and consequent reduction of flow resistance (i.e., lower injection pressure) [2, 18]
- elimination of burn marks, due to the diesel effect, especially on knit lines and in corners or flanges opposite to the gate [19]
- absence of air bubbles and voids in the moldings [12, 20]
- better mechanical properties, especially at weld lines, where the interference of air with the two welding melts can reduce the part strength while also forming v-notches on the surface of the molded part that act as a stress concentration during the part's end-use [1, 2, 16]
- improved process control, e.g., shorter filling and cooling times, lower packing pressure [15]

Moreover, evacuation of air from the cavity before injection of the polymer melt can also yield some benefits for the tool [16]:

- reduction of tool erosion and consequent reduction of mold maintenance
- less buildups of volatiles in the cavities
- no need for periodical cleaning of venting slots

The limitations and issues arising when adopting vacuum-assisted micro injection molding can be summarized as follows:

- need to implement an active venting system, which means higher investments, increased complexity for process control, and maintenance difficulty [19]
- increased cycle time [1]
- decrease of cavity air compression and consequent reduction of mold temperature [17]

8.3 Equipment and Design Solutions

8.3.1 Active Venting

The evacuation of air from the cavity in vacuum-assisted micro injection molding is performed by means of an active venting circuit [1]. Compared to conventional injection molding, air is not displaced by the advancing melt flow front pushing the air out of the cavity from either the parting plane, the ejector pins, or some slots machined for the purpose (i.e., passive venting). Instead, a pressure gradient is artificially created by means of a vacuum system connected to the molding tool.

The structure of an active vacuum system is shown schematically in Figure 8.2. The main elements of the circuit are the vacuum pump, the connecting/inlet elements, and the vacuum measuring device [1]. Vacuum pumps used for vacuum-assisted micro injection molding are usually mechanical systems that exploit a rotary device to remove gas molecules from the sealed volume. They usually work with nominal pumping speeds up to 1600 $m^3 h^{-1}$, and the ultimate cavity pressure they could yield ranges from 0.05 to 20 mbar [21].

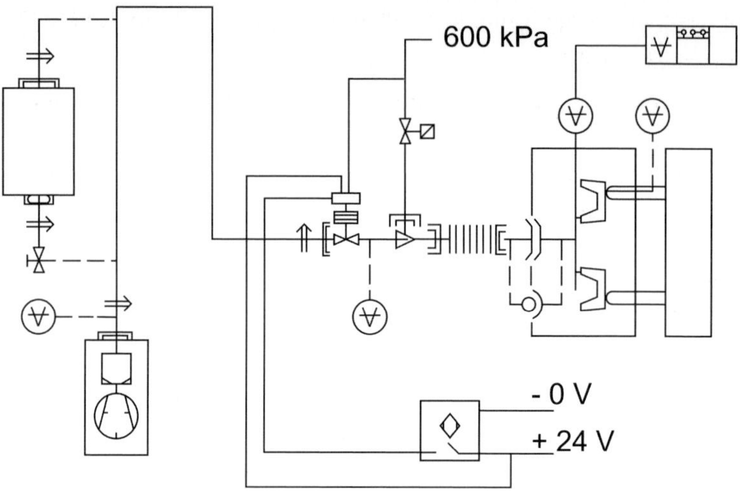

Figure 8.2 Vacuum system schematics, adapted from [1]

8.3.2 Mold Design for Vacuum-Assisted Micro Injection Molding

In vacuum-assisted micro injection molding, when designing the mold, the presence of some elements that allow the active evacuation of air from the cavity needs to be considered. The main elements that a mold for vacuum-assisted micro injection molding should include are the following:

- connecting elements for the vacuum pump
- venting channels to aspirate the air from the cavity
- vacuum inlets, placed at the end of the venting channel in the cavity or close to it
- sealing elements to allow the achievement of vacuum conditions inside the mold cavity

A typical design of molds for manufacturing micro- or nano-structured surfaces by micro injection molding is characterized by having the cavity in the moving half and the micro- or nano-structured interchangeable insert in the fixed mold half. For this reason, the main elements used to realize vacuum venting are included in the moving half. However, different design solutions can be adopted to place the venting channels into the mold and to create a vacuum in the cavity through the inlet elements.

Venting channels can be included in the mold assembly by creating circular holes through the assembly, which reach the parting plane where they are connected to the venting inlets. With this design solution, some slots are usually machined in the cavity, as opposed to the injection location, as shown in Figure 8.3(a). When applying vacuum venting, the evacuating air flows outside from the cavity through these slots, which convey it into some channels that eventually lead to the venting channels (Figure 8.4(a)).

Alternatively, the elements connecting the vacuum pump and the venting inlets can be directly located inside a protrusion of the main cavity, as indicated in Figure 8.3(b). This is significantly thinner compared to the cavity substrate, and can then be easily trimmed from the micro-molded part. Despite being technically easier, this solution might not be acceptable for parts demanding high quality. However, as shown in Figure 8.4(b), the venting channels are significantly smaller and minor modifications of the conventional micro injection molding mold allow the application of vacuum venting.

In both cases, when designing the mold, the presence of an O-ring to seal the parting plane should be contemplated. The venting inlets are placed within the sealed area, thus are able to efficiently evacuate air from the cavity. However, when the venting channels go through the moving half of the mold assembly, more sealing elements should be located between each plate to guarantee a complete isolation of the venting inlets.

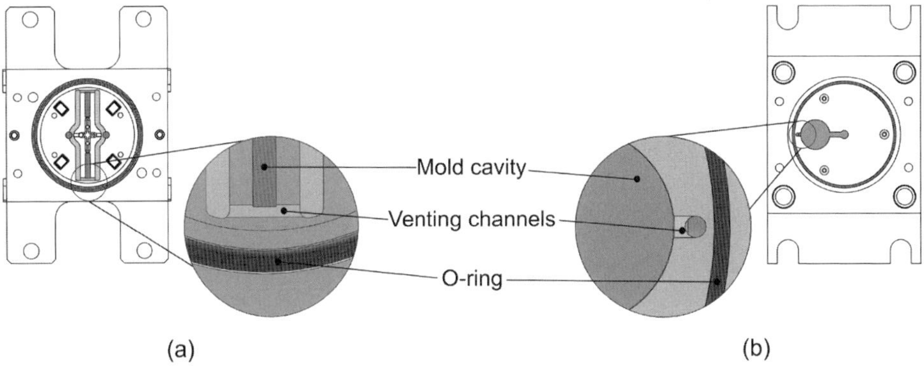

Figure 8.3 Venting channels machined (a) in the mold assembly and (b) in the cavity plate

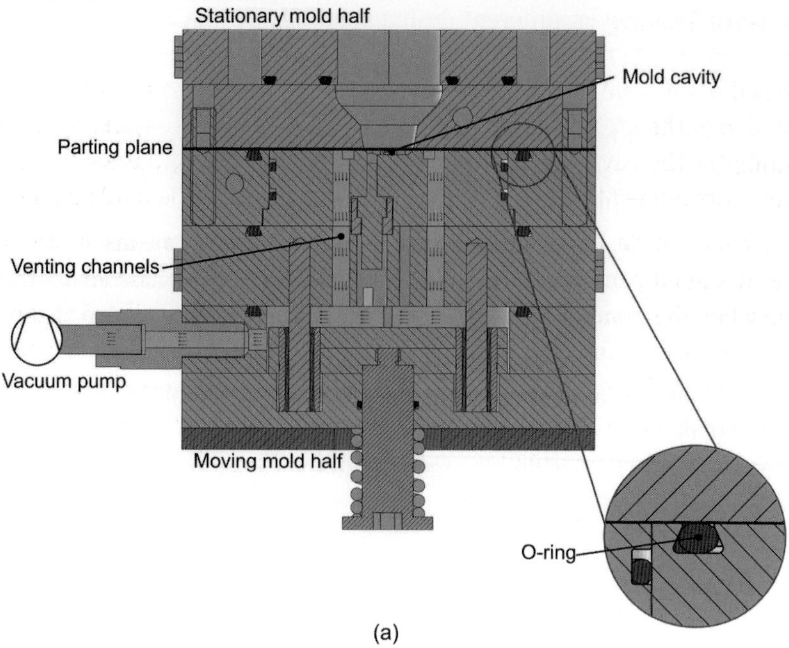

(a)

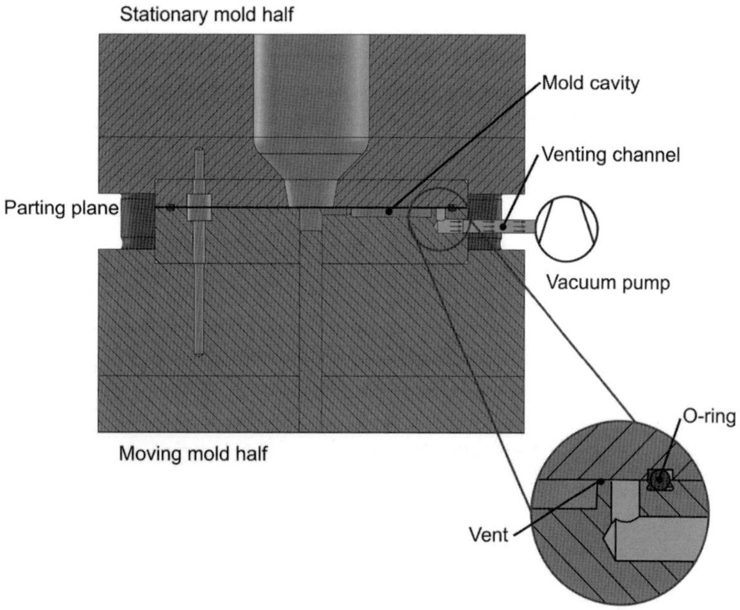

(b)

Figure 8.4 Mold design for vacuum-assisted micro injection molding with venting channels machined (a) in the mold assembly and (b) in the cavity plate. Different solutions for O-ring positioning are also shown

8.3.3 Cavity Sealing

The removal of air from the mold cavity is efficient only if the mold is sealed off. However, due to the presence of typical moving elements (e.g., ejectors), the complete sealing of the cavity is virtually impossible to achieve, unless the mold is completely surrounded by a closed jack or box that has just one parting line [1].

The most common design solutions realize the sealing by means of an O-ring located at the mold parting plane, which surrounds the cavity and seals the parting plane when the mold is clamped [7, 17, 22, 23]. A gasket is placed at the mold parting plane by machining a channel in one or both mold halves, as shown in Figure 8.4. When the mold closes, the action of the clamping force deforms the O-ring and creates an efficient sealing. O-rings for mold sealing applications are usually selected by considering the required mold temperature. Neoprene O-rings are used for mold temperatures up to 70 °C, while for higher temperatures, Vyton O-rings should be selected.

The air is removed from the cavity through a vent, placed within the sealed area, where the vacuum pump is connected. In general, the efficiency of the active venting is higher when the vent is placed at the end of the cavity as opposed to the injection location.

8.3.4 Vacuum Control

The main controllable parameters of an active venting system for micro injection molding are the evacuation time and the pressure. The former is generally controlled by means of the control unit of the injection molding machine, while the latter is controlled by the vacuum pump. In general, the optimization of vacuum control should guarantee the achievement of a critical vacuum degree, which allows the complete filling of micro cavities and the absence of morphological defects [24].

In order to maximize its positive effect on the reduction of the melt flow resistance, the active evacuation of air from the mold cavity should start 10–15 seconds before the beginning of the injection phase and last for its whole duration [20, 23]. In fact, the gas generated by the advancing melt flow front is not able to escape during the filling phase if the vacuum valve is closed. Thus, running the vacuum pump for the duration of the filling phase leads to a reduction of the pressure of the compressed gases and to an improved replication degree of the micro features [20].

The pressure of the air inside the cavity is controlled on the vacuum pump by means of a gauge. In order to maximize the effects on the replication, the pressure of the air in the cavity before the injection phase should be about 1–10 mbar [2, 5, 7]. However, the lower the cavity air pressure, the further the improvements of the filling depth in the micro cavities [23].

8.4 Effects on Replication

The effect of vacuum venting on micro feature replication has been increasingly reported in the literature as a possible technological improvement for the micro injection molding process. However, the results are conflicting as this auxiliary technology is deeply affected by other factors, such as process parameters, polymer properties, tool properties, and feature size. Thus, while trying to understand the effect of vacuum venting, complex interactions occurring during the micro injection molding process need to be considered.

The replication degree of a micro- or nano-structured surface manufactured by micro injection molding is generally evaluated by considering both the height (i. e., depth ratio between replicated features' height and tool depth) and the definition of the micro or nano features. Hence, while trying to understand the effects of vacuum venting, its effect on both these quality parameters should be considered.

8.4.1 Height of Replicated Features

The effect of vacuum venting on the reduction of the melt flow resistance, and the consequent increase of filling length, has been widely investigated in the literature. Table 8.1 summarizes the main results, showing how the effect of vacuum venting is not always consistent when changing a feature's dimensions and the injected polymer.

Vacuum venting was proven to significantly enhance the height of replicated micro features [2, 24]. However, significant differences were observed when molding micro- and nano-structured surfaces characterized by different aspect ratios [22]. In particular, when the ratio between a feature's width and depth was between 0.5 and 1, the depth ratio was observed to increase by a few percent (i. e., 1–2%) [5, 20, 22, 26–28]. Conversely, the benefits of vacuum-assisted micro injection molding are greater for higher aspect ratios [22, 29].

For low aspect ratio micro features, the optimization of micro injection molding process variables left little room, or even none, for further improvements when applying vacuum venting [4, 30]. In this case, the adoption of rapid heat cycle molding or high settings of process parameters are better alternatives to promote replication during both the filling and packing phases [2].

When increasing the aspect ratio of micro features, the achievement of complete replication by micro injection molding is more complex [21]. For such features, the adoption of vacuum venting can lead to significant improvements [22]. However, despite the advantages, the successful application of vacuum-assisted micro injection molding requires the optimization of other variables, which affect the quality of the replication, such as the process parameters and the selection of the molding polymer.

Table 8.1 Width and Aspect Ratio of Replicated Features for Published Research in Vacuum-Assisted Micro Injection Molding

Feature width [μm]	Aspect ratio	Polymer	ΔDR [%]	Ref.
100	1	ABS	>0	[2]
2–20	12	PMMA, PP, HDPE	/	[31]
8	0.5	PMMA	<1	[25]
60	4	PP	/	[24]
80	10	POM	4.6–5.4	[29]
3	5	PS	–16	[21]
4	1.25	PS	0	[4]

8.4.2 Feature Definition

Vacuum venting has been observed to be effective in enhancing the definition of replicated micro- and nano-structured surfaces. Even for low aspect ratio micro features, for which the increase of the replicated height can be small (about 1–2%), improvements in terms of feature definition can be significant [22], as shown in Figure 8.5.

The micro and nano features that often characterize the surface of micro injection molded parts are crucial for their functionality. For instance, in the case of multi-layer microfluidic devices, the coupling between the different layers needs to be very accurate in order to avoid leakage or malfunctions [32, 33]. For this reason, the replicated micro features must be characterized by precise replication of the tool cavities.

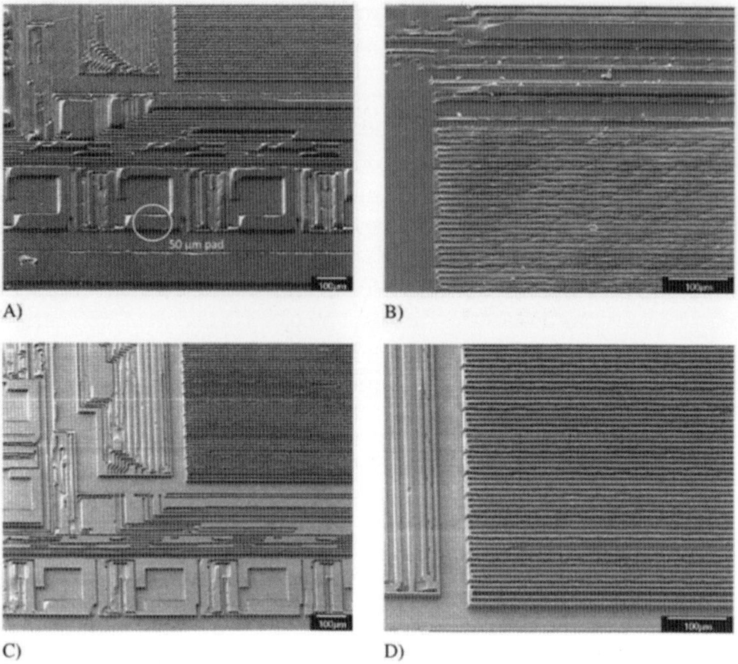

Figure 8.5 SEM images of replicated polymer micro features: (A, B) without vacuum venting and (C, D) with vacuum venting [22]

Feature definition has to be evaluated considering the following aspects in the polymeric replicated parts:

- sharpness of the edges
- absence of distortions
- absence of internal porosities and bubbles
- roughness of the replicated surfaces
- roundness and cleanness of the edges

The positive effects on feature definition consequent to the adoption of vacuum-assisted micro injection molding have been investigated with reports of significant improvements. Yoon et al. analyzed the definition of replicated lines and pads of different dimensions (4–50 µm) [22]. They observed that when air was evacuated from the cavity, the edges of polystyrene replicated micro features were sharper (Figure 8.5). Moreover, the smaller lines were more distorted and rougher when vacuum venting was not applied.

Chang et al. showed that the evacuation of the cavity air resulted in improved corner definition for PMMA channels characterized by a low aspect ratio (< 0.1 : 1) [34]. They indicated that the effect is more evident for higher-aspect-ratio positive features on the tooling. Similarly, Sha et al. indicated that the advancement of the melt flow in the corners can be hindered by the expansion, at high temperature, of the residual air [30].

Ong et al. studied the manufacture of high aspect ratio POM micro rods (diameter: 80–140 µm, depth: 800–1000 µm), indicating that vacuum venting could be exploited to enhance the physical appearance of the replicated polymer features [29]. In particular, when vacuum-assisted micro injection molding was adopted, less shrinkage, a better surface finish, and better repeatability in the micro parts were achieved [24].

The effect of vacuum venting on feature definition can be modified by different selections of process parameters. In particular, as the injection speed increases, venting becomes more difficult and the residual air can cause bubbles and voids in the micro parts [35]. Furthermore, edge sharpness can be improved by increasing the barrel temperature, which leads to lower viscosity of the polymer melt and, consequently, easier evacuation of air from the micro grooves [36].

8.4.3 Part Morphology

Vacuum-assisted micro injection molding can also be exploited to improve the morphology of molded micro parts. The residual air in the cavity during the injection phase can be embedded within the melt polymer, resulting in internal bubbles and voids in the final micro parts, as shown in Figure 8.6. This is accentuated by the critical processing conditions typical of the micro injection molding process, such as the high mold temperature, which is required to guarantee polymer flowability. The gasification of non-aqueous volatiles, occurring at high temperatures, leads to the formation of residual cavities within the molded micro walls and at their base, which eventually causes localized bulges [31]. The presence of these morphological defects is critical, due to the reduced dimensions of the polymer parts, thus constituting a critical issue for their mechanical resistance during functioning [24].

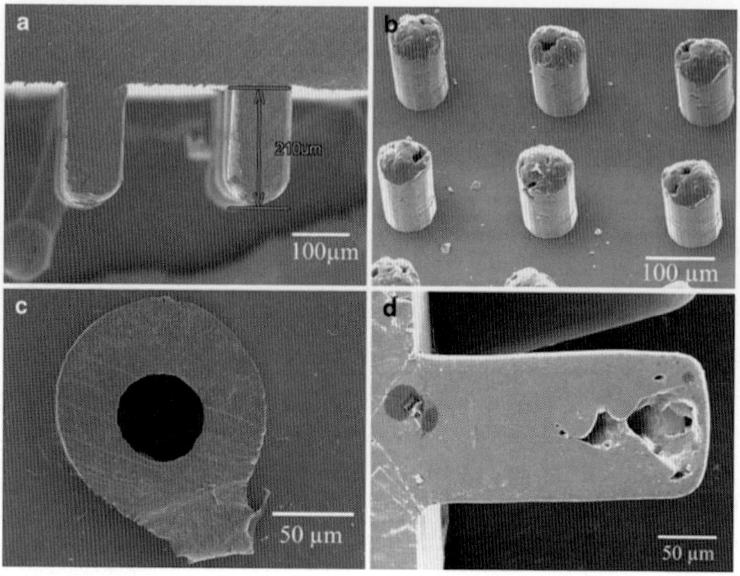

Figure 8.6 Effect of vacuum-assisted micro injection molding on the morphology of the micro parts: (a) short filling, (b) rough surface, and (c, d) hollow [24]

8.5 Venting Optimization

The successful enhancement of feature replication and definition for micro- and nano-structured surfaces by means of vacuum-assisted micro injection molding is dependent on the optimization of process variables and polymer selection. Evacuation of air from the cavity could lead to improvements in the quality of the molded micro parts [2, 18], but could also be irrelevant [25] or even negative [21]. Hence, when approaching the design of the vacuum-assisted micro injection molding process, its optimization is crucial to guarantee the achievement of a proper replication degree in the molded micro parts.

8.5.1 Effect of Micro Injection Molding Process Parameters

In micro injection molding applications, mold temperature is the main parameter responsible for the achievement of good replication of micro- and nano-structured surfaces [4]. A high value of this parameter can help to counteract the short freezing time, typical of narrow cavities, and thus it allows the achievement of markedly higher filling length. Vacuum-assisted micro injection molding can negatively affect the process by modifying the thermal boundary conditions at the cavity walls. The evacuation of air from the cavity could result in a decrease of the mold temperature just before the beginning of the injection phase, compromising the achievement of good replication [36].

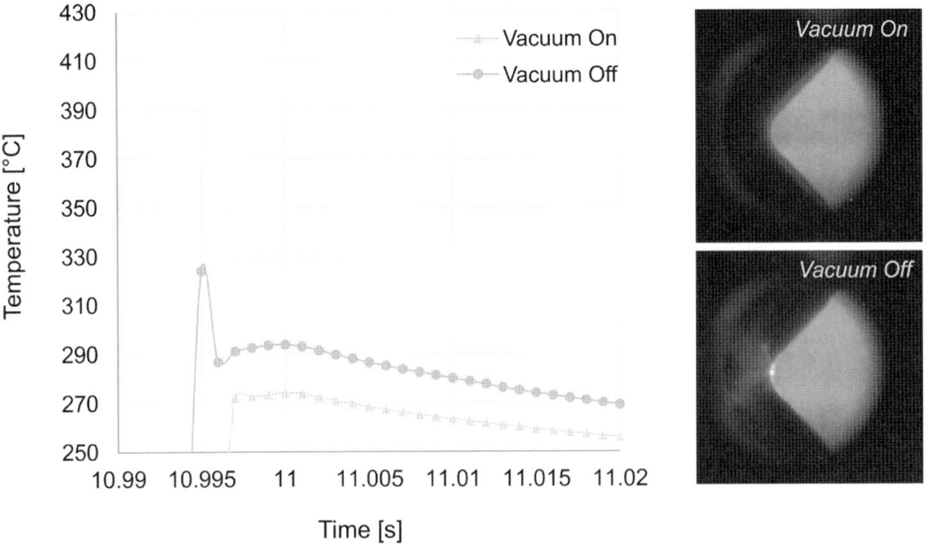

(a)

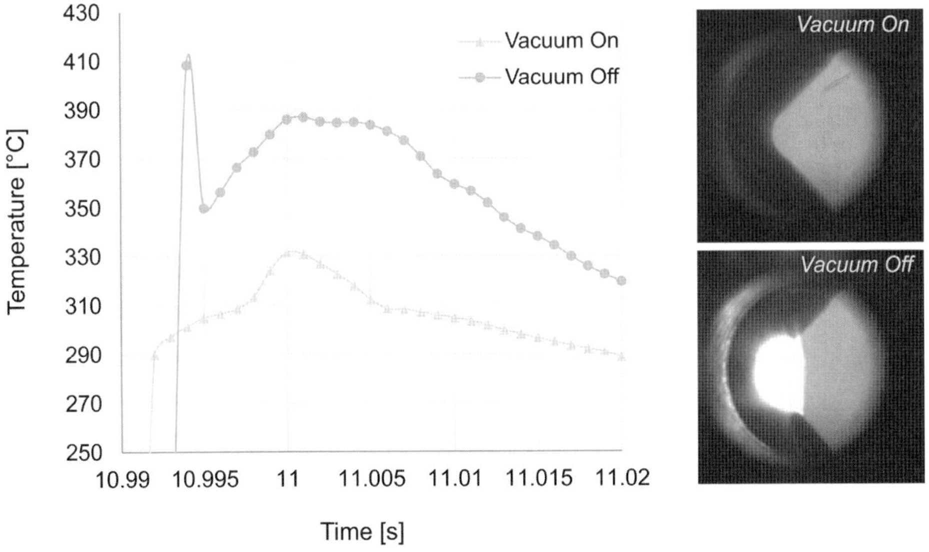

(b)

Figure 8.7 Flow front temperature data for different molding conditions at (a) 80 °C and (b) 140 °C. Adapted from [17]

Lucchetta et al. observed that, when manufacturing high aspect ratio micro features (i.e., micro grooves with a width of 3 μm and aspect ratio of 5), the application of vacuum venting results in a significant reduction of the replicated depth ratio (i.e., about 16% smaller) [21]. This negative aspect of vacuum-assisted micro injection molding was proven to be related to the removal of warm air from the cavity before the beginning of the injection phase. Sorgato et al. performed direct measurements of the mold/polymer temperature, using a high-speed infrared camera and a sapphire window (Figure 8.7), to show that when evacuating the air from the cavity, the temperature decreases significantly (i.e., about 7%), due to convection cooling of the cavity [17].

The molten polymer injected at high speed compresses the air trapped inside the cavity, causing its rapid compression against the cavity walls. The heat produced by the compression increases both the temperature of the flow front and the mold temperature. In the case of vacuum-assisted micro injection molding, a reduction of the melt and mold temperature occurs, and the thermal gradient between the polymer and the tool is affected. This modification of the thermal conditions during the filling phase is particularly significant for micro injection molding applications, where the successful replication of micro and nano features is strictly dependent on the ability of the polymer to flow in narrow cavities. In this sense, a lower temperature means higher melt viscosity, thus a reduction of the flowability of the polymer and the possibility to achieve complete replication, even when applying high values of the packing pressure.

The effect of vacuum venting on the reduction of mold surface temperature is modified by different selections of micro injection molding process variables. Specifically, when adopting high values of the mold temperature (i.e., rapid heat cycle molding to promote replication during the filling phase), the removal of air from the cavity results in a more marked drop of the flow front temperature (Figure 8.7(b)). Similarly, when injecting at a higher speed, the increase of the temperature is greater and thus is the reduction when applying vacuum venting.

8.5.2 Effect of Polymer Selection

The application of vacuum venting can interact with the selection of the molding polymer, and its effect on replication could be modified. The following properties of the molding polymer can be regarded as the most significant factors to be considered when approaching the optimization of the vacuum-assisted micro injection molding process:

- The viscosity of a molten polymer represents its ability to flow when subjected to a shearing flow. During the micro injection molding process, the polymer, heated above its melting temperature, is injected at high speed through a smaller section causing an increase of shear rate and shear stress, which are inversely proportional to cavity thickness [37, 38]. Thus, the lower the viscosity, the better the flowability of the injected polymer.

- The ability of a liquid (e.g., molten polymer) to maintain contact with a solid surface (e.g., cavity walls) as a result of the intermolecular interactions upon contact is defined as wetting [39]. The balance between adhesive and cohesive forces determines the wettability (i.e., degree of wetting) of a liquid-solid contact. In micro injection molding applications, this represents a critical property of the polymer-tool interface and can be used to understand the adherence of two materials [40]. The optimization of the interfacial effects between the polymer and the mold surface can be exploited to increase surface interaction, and thus replication [41].

These properties of the selected molding polymer allow vacuum venting to be more or less significant by affecting the filling behavior in the micro and nano cavities [22]. In general, a combination of higher temperature dependence of the viscosity and higher wetting properties facilitate replication and allow vacuum venting to significantly enhance the filling ratio [7].

When the set mold temperature is lower than the polymer glass transition temperature, the replication degree is not affected by the application of vacuum venting. Conversely, low melt viscosity can promote the replication when the residual air is evacuated from the cavity before and during the injection phase. This behavior is explained by the significant reduction of viscosity that follows an increase of mold temperature: a combination of low melt viscosity and vacuum venting can effectively increase the replication degree of the micro features [36].

Moreover, different polymers can have different sensitivity to cavity air evacuation. Sorgato et al. in their experiments compared the effects of vacuum venting when using two different polymers (i.e., PS and COC) [7]. The two polymers processed by vacuum-assisted micro injection molding yielded different replication degrees, due to their different wettability properties [22]. In fact, the higher the wettability of the polymer melt over the mold surface, the higher the interfacial interactions, and, as a consequence, the higher the replication degree after evacuating the air from the cavity. Conversely, without vacuum venting, the replication decreases due to the presence of residual air in the micro cavities that, despite the favorable wetting conditions, counteracts the replication.

In general, vacuum-assisted micro injection molding can be an efficient solution when molding polymers that are particularly sensitive to degradation at high temperatures [22]. Moreover, it can be applied when high settings of process parameters (e.g., high values of mold temperature and injection speed) are not feasible.

8.6 Concluding Remarks

Vacuum-assisted micro injection molding has been established as a common practice for the manufacturing of micro- and nano-structured surfaces, where the adoption of conventional passive evacuation setups (i.e., machined slots) is not possible. The application of vacuum venting is introduced in the conventional process as an auxiliary technology that allows the improvement of the replication degree (i.e., filling length, geometrical accuracy, and feature definition) by reducing the counteracting pressure generated by the air entrapped in micro features.

The evacuation of air from the cavity (before and during the injection phase) has been reportedly observed to provide significant advantages for the quality of the molded micro parts, the integrity of the tool (particularly prone to damage in micro molding applications), and the controllability of the process. However, the main disadvantages could be the increased cycle time and the increased costs for the required active venting system.

The effects on part quality resulting from the application of vacuum venting are evident considering the definition of the replicated polymer features (i.e., improved sharpness and reduced surface roughness) and their morphology (i.e., elimination of internal bubbles and voids). However, the effect on the replication degree of micro features is more complex and during its optimization other variables, such as the selection of different process parameters and of the molding polymer, should be considered.

In general, vacuum-assisted micro injection molding is well exploited for applications where the achievement of a complete replication through the optimization of process parameters is not possible. In these cases, cavity air evacuation can support the filling, which occurs in non-favorable conditions (e.g., low mold temperature, low injection speed, and micro features with high aspect ratios).

References

[1] Menges, G., Michaeli, W., Mohren, P., *How to make injection molds* (2001) Hanser, Munich

[2] Griffiths, C.A., Dimov, S.S., Scholz, S., Tosello, G., *J. Eng. Ind.* (2011) 133(1), p. 11006

[3] LeFan, J., Saint-Gobain Performance Plastics (2011), retrieved from http://www.medical.saint-gobain.com

[4] Masato, D., Sorgato, M., Lucchetta, G., *Mater. Des.* (2016) 95, pp. 219–224

[5] Chien, R.D., *Sens. Actuators, A* (2016) 128(2), pp. 238–247

[6] Lucchetta, G., Sorgato, M., Zanchetta, E., Brusatin, G., Guidi, E., Di Liddo, R., Conconi, M.T., *eXPRESS Polym. Lett.* (2015) 9(4), pp. 354–361

[7] Sorgato, M., Masato, D., Lucchetta, G., *Microsyst. Technol.* (2016) pp. 1–10, doi: 10.1007/s00542-016-3038-5

[8] Piotter, V., Hanemann, T., Ruprecht, R., Hausselt, J., *Microsyst. Technol.* (1997) 3(3), pp. 129–133

[9] Giboz, J., Copponnex, T., Mélé, P., *J. Micromech. Microeng.* (2007) 17(17), pp. 96–109

[10] Ebnesajjad, S., *Fluoroplastics, Volume 2: Melt Processible Fluoroplastics: The Definitive User's Guide* (2003) William Andrew Publishing/Plastics Design Library, Norwich, NY

[11] Bendada, A., Derdouri, A., Lamontagne, M., Simard, Y., *Appl. Therm. Eng.* (2004) 24(14–15), pp. 2029–2040

[12] Su, Y.-C., Shah, J., Lin, L., *J. Micromech. Microeng.* (2004) 14(3), pp. 415–422

[13] Yokoi, H., Han, X., Takahashi, T., Kim, W.K., *Polym. Eng. Sci.* (2006) 46(9), pp. 1140–1146

[14] Debondue, E., Fournier, J.-E., Lacrampe, M.-F., Krawczak, P., *J. Appl. Polym. Sci.* (2004) 93(2), pp. 644–650

[15] Rosen, M., *Plastics Eng.* (2015) 71(6), pp. 34–39

[16] Kazmer, D.O., *Injection Mold Design Engineering* (2007) Hanser, Munich

[17] Sorgato, M., Babenko, M., Lucchetta, G., Whiteside, B., *Int. J. Adv. Manuf. Technol.* (2017) 88(1–4), pp. 547–555

[18] Shen, Y.K., Chang, C.Y., Shen, Y.S., Hsu, S.C., Wu, M.W., *Int. Commun. Heat Mass Transfer* (2008) 35(6), pp. 723–727

[19] Yang, C., Yin, X.-H., Cheng, G.-M., *J. Micromech. Microeng.* (2013) 23(9), p. 93001

[20] Yokoi, H., Masuda, N., Mitsuhata, H., *J. Mater. Process. Technol.* (2002) 130–131, pp. 328–333

[21] Lucchetta, G., Sorgato, M., Carmignato, S., Savio, E., *CIRP Ann. Manuf. Technol.* (2014) 63(1), pp. 521–524

[22] Yoon, S.-H., Padmanabha, P., Cha, N.-G., Mead, J.L., Barry, C.M.F., *Int. Polym. Process.* (2011) 26(4), pp. 346–353

[23] Ying Choi, S., Zhang, N., Toner, J.P., Dunne, G., Gilchrist, M.D., *J. Micro and Nano Manuf.* (2016) 4(2), p. 21005

[24] Lu, Z., Zhang, K.F., *Int. J. Adv. Manuf. Technol.* (2009) 40(5–6), pp. 490–496

[25] Lin, H.Y., Young, W.B., *Appl. Math. Modell.* (2009) 33(9), pp. 3746–3755

[26] Chen, G., Yek, A., Bay, T., Wang, Z., Precision Micro Feature Moulding Using Vacuum Assisted Moulding Technique For Polymeric Microfluidic Chip Applications. *SPE ANTEC Tech. Papers* (2016) pp. 1236–1240

[27] Thiruvenkataswamy, R., *SPE ANTEC Tech. Papers* (2008) pp. 2463–2467

[28] Kalima, V., Pietarinen, J., Siitonen, S., Immonen, J., Suvanto, M., Kuittinen, M., Pakkanen, T.T., *Opt. Mater.* (2007) 30(2), pp. 285–291

[29] Ong, N.S., Zhang, H., Woo, W.H., *Mater. Manuf. Processes* (2006) 21(8), pp. 824–831

[30] Sha, B., Dimov, S., Pham, D., *Proc. of 4M2005 Conf.* (2005), pp. 1–4

[31] Liou, A.C., Chen, R.H., *Int. J. Adv. Manuf. Technol.* (2006) 28(11–12), pp. 1097–1103

[32] Masato, D., Sorgato, M., Parenti, P., Annoni, M., Lucchetta, G., *J. Mater. Process. Technol.* (2017) 246, pp. 211–223

[33] Marson, S., Attia, U.M., Lucchetta, G., Wilson, A., Alcock, J.R., Allen, D.M., *J. Micromech. Microeng.* (2011,) 21, pp. 1–9

[34] Chang, P.-C., Hwang, S.-J., Lee, H.-H., Huang, D.-Y., *J. Mater. Process. Technol.* (2007) 184(1–3), pp. 163–172

[35] Belofsky, H., *Plastic: Product Design and Process Engineering* (1995) Hanser, New York

[36] Sha, B., Dimov, S., Griffiths, C., Packianather, M., *J. Mater. Process. Technol.* (2007) 183, pp. 284–296

[37] Barnes, H.A., Hutton, J.F., Walters, K., *An introduction to rheology* (vol. 3) (1989) Elsevier, New York

[38] Macosko, C.W., Larson, R.G., *Rheology: principles, measurements, and applications* (1994) Wiley, New York

[39] Geoghegan, M., Krausch, G., *Prog. Polym. Sci.* (2003) 28(2), pp. 261–302

[40] Zitzenbacher, G., Huang, Z., Längauer, M., Forsich, C., Holzer, C., *J. Appl. Polym. Sci.* (2016) 133(21), pp. 43469(1–10)

[41] Rytka, C., Kristiansen, P. M., Neyer, A., *J. Micromech. Microeng.* (2015) 25(6), p. 065008

9 Modeling and Simulation of Micro Injection Molding

Guido Tosello, David Maximilian Marhöfer

9.1 Introduction

9.1.1 The Micro Injection Molding Process

The injection molding process is a widely applied and well-established production technology in industry and accounts for most of today's plastic parts. Due to its widespread use, it is naturally affected by the constant trend of miniaturization toward smaller parts incorporating increasingly more functionality. As a result, the micro injection molding process is commonly seen as one of the key technologies of the 21st century for the large-scale and highly automated production of high precision micro structured and truly three-dimensional parts. The process is capable of yielding net-shaped multimaterial plastic parts with a complex geometry and diverse functionality. Due to the continuous trend of miniaturization of parts and reduction of weight and costs, the field is constantly growing and is established for the production of microproducts made out of thermoplastics. Typical application fields for the micro injection molding process are the medical industry, the telecommunications industry, and optics [1–8]. The miniaturized version of the injection molding process exhibits the same basic process cycle with the following steps [5, 9]:

- the melting of the plastic granules,
- the metering of the molten plastic,
- the injection of the material into the cavity of the mold (filling phase),
- the packing of the part to compensate for shrinkage,
- the cooling to solidification, and
- the ejection from the mold.

For more information on the process, characteristics, and equipment used in micro injection molding, please refer to the other relevant chapters in this book.

9.1.2 Why Simulate the Injection Molding Process?

> "A computer simulation is any computer-implemented method for exploring the properties of mathematical models where analytic methods are unavailable" [10].

In other words, computer simulations can be described, in general, as the numerical computation of the solution of equations and their related boundary conditions, which govern a mathematical model. They are applied to these mathematical models if an analytic solution is not available or is too complex. As physical and mathematical models are available for many issues in engineering and technology, simulations are consequently widely applied in any field of engineering. Simulations cover areas like structural analyses, meteorology processes, fluid dynamics, logistics, and electric and electronic circuits to name a few.

The following general motives for performing simulations were identified as being relevant and appropriate in the area of injection and micro injection molding [11]:

- Simulations enable us to investigate systems in detail when observations in reality are impossible for pragmatic reasons (too fast, too slow, inaccessible parts, system does not yet exist, and so on) as in the case of climate change or the filling phase in injection molding.
- Simulations enable us to quickly and easily manipulate the investigated system when these changes are too costly, too laborious, or simply impossible in reality as in the case of material exchange in constructions or newly developed plastic parts.
- Simulations enable us to visualize and improve the understanding of the underlying system when the visualization or implementation of the system in reality might be too costly or too dangerous as in the case of combustion inside an engine or plastic flow inside a steel mold.

These reasons can be specifically applied to and transferred to the area of injection molding. In particular, the following goals can thus be derived for simulations of conventional and micro injection molding processes [6, 12–21]:

- shortening the development time and the time-to-market for a new product,
- assisting the part design, especially at an early stage in the design process when the molds or tools do not exist yet,
- avoiding design errors in the costly mold and, subsequently, the necessary re-engineering actions to fix these errors,
- assisting material selection,
- eliminating the need or reducing the amount of physical prototyping and experiments,
- optimizing the part and mold design as well as the parameter settings of the manufacturing process,

- predicting and improving the part quality, for example morphology and properties,
- gaining a better understanding of the process characteristics, and
- visualizing the process and inspecting the final product for, for example, possible air entrapment, warpage and shrinkage, weld lines, and so on.

In conclusion, simulations are widely applied and accepted in the plastics industry as helpful and efficient tools. Due to the complexity of the injection molding cycle, the equipment, and the materials used, as well as the requirement of an increase in the quality of the parts, process simulations of injection molding are indeed very attractive and relevant, and are gaining more attention. Process simulations in the plastic injection molding area have, in fact, been established for decades and are common practice in the plastics industry. Because of the rising number of applications of micro injection molded components, the interest and demand for reliable and supportive simulations in this area are increasing rapidly. Hence, simulation tools and related modeling must be addressed to fully develop the potential of micro injection molding [20, 22–25].

Nonetheless, carrying out simulations also comes with some drawbacks. It must not be forgotten that the underlying models usually incorporate idealizations, approximations, and assumptions—no matter how realistic or accurate the results might appear. In addition, the underlying model must be understood and there must be evidence to believe in the applicability and validity of the model independent of how good the real data are reproduced [11].

This chapter will give some insight into the modeling and simulation of the micro injection molding process, into what is currently feasible and investigated in this area, and into where possible limitations and pitfalls arise.

9.2 Mathematical Background

9.2.1 Viscosity of Plastics

During the filling phase, the plastic flows into the cavity driven by the applied pressure of the injection molding machine. Hence, the plastic's viscosity is of major importance during the filling. The viscosity mathematically links the applied shear rate to the shear stress that the fluid experiences.

Newtonian fluids show a strictly linear behavior, meaning the viscosity is independent of the shear rate. In contrast, polymers are non-Newtonian fluids and, therefore, the viscosity is a function of the applied shear rate. When polymers experience shear, two effects come into play, leading to a decrease in viscosity, the so-called shear thinning effect: heat dissipation and the alignment of the polymer chains [26].

In order to mathematically describe the viscosity, the power-law model was introduced in the 1920s [18]. It is one of the simplest and most common models for non-Newtonian fluids. According to the power-law model, the viscosity η of a plastic melt depends exponentially on the shear stress γ and is given as [27]

$$\eta(\dot{\gamma}) = m\dot{\gamma}^{n-1} \tag{9.1}$$

where m is the viscosity at a shear rate of one reciprocal second and n is the power-law index, which are both data-fitted constants. Generally, $n \leq 1$ for plastic melts, which reflects the shear thinning effect. This model describes the viscosity of plastics well at high shear rates. Nonetheless, injection molding is not only a high shear rate process. The shear rate is highest at the wall of the mold and decreases toward the center of the fountain flow where the shear actually vanishes completely, and the polymer melt exhibits the so-called zero shear rate viscosity.

Consequently, the power-law model is insufficient. Instead, the Cross model is widely applied and is part of most simulation software [14]. This model describes the shear rate dependency of the viscosity over a wide range of shear rates very well, because it combines a Newtonian region at low shear rates with a power-law shear thinning region at higher shear rates [18]. The viscosity is mathematically given by the Cross model as [26, 28]

$$\eta(\dot{\gamma}) = \frac{\eta_0}{1 + \left(\frac{\eta_0 \dot{\gamma}}{\tau^*}\right)^{1-n}} \tag{9.2}$$

where η_0 denotes a zero shear rate viscosity and τ^* is the critical stress level at the transition between the Newtonian plateau and the power-law region, which is determined by curve-fitting. The term $1 - n$ describes the slope of the viscosity curve in the power-law region when logarithmically plotted versus the shear rate, with n being equal to the power-law index.

The applied viscosity model in the simulation software must also account for the temperature changes happening during the entire injection molding cycle. The concept of time–temperature superposition can also be applied to the viscosity. The core of the concept is to shift the known viscosity curve at a reference temperature along the time axis in order to gain the viscosity curve at any temperature that differs from the reference temperature. The shift factor is given by the Williams–Landel–Ferry (WLF) equation, which can also be applied to the zero shear rate viscosity of the Cross model to introduce the temperature. This step yields the Cross–WLF model with its zero shear rate viscosity given as [14, 28]

$$\eta_0(T,p) = D_1 \exp\left[-\frac{A_1(T - D_2 - D_3 p)}{A_2 + T - D_2}\right] \quad (9.3)$$

where A_1, A_2, D_1, and D_3 are constants, T is the temperature, and $D_2 = T_0$ is the reference temperature. The viscosity is also dependent on the pressure as given in Eq. 9.4. In many simulations or material models, this pressure dependency is commonly neglected by assuming that $D_3 = 0$. The pressure dependency is, nevertheless, quite relevant, because the injection molding process often exhibits high injection pressures [14]. More models exist to describe the viscosity of plastics. Common examples, which will not be discussed in more detail in this work, are the simple Cross, Carreau, and the Carreau-Yasuda models, which are described in other literature [27, 29]. Microscale viscosity is possibly an important aspect for the accurate modeling of micro injection molding. More details on this phenomenon and its relevance are described in Section 9.4.3. However, the currently available simulation tools do not consider microscale viscosity.

Moreover, semi-crystalline polymers show a sudden and significant increase in viscosity at their melting temperature due to the incipient formation of crystalline domains inside the materials. The aforementioned models do not undergo this viscosity increase due to crystallization [18]. Additionally, the micro structure of the final micro part might differ from a macro part, as micro injection molding causes higher shear rates and a different cooling and solidification behavior of the melt. The modeling of crystallization, and thus its impact on the mechanical characteristics of plastic parts, is a current topic of research. It is not yet fully integrated in any commercially available simulation software.

9.2.2 Thermodynamics

Besides the viscosity, information about the thermodynamic properties of materials is necessary for computer simulations of the injection molding process. This can be derived from the equation of the state of the material, which links the pressure, the specific volume (the inverse of the density), and the temperature. It is, therefore, also called pvT data [18].

For plastics, the data are usually provided as pvT diagrams, which show the specific volume v as a function of the pressure p and the absolute temperature T. The commonly used model for describing the curves of such a diagram is the two-domain Tait model, which is mathematically given as [30, 31]

$$v(p,T) = v_0(T) \cdot \left[1 - C \ln\left(1 + \frac{p}{B(T)}\right)\right] + v_t(p,T) \quad (9.4)$$

where $C = 0.0894$ is a universal constant, and the function $B(T)$ denotes the pressure sensitivity of the material. The two domains are equal to the solid and molten phase of the plastic, which are separated by the transition temperature T_t. The transition temperature is given as

$$T_t = b_5 + b_6\, p \tag{9.5}$$

In the lower temperature region, meaning a solid phase with $T \leq T_t$, the missing functions of Eq. 9.4 are given as

$$v_0(T) = b_{1s} + b_{2s}(T - b_5)$$

$$B(T) = b_{3s}\, \exp\left[-b_{4s}(T - b_5)\right]$$

$$v_t(p,T) = b_7\, \exp\left[b_8(T - b_5) - b_9\, p\right] \quad \text{if semi-crystalline}$$

$$v_t(p,T) = 0 \quad \text{if amorphous} \tag{9.6}$$

and the upper temperature region, meaning a molten phase with $T \geq T_t$, is respectively described by

$$v_0(T) = b_{1m} + b_{2m}(T - b_5)$$

$$B(T) = b_{3m}\, \exp\left[-b_{4m}(T - b_5)\right]$$

$$v_t(p,T) = 0 \tag{9.7}$$

where the unknowns b_{1s}, b_{2s}, b_{3s}, b_{4s}, b_{1m}, b_{2m}, b_{3m}, b_{4m}, b_5, b_6, b_7, b_8, and b_9 in Eq. 9.6 and 9.7 are all data-fitted coefficients. The data are usually collected from measurements with rheometers. Because of the ease of use and the ability to reproduce high shear rates, high-pressure capillary rheometers are often deployed [28].

9.2.3 Flow of Plastics

The flow of molten plastics is of an anisothermal, compressible, and viscous nature and can consequently be described by Navier–Stokes equations. The Navier–Stokes equations are a system of non-linear partial differential equations. They are the fundamental mathematical model in general fluid dynamics and consist of the following:

- The continuity equation, which describes the conservation of mass. With the density ρ, and the velocity $\mathbf{u}$, the continuity equation is commonly given by [32]

$$\frac{\partial \rho}{\partial t} + \nabla \cdot (\rho \boldsymbol{u}) = 0 \tag{9.8}$$

- The momentum equation, which describes the conservation of momentum. The momentum equation is ordinarily formulated with the pressure p, the stress σ, and the gravity $\mathbf{g}$ as [18]

$$\rho \frac{\partial \boldsymbol{u}}{\partial t} = -\nabla p + \nabla \cdot \boldsymbol{\sigma} + \rho \boldsymbol{g} \tag{9.9}$$

- The energy equation, which describes the conservation of energy. With the specific heat capacity c_p at constant pressure, the absolute temperature T, the coefficient of volume expansion β, and the thermal conductivity κ, the energy equation, sometimes also referred to as heat-transfer equation, is commonly given as [18]

$$\rho c_p \frac{DT}{Dt} = \beta T \frac{Dp}{Dt} + p \nabla \cdot \boldsymbol{u} + \nabla \cdot (\kappa \nabla T) + \boldsymbol{\sigma} : \nabla \boldsymbol{u} \tag{9.10}$$

Current commercially available software tools use the Navier–Stokes equations to describe the flow of plastic during the injection molding process. The contributions of gravity and inertia in the flow are considered in simulation software packages, and options to switch the effects on or off are often offered to the user. It is a commonly accepted aspect that gravity is negligible when it comes to the processing and simulation of micro molded parts, because of their small spatial dimensions and the small influence of body forces at the microscale [27, 28, 30, 33, 34]. The significance of inertial forces is often assumed to be inferior to viscous forces or surface tension in conventional and micro injection molding. Yet, this issue is discussed controversially by the scientific community, especially in the cases of filled plastics and powder feedstocks due to their heavier masses [28, 31, 34–37]. The question of the significance of surface tension in micro injection molding and simulations also remains unanswered. Contradictory results were found in different studies; with some researchers concluding it is negligible, others claiming it certainly is important [13, 38–41]. Some researchers state that surface tension becomes relevant for very small channels in the sub-micrometer range [5, 36]. The effect of surface tension is not implemented in any way in commercial software codes.

More details on the underlying theory and on the mathematical models governing the simulations are not introduced here. Many aspects can be found in dedicated, more specific literature, for instance about computer modeling and polymer rheology. Further reading comprises comprehensive works such as [18, 29, 31] or specific scientific publications such as [4, 42–45].

9.3 State-Of-The-Art and Challenges of Micro Injection Molding Simulations

In the field of conventional and micro injection molding, process simulations are based on the method of finite element analysis (FEA). In particular, the finite difference method (FDM) and the finite element method (FEM) are predominantly employed in commercial software. Sometimes, the finite volume method (FVM) is applied also. A combination of these methods can also be found, so that, for example, the fluid flow might be computed by FEM and the temperature distribution by FDM [23].

Despite their similarity, micro injection molding is not just simply scaling down the conventional injection molding process. Significant differences between the two technologies, the physics behind them, and the applied equipment, exist. These differences also influence the simulation accuracy of software primarily developed for macroscopic thermoplastic parts [5, 7, 9, 13, 15, 22, 25, 31, 46]:

- Relevant microscopic effects such as wall slip, surface tension, or microscale viscosity are not or insufficiently implemented in today's commercially available simulation packages.
- Dedicated micro injection molding machines for improved metering and more precise injection volumes and control exist. These machines can show different characteristics to conventional machines, such as electric drives or plunger mechanics. However, conventional machines are the default machine type in obtainable software tools and micro machines are not earmarked for use.
- Due to the larger surface-to-volume ratio and the smaller thickness, the surface roughness and the altering heat transfer coefficient are no longer negligible. Current simulation software, nonetheless, provides only very limited possibilities to account for changing the heat transfer coefficient, not to mention the implementation of surface roughness.
- The smaller cavity size in micro cavities causes considerably increased shear rates and shear heating. The material data offered might only be available or valid over a limited range of temperatures or shear rates and are perhaps also outdated, leading to a demand for the extrapolation of material data or the acquisition of such data by non-standard methods.

The aforementioned major differences to conventional injection molding thus significantly influence the accuracy of the simulations of micro molded parts. The drawback is that applying these simulation packages to micro injection molding often still yields improper quantitative results. The programs have, especially, difficulties with micro cavities exhibiting high aspect ratios. In order to investigate the differences of micro injection molding and possibly improve the simulations results, a multitude of studies on the process and on simulation validations have been performed in science and research in recent years.

The reported investigations are often based on simple test geometries, enabling the use of mid-plane or dual-domain meshes, thus reducing the complexity of the model and the computation time. However, mid-plane and dual-domain meshes were found to be insufficient for the accurate simulation of micro parts by several groups of researchers. More advanced 3D models were recommended for superior performance. Additionally, although the simple geometries are suitable for scientific examinations, industrial products have a much more complex nature [6, 24, 26, 36, 40, 47].

Further comparative studies have shown the suitability of simulations to identify the most important process parameters linked to the part quality and to defects like air entrapment [4, 26, 47, 48].

In-house code is often applied for the simulation of micro injection molding at universities. The advantages, such as implementation flexibility, overcoming software limitations, and achieving improved results, come, however, at a price. The reproducibility and validation of the results are significantly reduced. Furthermore, the in-house code is specific and is not available to industry. Consequently, more general approaches with commercially obtainable software are preferred [36, 49].

Best practice strategies with validation experiments on dedicated micro molding machines and a simple micro test part have been published [13]. The researchers pointed out several vital aspects, such as the importance of the right meshing parameters when creating the simulation model, the visualization of the flow front as a supportive validation method, and improved pressure prediction by the implementation of the nozzle of the injection molding machine.

Additionally, validation studies with time-controlled and pressure-controlled simulations, and a model comprising the part and the feed system, were reported [46, 50], yielding the right trends, but still underestimating the real injection time and injection pressure. Furthermore, the molding and experimental short shots were monitored to gain input data for the simulations, which yielded good predictions of the injection time, pressure, part mass, and flow length.

Proper boundary conditions of the simulations are essential as emphasized by the research community [22, 46, 51, 52]. The focus of the first study was the necessity for extensive material characterization and the modification of the deployed software for the incorporation of relevant microscopic effects, similar to the aforementioned in-house approaches.

The latter studies dealt with the adaptation of the heat transfer coefficient (HTC) in micro injection molding. The HTC generally depends on the surface roughness, the contact pressure, the materials' thermal conductivities, the interstitial fluid, and the transient behavior of the heat transfer [53]. In a reverse engineering approach, the HTC was investigated for varying cavity sizes in the order of a few 100 μm and varying process parameters [51]. The HTC increased with decreasing cavity thickness as typically assumed for micro cavities, but decreased also with increasing injection speed, leading to lower values at typical speeds for micro injection molding. The determined HTC values covered a large range from adiabatic up to 25,000 W m^{-2} K^{-1}.

In another study, the heat flux in micro cavities of less than 10 μm was determined as being considerably lower than 5000 W m^{-2} K^{-1} [54], a typical value in conventional injection molding and a default value in commercial simulation software. The results are, hence, in contrast to the frequent assumption of a higher HTC in micro molding. Furthermore, successful simulations and flow validations with a 1-mm-thick test part were run with the default values of the HTC [46].

In conclusion, the necessity of the adaptation and change of software default values of the HTC appears to depend on the exact size of the micro cavities. The topic of the HTC in micro molding is still controversially discussed in the research community.

9.4 Best Practice Strategies for Micro Molded Component Simulations

New opportunities for improved accuracy have opened up due to the current and quick developments of simulation technology. With every new release of their software packages, software manufacturers provide new tools and improved features. Keeping up with this evolution has the objective of achieving more accurate and more reliable simulation results for micro injection molding. Consequently, new and comprehensive strategies for simulation models in micro injection molding are necessary for full benefit. The implementation methodology and model of the process simulations are essential for improving the results for micro plastic parts. This section, therefore, focusses on some recommendations on how to create the model when micro molded parts are involved.

9.4.1 Modeling

The implementation strategy of the simulation model of a typical micro plastic part is recommended to be based on the analysis and incorporation of the whole injection molding system including the actual part, the feed system, the inserts, the mold block, the cooling or heating system, the deployed material, and the deployed injection molding machine [13, 55]. A proper implementation strategy for establishing the simulation model increases the result quality (accuracy) [39, 56]. Furthermore, the principle "garbage in, garbage out" applies to simulations [57], meaning accurate simulation results can only be achieved with accurate and sound data on the properties and behavior of the components used. An example of such a comprehensive approach, modeling all components of the injection molding system, is depicted in Figure 9.1.

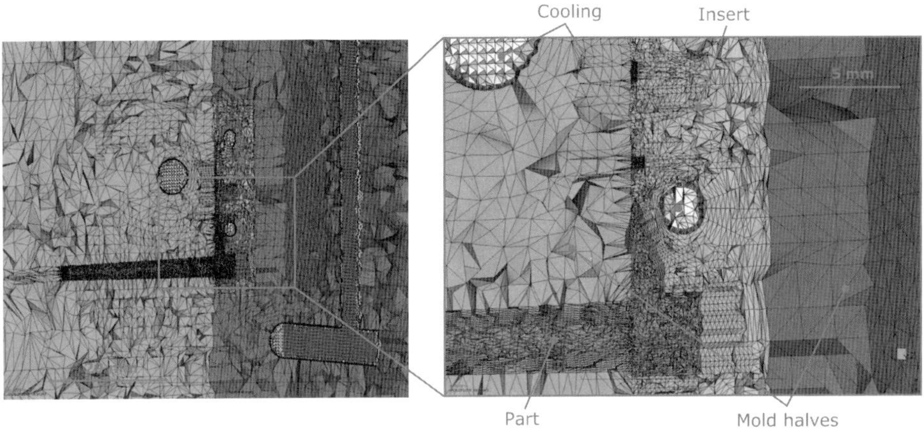

Figure 9.1 Comprehensive model with meshed components of the injection molding system: actual plastic part, cooling channels in mold and insert, mold insert, and mold block consisting of the stationary and movable halves

Since the simulations should, on the other hand, be economical and practically feasible, components like the mold and cooling system can be simplified to reduce the necessary computation time. Moreover, less sophisticated simulations with the part only can still be run to support major design decisions at an early design stage. Regarding newly developed components, the mold might not yet be designed, or the molding machine has yet to be decided. The accuracy can, however, be compromised.

The investigated plastic part should also include a complete feed system [13, 55]. With regard to micro parts, the feed system contributes to a large extent of the total volume and flow length. Including the feed system is thus inevitable for micro parts in order to predict the injection phase correctly. However, attention must be paid to the warpage prediction, since the feed system mechanically constrains the part. The feed system should, consequently, be excluded from the warpage computation.

The implementation of the characteristics of the injection molding machine is important [46, 55]. The injection happens very quickly for micro components and, especially, conventional hydraulic machines influence the filling because of limited response, acceleration, and positioning accuracy. Besides, special micro injection molding machines and units do exist. These machines differ from the conventional machines not only by structure, but they also have a fully electric drive enabling faster and more precise injection and response. Simulation software assumes conventional machines by default though. The technical specifications of the machine are, hence, supposed to be implemented in the machine model too. Furthermore, the geometry of the machine should be implemented in addition to the part and feed system, as shown in Figure 9.2. By doing so, the material compressibility and pressure drop over the nozzle and barrel are taken into account, which can contribute, in micro injection molding, to a considerable share of the total injection pressure [14, 58, 59].

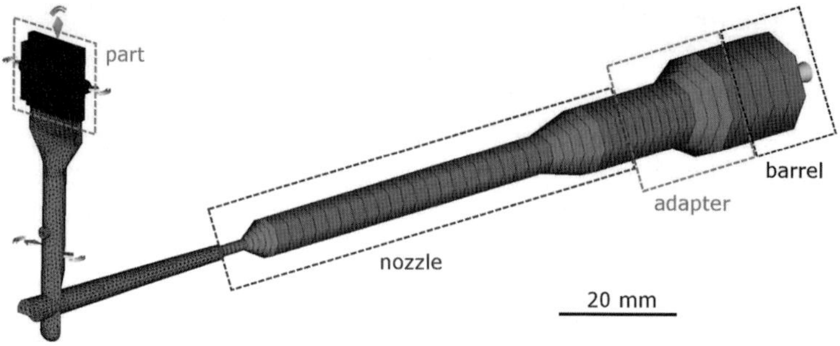

Figure 9.2 Model of the actual plastic part (micro structured test part with a footprint of approximately 12 mm x 12 mm) with feed system, additional machine geometry being modeled and resembling the actual nozzle and the used volume of the barrel of the machine [60]

The injection velocity is one of the most crucial parameters influencing, for instance, the injection time, the flow length, the maximum shear rate occurring, and the melt temperature. It is also one of the most influential parameters in the simulations and in predicting the values of the aforementioned process parameters. In order to improve simulation accuracy, it has been proven, so therefore it is recommended, that the true speed profile in the filling phase of the deployed injection molding machine should be implemented. Possible implementation approaches are the monitoring and institution of the ram speed versus time or versus ram position [39, 50, 55].

The high injection speeds and thus very short injection times make it necessary to adjust the solver settings of commercial software. The default time steps are quite large in comparison to the injection times and the temporal resolution of the simulation analysis can be compromised. [6]

9.4.2 Meshing

The mesh is the spatial discretization of the investigated geometry. Hence, the mesh is of significant importance for the accuracy of the simulation and the spatial resolution of the model [29]. It influences how well the original geometry is reproduced, the accuracy of the numerical results, and the necessary computation time. In general, the numerical accuracy and the computation time increase with decreasing specific element edge length or increasing mesh density. The mesh density is the number of elements per volume or surface area. For one and the same geometry, the resulting mesh can look very different, depending on the settings of the numerous meshing parameters.

As opposed to conventional injection molding, the commonly used 2D modelling or 2.5D Hele-Shaw flow approximation might not be valid, and full 3D modeling is strongly recommended for micro parts [20, 24, 25, 36]. The modeling in 2D or 2.5D introduces errors such as the missing detection of filling imbalances in runner systems. This is due to false assumptions and boundary conditions [18, 34, 61]. The more accurate 3D modeling, however, comes at the expense of longer computation times, as the number of elements is larger and the boundary conditions become more complex. The mesh elements and the mesh edge length thus must be chosen with special care.

Micro injection molding exhibits, in general, large variations in the size of the components involved. The actual micro plastic part has outer dimensions and feature sizes in the micron and sub-micron range, whereas machine and mold parts are of a conventional size in the order of centimeters (inches). The application of a multiscale mesh extending several orders of magnitude regarding the mesh size is, therefore, inevitable. In the multiscale mesh approach, the mesh density is adjusted locally according to the size of the meshed geometry and entity. The mesh applied to micro parts locally requires element edge lengths down to tens of microns at the actual part and even single-digit microns (0.0001 in) at thin gating sections [39, 62, 63]. The mesh size of the mold is typically in the millimeter range.

The multiscale approach ensures an appropriate mesh size for all components included, optimizing the required computation time and being a compromise between the necessary fine mesh for small entities such as micro plastic parts and an adequate coarse mesh for large entities such as mold plates. As mentioned, the element edge length influences the reproducibility of the original part geometry. Therefore, especially for round micro features, the element edge length must be very small to prevent the loss of geometrical accuracy.

Moreover, the degree of the mesh density plays an important role in identifying and indicating the hesitation effect of the polymer melt in very thin areas [29]. Because of the small dimensions of micro cavities, this effect is often seen in micro injection molding.

A comparison between the meshing approach with a homogeneous element edge length all over the model and the multiscale mesh approach is illustrated in Figure 9.3. The meshed plastic part, the cooling structure, and the mold block enclosing everything are shown. In this exemplary study, the total number of elements is comparable. Yet, the resolution of the mesh is in favor of the micro part in the case of the multiscale mesh, which shows a specific element edge length down to 40 µm (0.0016 in).

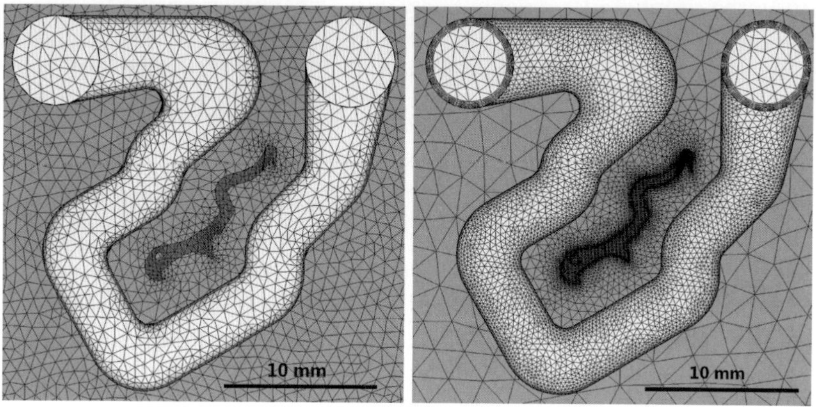

Figure 9.3 Comparison between a conventional (left) and multiscale (right) meshing approach applied to a micro part (center), cooling structure (around the part), and mold block (background). The multiscale mesh yields a finer resolution for the micro part despite a comparable number of elements and thus computation time [60]

9.4.3 Material Data

Special attention should be paid to the material data, because they are the essential foundation for the proper simulation of the plastic flow; without accurate and up-to-date material data, accurate simulations will not be possible. This is especially true for micro molding. Plenty of materials are usually included in the databases of simulation packages by default. However, the available material data might be limited or outdated [6, 60].

The research community has pointed out that an accurate and comprehensive set of material data is substantial [14, 22, 46, 51, 52]. Here, the focus is on having a detailed material characterization with pressure- and temperature-dependent data available. This is a vital aspect for all rheological and thermodynamic material properties, since the injection molding process cycle covers a larger temperature and pressure span, typically from ambient conditions to several 100 °C or 100 MPa, respectively. Due to the increased surface-to-volume ratio and heat loss in micro cavities, the injection often happens at much higher speeds than for conventional plastic parts. The high injection speeds and small cavity sizes also cause the polymer to experience high shear rates. As a consequence, the provision of high shear rate rheology data is recommended, yet it is often not available by default in commercial simulation software [13, 22, 39].

Also, the microscale viscosity can become important for very small cavity sizes [3, 5, 40, 64–66]. The use of micro rheological data is, therefore, recommended. Nonetheless, getting such data is very difficult. Specialized and standardized equipment for micro rheological measurements or guidelines for deriving size-dependent data from available macroscale data are not available.

The viscosity in micro channels was found to be lower than the viscosity measured by a conventional rheometer because of microscopic effects such as wall slip or shear heating. Two exemplary studies are shown in Figure 9.4. The viscosity drop is considered notable for polymer flows in channels with dimensions in the 150–200–300 μm range [56, 65].

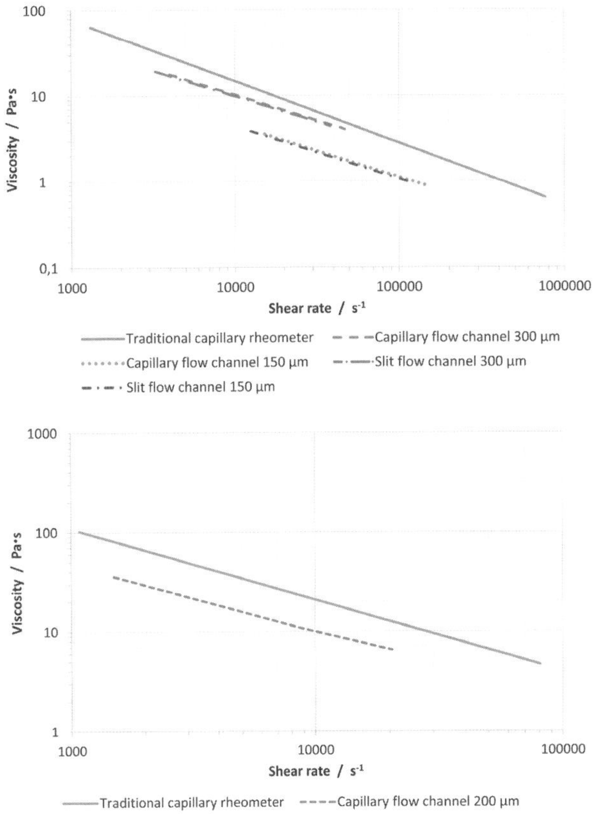

Figure 9.4 Exemplary measurement curves of the microscale viscosity of polystyrene (PS) (top: PS at 195 °C, data replotted from [65], bottom: PS at 200 °C, data replotted from [56]). The viscosity was found to be lower in micro channels than actually measured with conventional-sized capillaries of rheometers

The microscale viscosity is, nevertheless, controversially discussed. Contradictory evidence has been found by other researchers, the measurement methods are often not completely reliable, and well-developed equipment for such measurements has not been established yet [3, 67]. Current simulation tools do not provide any possibility to cope with the phenomenon of microscale viscosity. Successfully modeling and describing microscale viscosity remain the subject of current and future research.

9.4.4 Validation and Verification

Simulations in micro injection molding show promising results. Yet, the simulation software packages also show certain limitations. In addition, simulations are generally only valid in a certain context and the investigated materials and geometries are complex and change case by case. Consequently, the validation and verification of simulation results currently are, and will remain, inevitable when it comes to micro injection molding.

Validation of the simulations is based on qualitative or quantitative comparisons to experimental work. The common approach for the quantitative approach is measuring a typical process or part characteristics, such as the injection time, the injection pressure, or the part weight. The use of special and more precise micro molding machines or the conduction of multiple measurements is recommended to rule out or minimize the influence of process variance in the comparison [13].

The short shot method is the main qualitative validation method. In this technique, several short shots with varying stroke length are produced and then compared to the pattern of the flow front predicted by the simulation. The short shot method can also be employed for estimating the heat transfer coefficient in micro cavities for given processing conditions [51].

Several studies exist addressing or even combining both methods. The conducting researchers yielded that the short shot method is also a precious tool for micro parts [13, 60]. The short shot investigations showed good agreement between the simulated flow front pattern and the real one monitored by performing short shots. This agreement was found on the global part level as well as locally on the micro feature level of a micro patterned plastic part as illustrated in Figure 9.5 and Figure 9.6. The example shown is a microfluidic part. The molding was done in polypropylene on a conventional hydraulic injection molding machine with an 18-mm screw size. The corresponding simulation was performed in Autodesk Moldflow© with a multiscale mesh with element edge lengths in the range of 40 μm to 800 μm and about 2.8 million tetrahedral elements.

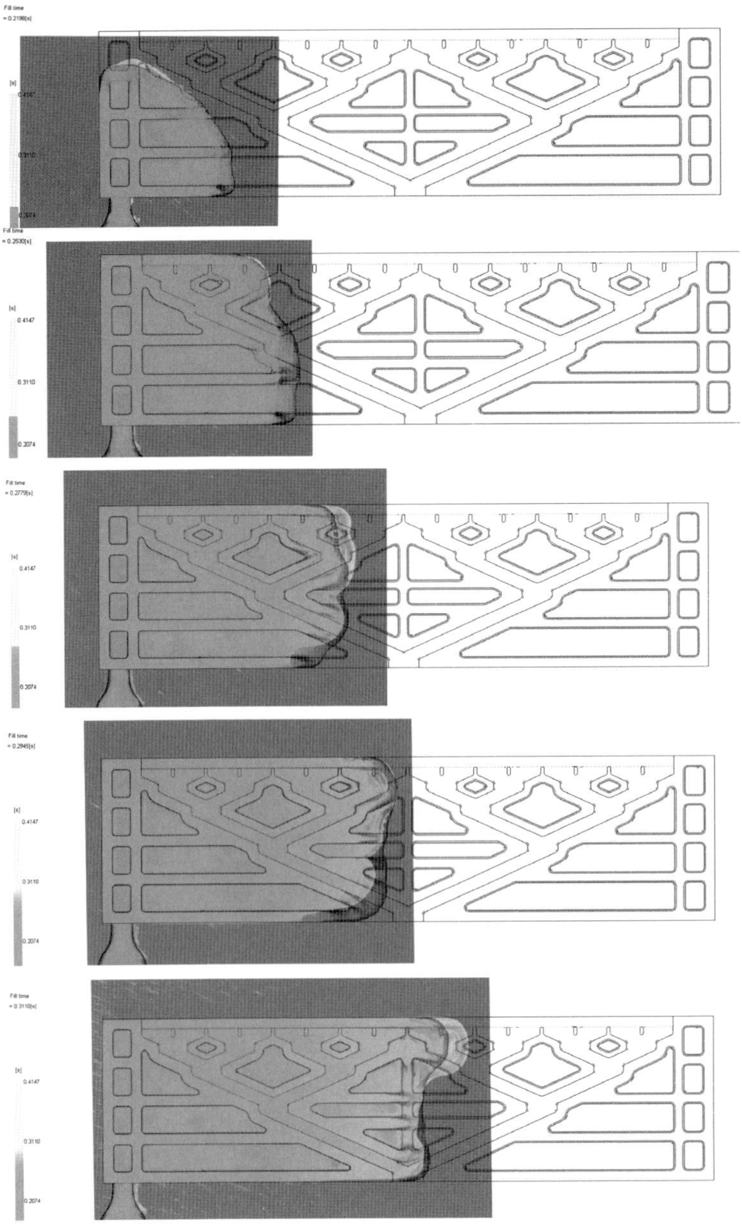

Figure 9.5 Comparison of molded short shots (pictures taken with a digital camera) and the simulated flow front pattern (pictures taken in Autodesk Moldflow©) on the global part level of a microfluidic part [60]. The figure is presented in the color supplement

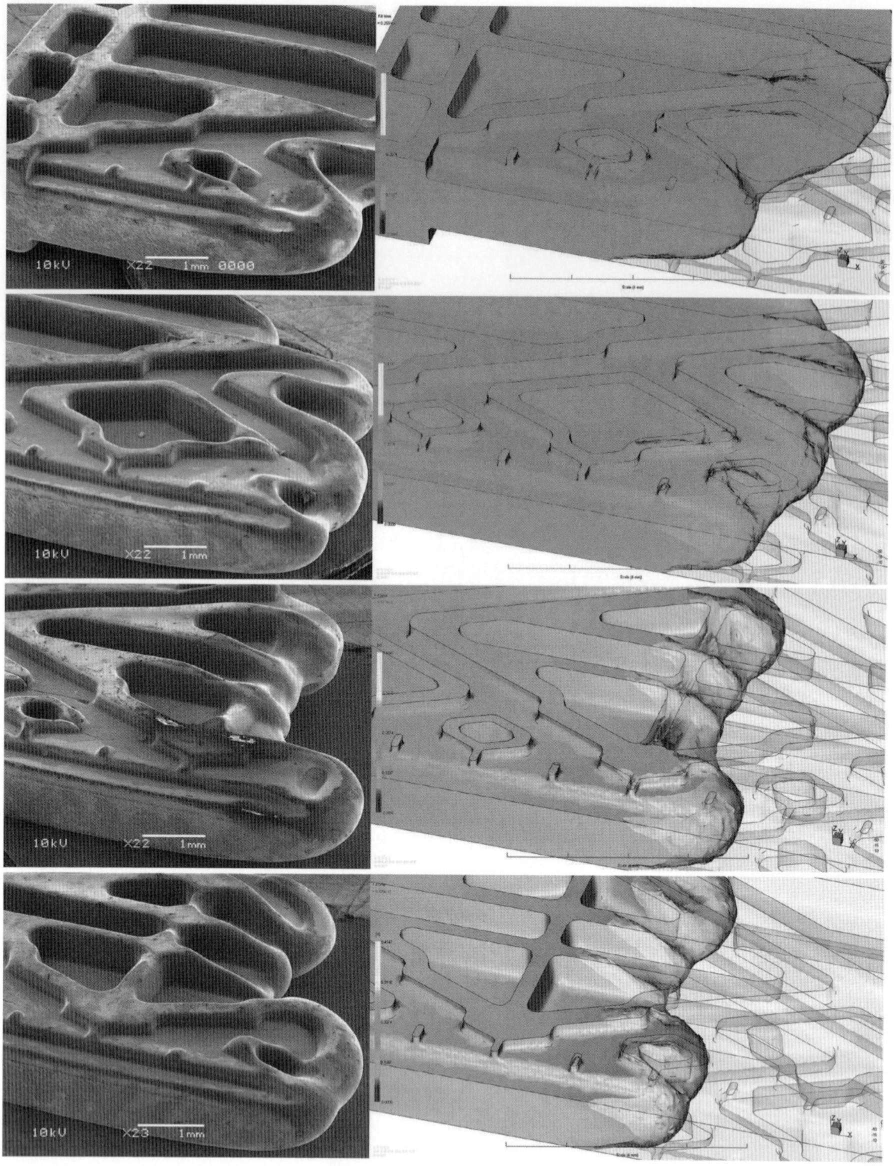

Figure 9.6 Comparison of molded short shots (pictures taken with a scanning electron microscope) and the simulated flow front pattern (pictures taken in Autodesk Moldflow©) on the local feature level of a microfluidic part [60]. The figure is presented in the color supplement

The short shot method was also exploited for experimental determination of the HTC in a reverse engineering procedure [51, 52]. In this approach, the flow length and the injection pressure of experiments with two test geometries with thicknesses in the sub-millimeter range were put into a relationship. The heat transfer coefficient was then adapted in the simulations in order to get a good fit between the simulation outcome and the experimental data, as depicted in Figure 9.7. The investigations revealed that the HTC varied from the default value. In contrast to the expected higher value as commonly assumed in micro molding, the HTC depended on the velocity and also showed lower values at high, although for micro molding typical, injection speeds. The short shot method was hence proven to serve as a beneficial tool in the determination of the HTC in micro cavities and its dependencies on injection speed, injection pressure, and cavity size.

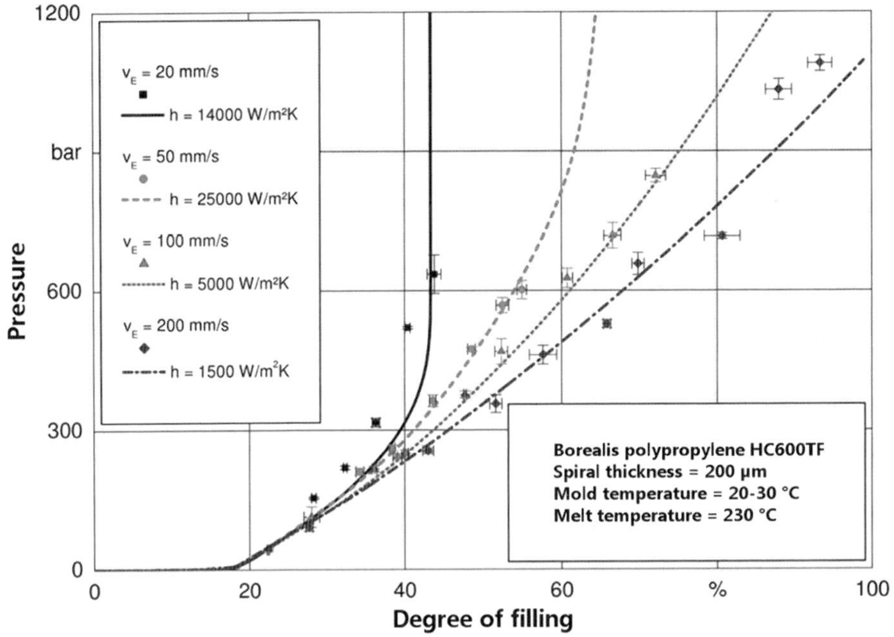

Figure 9.7 Mold filling curves with the degree of filling plotted over pressure for a micro test spiral with a thickness of 200 µm. The HTC in the simulation was adapted to fit the experimental data of short shots at different injection speeds [52]

9.5 Examples of Simulation-Aided Design and Simulated Phenomena

As mentioned before, simulations of micro injection molding are on the rise and are not entirely established as common tools in the development of micro plastic parts. This section is aimed at providing some insight into simulation studies done on a micro-structured part in order to outline the application opportunities and to increase awareness of micro molding simulations.

9.5.1 Gate Design Optimization

The design of the feed system and the gate of a plastic part is a crucial step in injection molding design. The feed system significantly influences, for instance, the cycle time, the injection pressure, the material consumption, and the plastic's flow pattern inside the cavity. In micro injection molding, the feed system often constitutes the majority of the injected volume or flow length and is much larger than the actual micro component [46, 55]. Hence, the feed system becomes even more influential and it is only natural that optimization of the feed system and gate is exceptionally beneficial. Since the feed system can become very complex, simulation-aided optimization of the design and dimensions is recommended in established injection molding literature [34, 68].

Nonetheless, there are hardly any design and optimization studies on micro plastic parts available. The reason here is probably the aforementioned limitations of the commercial software packages. However, it is believed that the application of simulations can support the layout of the feed systems of micro parts also, especially for micro-patterned components.

An example shall be introduced in the following. The investigated part was a microfluidic component with a footprint of 20 mm × 20 mm and a thickness of up to 2 mm. It could be classified, in fact, as a micro-patterned part because of the microfluidic features. The channel depths were in the range of 300 µm to 600 µm, the channel width less than 1 mm, and the wall thickness about 400 µm. The flow guiding micro pillars in the center of the part were 600 µm high, slightly conical for easier ejection, and 200 µm to 250 µm in diameter. The part weight was about 0.54 g.

Because of the sheet-like shape of the substrate exhibiting the micro structures, the part was gated with a film gate. The film gate thickness was varied: 0.28 mm, 0.42 mm, and 0.56 mm (equal to 20%, 30%, and 40% of the wall thickness where the gate was attached). The gate was, moreover, 1 mm long and 16 mm wide. The CAD model of the whole plastic part, including the cold runner feed system and the gate, is shown in Figure 9.8.

The part was meshed with the multiscale mesh approach described in Section 9.4.2. The element edge length of the mesh was adapted to the micro structures and hence set to values in the range of 65 µm to 800 µm. The result of the meshed part is depicted in Figure 9.9. The entire model comprised about 4.1 million tetrahedral elements.

The aim of the simulation-aided design optimization for this part was to find the right thickness for the film gate. The gate had to be removed from the part after molding, which was facilitated by a thin gate. However, the thin gate was expected to compromise the packaging, leading to high shrinkage and low dimensional accuracy. Moreover, the thin cross section was expected to promote excessive shear and thus polymer degradation, so that a thick gate would be favorable.

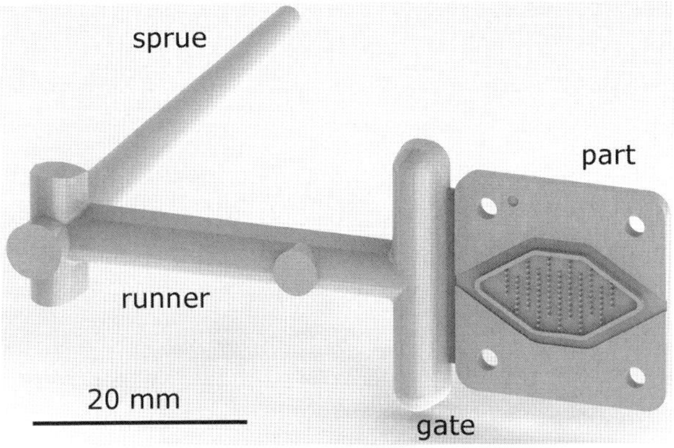

Figure 9.8 CAD model of a microfluidic part with a film gate and feed system [60]

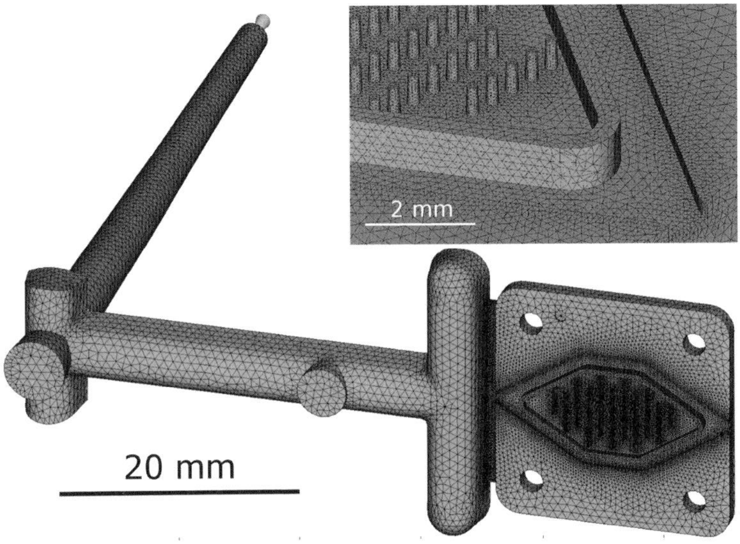

Figure 9.9 Microfluidic plastic part, meshed with the multiscale approach and local mesh densities, resulting in element edge lengths in the range of 65 μm to 800 μm [60]

The simulations of the proposed film gate proved that this type of gate was suitable for the plastic part. All three gate layouts enabled the polymer flow to enter the cavity uniformly across the entire gate width. Furthermore, short shots could not be detected for any of the chosen gate thicknesses.

Additionally, the shear rate could be investigated and was at an acceptable level for the thickest gate only, as shown in Figure 9.10, whereas for the small and medium gates, the risk of polymer degradation emerged.

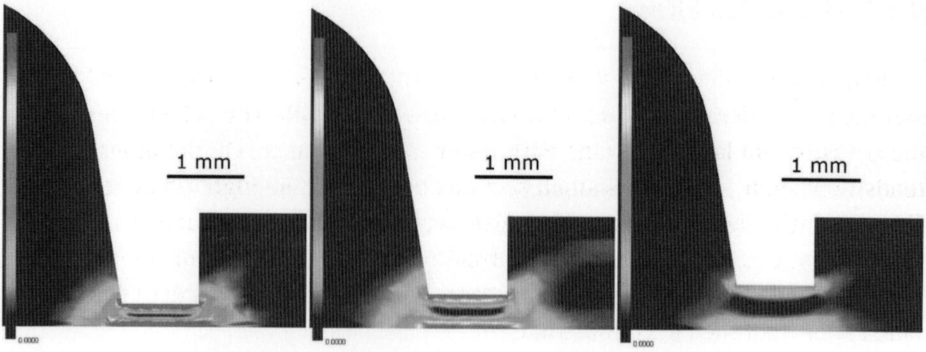

Figure 9.10 Predicted shear rate at the film gates of the microfluidic chip for different thicknesses (from left to right: 280 µm, 420 µm, 560 µm). Only the largest gate showed acceptable shear rates. Color scales from blue to red: shear rate from 0 s^{-1} to 40000 s^{-1} (allowed limit for used polymer) [60]. The figure is presented in the color supplement

The predicted average volumetric shrinkage for all three gates is depicted in Figure 9.11. The thickest gate achieved the lowest average volumetric shrinkage, the highest part mass (mass increase due to a larger gate thickness was negligible), and thus the best packing performance. In addition, the shrinkage was quite symmetrical for all gate configurations. Uniform and low shrinkage was important for the part's assembly after the molding.

In total, the film gate with a thickness of 560 µm showed the best performance and was, therefore, selected to be incorporated in the final design of the mold. The numerical simulations of the micro injection molding process were successfully applied in the gate design of the microfluidic device presented.

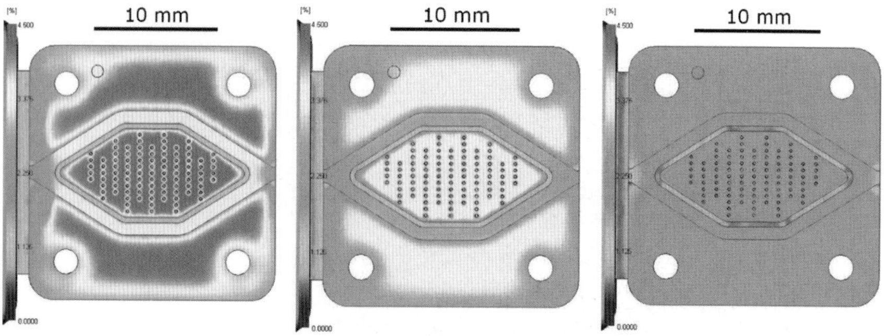

Figure 9.11 Predicted average volumetric shrinkage of the microfluidic chip for different film gate thicknesses (from left to right: 280 µm, 420 µm, 560 µm). The packing performance increases with gate thickness. Color scales from blue to red: shrinkage from 0% to 4.5% [60]. The figure is presented in the color supplement

9.5.2 Hesitation Effect

A phenomenon called hesitation effect frequently occurs when plastic parts exhibit significantly different wall or feature thicknesses [20, 40]. The polymer melt flows more easily into larger sections with lower flow resistance. On the other hand, it tends to hesitate as it enters small sections because of the higher flow resistance. Consequently, the hesitation effect also occurs in micro structured components that consist of a relatively thick substrate featuring superficial micro structures, such as microfluidic devices or optical arrays. In these cases, the micro features fill much later than the bulky substrate.

Numerical process simulations can identify and display the hesitation of polymer flow in micro structures, as illustrated in Figure 9.12. The part shown is the same microfluidic device as in the previous section, exhibiting micro pillars for flow guidance. In this particular case, the hesitation effect was successfully indicated prior to the experimental molding of the part. The hesitation led to the flow actually reaching the end of the cavity before the micro pillars were completely filled. This behavior had to be considered while setting up the packing phase of the molding process.

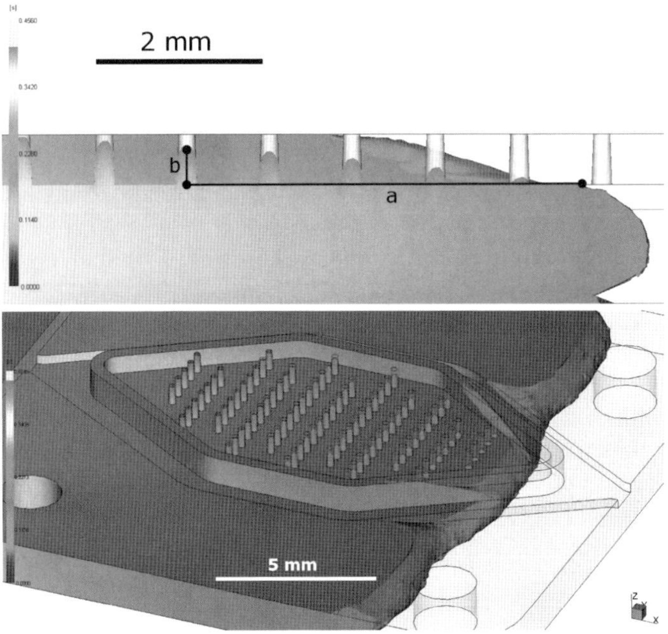

Figure 9.12 Cross-sectional (top [60]) and 3D view (bottom) of the flow inside the cavity of a micro structured part, displaying a distinct hesitation effect at the micro pillars: filling in the bulk direction is faster than in the direction of the micro pillars; in other words, the flow length, *a*, in the bulk direction is longer than the flow length, *b* , in the direction of the micro pillars. The figure is presented in the color supplement

9.6 Conclusions

Micro injection molding is established as the key technology for the large-scale fabrication of precision and micro plastic components. Furthermore, simulations are accepted and precious tools for the product development of plastic parts made by classical injection molding. Owing to the growing market and number of applications, the increasing demands for part quality and functionality, and the general entrepreneurial desire for shorter product development times, the need for reliable and economic process simulations is increasing noticeably in the field of micro injection molding.

Micro injection molding has different characteristics to conventional injection molding. Yet, process simulations can be employed to assist the product, tool, and process design of micro molded parts. So far, simulations have mostly supplied a qualitative input to design decisions. An example of the application of computer simulations in the design process and optimization of a feed system for a micro-patterned part was outlined. Also, the hesitation effect, often seen in injection molded parts with significant thickness variation and micro parts, can be indicated by the simulations.

Furthermore, implementation strategies and improvements are currently being developed in order to yield qualitatively more reliable results and push the state-of-the-art and applicability of micro molding simulations. The aspect of divergence between micro and conventional injection molding must not be forgotten when commercially available simulation tools are employed and when process simulations are set up. For improved simulation accuracy, the simulation model of micro parts should be as comprehensive as possible and should also include the mold, the feed system, and the machine geometry. Besides, the machine characteristics should be implemented in the simulation as well.

When it comes to meshing, full 3D modeling is required. The also quite common 2D or 2.5D approach is not sufficient, as true micro parts can often be considered as chunky parts. Hence, the assumptions of the Hele-Shaw flow approximation are not valid anymore. Additionally, the multiscale mesh approach is strongly recommended. It enables provision of the appropriate mesh resolution for the micro part on the one hand and the optimization of the required computation time and power on the other hand. Reliable, up-to-date, and comprehensive material data are, moreover, substantial for the accurate simulation of the micro injection molding process. The required extent and level of detail of the material characterization often goes beyond that which classical measurements or models can provide.

The necessity of the implementation of microscale viscosity or the changing heat transfer coefficient in micro injection molding is being discussed controversially. Researchers recommend considering microscale viscosity at very small cavity sizes. The heat transfer coefficient is commonly assumed to increase for micro cavities compared to the default values of classical injection molding. Studies have shown that the heat transfer coefficient exhibits deviations in both directions, as it also strongly depends on the processing conditions.

Due to the limitations of process simulations and of commercial code in micro injection molding, steady validation and further development of the simulation tools remain essential. The short shot method is one of the most important methods for the validation of simulation results. Moreover, the approach allows us to investigate the heat transfer coefficient in micro cavities, which varies from the one experienced in conventional injection molding.

Simulations will play an increasingly influential role in the product development of micro molded parts, and the designers and engineers shall acknowledge their potential and their restrictions equally.

References

[1] Attia, U.M., Alcock, J.R., A review of micro-powder injection moulding as a microfabrication technique. *J. Micromech. Microeng.* (2011) 21(4), p. 43001

[2] Giboz, J., Copponnex, T., Mélé, P., Microinjection molding of thermoplastic polymers: a review. *J. Micromech. Microeng.* (2007) 17(6), pp. R96–R109

[3] Zhiltsova, T.V., Oliveira, M.S.A., Ferreira, J.A., 2013, Integral approach for production of thermoplastics microparts by injection moulding. *J. Mater. Sci.* (2013) 48(1), pp. 81–94

[4] Tada, K., Fukuzawa, D., Watanabe, A., Ito, H., Numerical simulation for flow behaviour on micro- and nanomoulding. *Plast., Rubber Compos.* (2010) 39(7), pp. 321–326

[5] Yang, C., Yin, X.-H., Cheng, G.-M., 2013, Microinjection molding of microsystem components: new aspects in improving performance. *J. Micromech. Microeng.* (2013) 23(9), p. 93001

[6] Nguyen-Chung, T., Loser, C., Jüttner, G., Pham, T., Obadal, M., Gehde, M., Simulation of the micro-injection moulding process: effect of the thermo-rheological status on the morphology. *Proc. Inst. Mech. Eng., Part E* (2011) 225(4), pp. 224–238

[7] Drummer, D., Messingschlager, S., Ceramic injection molding material analysis, modeling and injection molding simulation. *Proceedings of the 29th International Conference of the Polymer Processing Society (PPS-29)* (2014), AIP Publishing, Melville, NY, US, pp. 582–586

[8] Griffiths, C.A., Dimov, S.S., Scholz, S., Hirshy, H., Tosello, G., Process Factors Influence on Cavity Pressure Behavior in Microinjection Moulding. *J. Manuf. Sci. Eng.* (2011) 133(3), p. 31007

[9] Drummer, D., Ehrenstein, G.W., Hopmann, C., Vetter, K., Meister, S., Fischer, T., Piotter, V., Prokop, J., Innovative Prozesstechnologien für die Herstellung thermoplastischer Mikrobauteile – Analyse und vergleichende Bewertung. *Zeitschrift Kunststofftechnik (J. Plast. Technol.)* (2012) 8(5), pp. 440–467

[10] Humphreys, P., Computer Simulations. *PSA Proc. Bienn. Meet. Philos. Sci. Assoc.* (1990) 2, pp. 497–506

[11] Hartmann, S., The World as a Process: Simulations in the Natural and Social Sciences, in *Modelling and Simulation in the Social Sciences from the Philosophy of Science Point of View*, Hegselmann, R., Mueller, U., Troitzsch, K.G. (Ed.)(1996) Springer Netherlands, Dordrecht, The Netherlands, pp. 77–100

[12] Thornagel, M., Simulating flow can help avoid mould mistakes. *Met. Powder Rep.* (2010) 65(3), pp. 26–29

[13] Costa, F.S., Tosello, G., Whiteside, B.R., Best practice strategies for validation of micro moulding process simulation. *Proceedings of the Polymer Process Engineering 2009 Conference* (2009) University of Bradford, UK, pp. 331–364

[14] Vietri, U., Sorrentino, A., Speranza, V., Pantani, R., 2011, Improving the predictions of injection molding simulation software. *Polym. Eng. Sci.* (2011) 51(12), pp. 2542–2551

[15] Khalilian, S.A., Park, S.S., Freiheit, T.I., The Application of Commercial Injection Molding Software to Micro-Component Design and Process Development. *Proceedings of the 8th International Conference on Micro Manufacturing (ICOMM 2013)*, Research publishing, Singapore (2013), pp. 50–56

[16] Tosello, G., Gava, A., Hansen, H.N., Reinecke, H., Lucchetta, G., Schoth, A., Experimental validation of micro molding simulations using different process setting conditions. *Proceedings of the 67th Annual Technical Conference of the Society of Plastics Engineers 2009 (ANTEC 2009)* (2009) Society of Plastics Engineers (SPE), Chigago, IL, USA, pp. 1787–1793

[17] Chen, L., Kietzmann, C., Astbury, D., Shao, L., Advanced cooling simulation technologies for the injection molding process, SPE China Proceedings, Shanghai (2013), p. 8

[18] Kennedy, P., Zheng, R., *Flow Analysis of Injection Molds* (2013) Carl Hanser Verlag, Munich, Germany

[19] Park, S., Ahn, S., Kang, T.G., Chung, S.-T., Kwon, Y.-S., Chung, S.H., Kim, S.-G., Kim, S., Atre, S.V., Lee, S., German, R.M., A review of computer simulations in powder injection molding. *Int. J. Powder Metall. (Princeton, N.J.)* (2010) 46(3), pp. 37–46

[20] Attia, U.M., Marson, S., Alcock, J.R., Micro-injection moulding of polymer microfluidic devices. *Microfluid. Nanofluid.* (2009) 7, pp. 1–28

[21] Marhöfer, D.M., Tosello, G., Islam, A., Hansen, H.N., Gate Design in Injection Molding of Microfluidic Components using Process Simulations. *J. Micro Nano-Manuf.* (2016) 4(2 (June)), p. 25001

[22] Cramer, A., Michaeli, W., Friesenbichler, W., Duretek, I., Simulation des Spritzgießprozesses von Mikrobauteilen. *Zeitschrift Kunststofftechnik (J. Plast. Technol.)* (2007) 3(1), pp. 1–26

[23] Kennedy, P., *Practical and Scientific Aspects of Injection Molding Simulation* (2008) Technical University of Eindhoven, The Netherlands

[24] Young, W.B., Simulation of the filling process in molding components with micro channels. *Microsyst. Technol.* (2005) 11(6), pp. 410–415

[25] Jeggy, C., Micro-Injection Moulding: From Process to Modelling (2004), Presses universitaires de Louvain, Leuven, Belgium

[26] Griffiths, C.A., Dimov, S.S., Brousseau, E.B.J.P., Packianather, M.S., The finite element analysis of melt flow behaviour in micro-injection moulding. *Proc. Inst. Mech. Eng., Part B* (2008) 222(9), pp. 1107–1118

[27] Dealy, J.M., Wang, J., *Melt Rheology and its Applications in the Plastics Industry* (2013) Springer Science+Business Media, Dordrecht, The Netherlands

[28] Osswald, T.A., Rudolph, N., *Polymer Rheology* (2014) Carl Hanser Verlag, Munich, Germany

[29] Zhou, H., *Computer Modeling for Injection Molding* (2013) John Wiley & Sons, Hoboken, NJ, USA

[30] Wang, J., *Some critical issues for injection molding* (2012) InTech, Rijeka, Croatia

[31] Löhe, D., Haußelt, J., *Microengineering of Metals and Ceramics. Part I: Design, Tooling and Injection Molding* (2005) Wiley-VCH Verlag, Weinheim, Germany

[32] Bruus, H., *Theoretical Microfluidics* (2008) Oxford University Press, Oxford, UK

[33] Gad-el-Hak, M., *MEMS – Introduction and Fundamentals* (2006) CRC Press/Taylor and Francis Group, Boca Raton, FL, USA

[34] Beaumont, J. P., *Runner and Gating Design Handbook – Tools for Successful Injection Molding* (2004) Carl Hanser Verlag, Munich

[35] Heaney, D. F., *Handbook of Metal Injection Molding* (2012) Woodhead Publishing, Cambridge, UK

[36] Choi, S. J., Kim, S. K., Multi-scale filling simulation of micro-injection molding process. *J. Mech. Sci. Technol.* (2011) 25(1), pp. 117–124

[37] Islam, A., Giannekas, N., Marhöfer, D. M., Tosello, G., Hansen, H. N., Experimental investigation on shrinkage and surface replication of injection moulded ceramic parts. *Proceedings of the 4th International Conference on Nanomanufacturing (nanoMan 2014)* (2014) Bremen, Germany

[38] Cao, W., Hassager, O., Wang, Y., Surface tension effect on micro injection molding. *Proceedings of the 24th Annual Meeting of the Polymer Processing Society (PPS2008)* (2008) Polymer Processing Society (PPS), Salerno, Italy, p. 141

[39] Qin, Y., *Micromanufacturing Engineering and Technology* (2015) Elsevier, William Andrew, Amsterdam, The Netherlands, Boston, USA

[40] Yao, D., Kim, B., Simulation of the filling process in micro channels for polymeric materials. *J. Micromech. Microeng.* (2002) 12(5), pp. 604–610

[41] Yao, D., Kim, B., Scaling Issues in Miniaturization of Injection Molded Parts. *J. Manuf. Sci. Eng.* (2004) 126, pp. 733–739

[42] Guerrier, P., Tosello, G., Kirstein, K., Jesper, N., Hattel, H., Nielsen, K. K., Hattel, J. H., Three-dimensional numerical modeling of an induction heated injection molding tool with flow visualization. *Int. J. Adv. Manuf. Technol.* (2015) 85, pp. 643–660

[43] Fan, B., Kazmer, D. O., Nageri, R., An Analytical Non-Newtonian and Nonisothermal Viscous Flow Simulation. *Polym. Plast. Technol. Eng.* (2006) 45(3), pp. 429–438

[44] Wang, F., Li, J., Yang, B., Modeling and simulation of fiber-reinforced polymer mold-filling with phase change. *J. Appl. Polym. Sci.* (2016) 133(3), pp. 1–18

[45] Azdast, T., Behravesh, A. H., Mazaheri, K., Darvishi, M. M., Numerical simulation and experimental validation of residual stress induced constrained shrinkage of injection molded parts. *Polimery* (2008) 53(4), pp. 304–310

[46] Tosello, G., Costa, F. S., Hansen, H. N., Micro Injection Moulding High Accuracy Three-Dimensional Simulations and Process Control. *Proceedings of Polymer Process Engineering 11: Enhanced Polymer Processing* (2011) Bradford, UK

[47] Yu, L., Koh, C. G., Lee, L. J., Koelling, K. W., Madou, M. J., Experimental investigation and numerical simulation of injection molding with micro-features. *Polym. Eng. Sci.* (2002) 42(5), pp. 871–888

[48] Tada, K., Fukuzawa, D., Watanabe, A., Ito, H., Computer Simulation for Flow Behavior on Micro- and Nano-Molding. *Proceedings of the 67th Annual Technical Conference of the Society of Plastics Engineers 2009 (ANTEC 2009)* (2009) Society of Plastics Engineers (SPE), Chigago, IL, USA, pp. 1794–1798

[49] Heldele, R., Schulz, M., Kauzlaric, D., Korvink, J. G., Haußelt, J., Micro powder injection molding: process characterization and modeling. *Microsyst. Technol.* (2006) 12(10–11), pp. 941–946

[50] Tosello, G., Schoth, A., Hansen, H. N., Implementation strategies for the optimization of micro injection moulding simulations. *Proceedings of the 4th International Conference on Multi-Material Micro Manufacture (4M 2008)*, Dimov, S. S., Menz, W. (Ed.) (2008) Whittles Publishing, Cardiff, UK

[51] Nguyen-Chung, T., Jüttner, G., Löser, C., Pham, T., Gehde, M., Determination of the heat transfer coefficient from short-shots studies and precise simulation of microinjection molding. *Polym. Eng. Sci.* (2010) 50(1), pp. 165–173

[52] Nguyen-Chung, T., Jüttner, G., Pham, T., Mennig, G., Gehde, M., Die Bedeutung präziser Randbedingungen für die Simulation des Mikrospritzgießens. *Zeitschrift Kunststofftechnik (J. Plast. Technol.)* (2008) 4(6), pp. 1–25

[53] Heinle, M., Drummer, D., Heat Transfer Coefficient in Injection Molding of Polymers. *Int. Polym. Process.* (2015) 30(4), pp. 434–441

[54] Nakao, M., Tsuchiya, K., Sadamitsu, T., Ichikohara, Y., Ohba, T., Ooi, T., Heat transfer in injection molding for reproduction of sub-micron-sized features. *Int. J. Adv. Manuf. Technol.* (2008) 38(3–4), pp. 426–432

[55] Marhöfer, D. M., Tosello, G., Hansen, H. N., Islam, A., Advancements on the Simulation of the Micro Injection Moulding Process. *Proceedings of the 10th International Conference on Multi-Material Micro Manufacture (4M)* (2013) Research Publishing Services, San Sebastian, Spain, pp. 77–81

[56] Tosello, G., *Precision Moulding of Polymer Micro Components* (2008) Technical University of Denmark (DTU), Denmark

[57] Lidwell, W., Holden, K., Butler, J., *Universal Principles of Design* (2010) Rockport Publishers, Beverly, MA, USA

[58] Guerrier, P., Tosello, G., Kirstein, K., Jesper, N., Hattel, H., Three-dimensional numerical modeling of an induction heated injection molding tool with flow visualization. *Int. J. Adv. Manuf. Technol.* (2016) 85, pp. 643–660

[59] Guerrier, P., Tosello, G., Hattel, J. H., *Analysis of cavity pressure and warpage of polyoxymethylene thin walled injection molded parts: Experiments and simulations* (2015) AIP Conference Proceedings, AIP Publishing, Cleveland, OH, USA, p. 110006

[60] Marhöfer, D. M., *Design and Manufacture of Molded Micro Products Using Concurrent Engineering* (2016) Technical University of Denmark (DTU), Denmark

[61] Minin, I. V., Minin, O. V., *Computational Fluid Dynamics Technologies and Applications* (2011) InTech, Rijeka, Croatia

[62] Bibber, D., Secrets of success in micro molding. *Plast. Technol.* (2012) 58(3), pp. 33–38

[63] Bibber, D., Micro Molding Thin-Walled Devices. *Manuf. Eng.* (2013) May, pp. 101–109

[64] Chien, R.-D., Jong, W.-R., Chen, S. C., Study on rheological behavior of polymer melt flowing through micro-channels considering the wall-slip effect. *J. Micromech. Microeng.* (2005) 15, pp. 1389–1396

[65] Chen, C. S., Chen, S. C., Liaw, W. L., Chien, R.-D., Rheological behavior of POM polymer melt flowing through micro-channels. *Eur. Polym. J.* (2008) 44(6), pp. 1891–1898

[66] Chen, S. C., Tsai, R. I., Chien, R.-D., Lin, T. K., Preliminary study of polymer melt rheological behavior flowing through micro-channels. *Int. Commun. Heat Mass Transfer* (2005), 32(3–4), pp. 501–510

[67] Ito, H., Saito, T., Yasuhara, T., Kikutani, T., Yamagiwa, Y., Muroi, K., Evaluation of Material Properties and Processability Using The Intelligent Micro-scale Polymer Processing System. *Proceedings of the 63rd Annual Technical Conference of the Society of Plastics Engineers 2005 (ANTEC 2005)* (2005) Society of Plastics Engineers (SPE), Boston, MA, USA, pp. 693–697

[68] Kazmer, D. O., *Injection Mold Design Engineering* (2007) Carl Hanser Verlag, Munich, Germany

10 Metrological Quality Assurance in Micro Injection Molding

Danilo Quagliotti, Matteo Calaon, Guido Tosello

10.1 Introduction

In the last two decades, the manufacturing industry has been the scene of a technological revolution. The opportunity to reduce the dimensions of the products rapidly allowed the realization of new advanced micro and nano systems that were unthinkable just a few years earlier. Molding technologies are still the main players in this change in the mass production of polymer micro parts and parts with micro and nano surfaces [1]. As widely discussed in Chapter 4 (see also [2]), the achievement of a full surface replication of the tool insert component when molding the polymer melt is essential in micro injection molding technology. A replication process requires reproducing a master geometry by conveying it to a substrate material. It is typically induced by means of different energy sources (usually heat and force) and direct physical contact between master and substrate. Consequently, the accuracy required for the manufacture of micro molded components must be ensured by means of a metrological approach to surface replication and dimensional control of both master geometry and replicated substrate [3, 4].

Nowadays, micro injection molding of thermoplastic polymers allows the replication of parts with an overall size below (3–5) mm down to less than 1 mm and a mass of less than 1 g (e.g., microfluidic systems) down to (0.5–1) mg (e.g., micro gears). Furthermore, such components have features in the order of 100 μm down to less than 1 μm, and dimensional tolerances between (10–20) μm down to (10–20) nm (see Figure 10.1). Concepts of micro and nanostructured components include a large variety of materials, different dimensional scales, increasingly small dimensions and dramatic growing of the complexity of the surfaces, which are natural interfaces of the products to the environment. On the one hand, this trend improves many products we use in everyday life, allowing the incorporation of more and more functionalities in one single micro- or nano-structured part; on the other hand, it poses new challenges in assessing the quality of the production. It means that a detailed knowledge of absolute dimensions and of geometrical quantities must include the establishment of the traceability and an adequate assessment of the measurement uncertainty, which is a decisive parameter to deal with the quality assurance of micro and nano manufactured components [5].

The methods for uncertainty evaluation currently available in manufacturing are based on the upper bound strategy [6]. At the same time, in micro manufacturing, several approaches are suggested (see Section 10.5), but there are no shared methodologies, yet, to deal with the uncertainty evaluation of features of size at the sub-millimeter scale. Hence, considering the incessant scaling down of critical dimensions [2–4, 7], traditional techniques appear to be outdated in micro and nano manufacturing engineering and not adequate any longer for an assessment of the measurement uncertainty, which is required to be kept under reasonable limits [8, 9]. Furthermore, as the technology progresses, the interest in three-dimensional areal characterization also increases and, in this context, optical instruments appear more adequate because they perform relatively fast acquisitions without damaging the surface of the specimen under measurement (e.g., as is the case with the stylus of a contact instrument on soft materials such as polymers) [5, 10, 11]. Therefore, alongside traditional methods, in the following, importance will be given to areal techniques mostly based on optical measurements. Beforehand, in the next section, some basic concepts of metrology, which are of central importance hereinafter, are also given.

■ 10.2 Quality of the Measurement Process: Calibration and Traceability

Calibration and traceability (see also Chapter 4) demand essential actions to assure consistency in the measurement results and, consequently, they are fundamental in the quality assurance of the micro manufactured production.

Calibration is an "operation that, under specified conditions, in a first step, establishes a relation between the quantity values with measurement uncertainties provided by measurement standards and corresponding indications with associated measurement uncertainties and, in a second step, uses this information to establish a relation for obtaining a measurement result from an indication" [12].

In close connection with calibration, the metrological *traceability* is the "property of a measurement result whereby the result can be related to a reference through a documented unbroken chain of calibrations, each contributing to the measurement uncertainty" [12], with the final purpose of relating the measurement results to the International System of Units (SI) [13]. Hence, the measurement uncertainty is an essential concept for calibration and traceability.

Measurement instruments may also require an *adjustment*, i.e., a "set of operations carried out on a measuring system so that it provides prescribed indications corresponding to given values of a quantity to be measured" [12]. The adjustment can be a mechanical or a software change in the instrument and consists of tuning some parameters in order to provide an indication that is closer to a reference value. An adjustment does not require any evaluation of the measurement uncertainty. Therefore, it must not be confused with the calibration.

In the following, an example of a calibration chain establishes the metrological traceability of a micrometer to the SI unit "through a documented unbroken chain of calibrations", consisting of the following steps [14, 15]:

- The micrometer is calibrated against a gauge block.
- The gauge block was previously calibrated by a mechanical length comparator. An accredited laboratory or primary institute issued a relative calibration certificate[1].
- The comparator was also calibrated by another gauge block that, in turn, was calibrated using an optical interferometer with a laser source. An accredited laboratory or primary institute issued relative calibration certificates.
- The laser source was initially calibrated by an iodine-stabilized laser that realizes the definition of the meter in a primary institute (e.g., PTB in Germany, NIST in the USA, NPL in the UK, etc.).

The steps are summarized in Figure 10.1. At each step of the unbroken chain, the calibration is performed by another instrument or by a calibrated artifact (also called transfer standard or measurement standards or material measures [17]) and the uncertainty is propagated from the calibrated standard to the measurement system, up to the final measurement result. Hence, the final uncertainty increases with the number of steps. A short chain of calibrations is desirable to keep the uncertainty related to the traceability process small. Furthermore, a smaller calibration uncertainty is expected at the beginning of the calibration chain.

1 Additional information about national and international metrology and relative terminology can be found elsewhere [16].

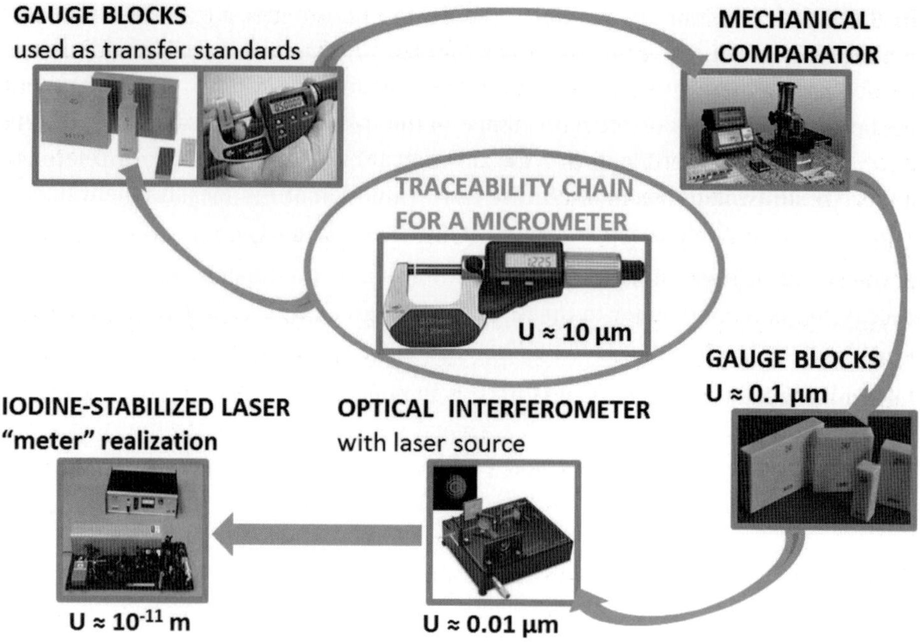

Figure 10.1 Traceability chain for a micrometer through a primary meter, primary gauge block, and workshop gauge block [15]

10.2.1 Accuracy and Precision

Accuracy and precision are important attributes of the measurement. The official definition of *accuracy*, according to the International Vocabulary of Metrology (VIM) [12], states that the accuracy of a measurement is the "closeness of agreement between a measured quantity value and a true quantity value of a measurand".

Following the definition, a common interpretation explains measurements for which traceability is established as accurate [11]. Nevertheless, another interpretation [18] points out that "the notion of accuracy is not meaningful when reporting the result of a measurement" because the true value is not known and "proven traceability does not ensure accuracy". The definition of traceability, in fact, refers to the measurement uncertainty and not at all to the accuracy. The VIM, in addition, specifies that the "measurement accuracy is not a quantity and is not given a numerical quantity value. A measurement is said to be more accurate when it offers a smaller measurement error" [12]. However, being the result of a measurement always expressed by an uncertainty, the "true value" is believed to be within the uncertainty interval with a certain probability, corresponding to the expanded uncertainty (see Section 10.5). Hence, the measurement error cannot be quantified.

In the following, the word accuracy is used to indicate measurement systems for which the uncertainty result of a traceability process is larger or smaller with respect to other corresponding measurement systems, without giving an absolute meaning to the word accuracy. In the example in Figure 10.1, for example, "the micrometer is calibrated against a gauge block, which was previously calibrated by a more accurate instrument, etc."

Moreover, in place of "true value", the terms "reference value" or "nominal value" are used according to the context.

A more tangible property of measurement is the *precision*, which is specified as a "closeness of agreement between indications or measured quantity values obtained by replicate measurements on the same or similar objects under specified conditions" [12]. In other words, the precision quantifies the variability of a session of repeated measurements in specified conditions. The most common conditions are regarded as repeatability and reproducibility.

10.2.1.1 Repeatability and Reproducibility

Repeatability and *reproducibility* are expressed as uncertainty contributors and both specify agreement (or disagreement) of the measurement results, generically indicated as precision of a measurement system, or of a measurement session. Nonetheless, they are two different uncertainty contributors because they are evaluated in different conditions. In particular, repeatability refers to the variability of the results when the measurements are performed in unchanged measurement conditions (best-way undertaken measurements). Therefore, repeatability can be considered a lower limit with respect to the reproducibility. Correspondingly, the reproducibility refers to the variability of the results when at least one of the measurement conditions is changed. Hence, it normally takes into account the repeatability too.

As soon as the micro injection molding process (or an experiment in general) is replicated, information about the production and the process is obtained by analyzing the results of one or more sessions of measurements. Hence, repeatability (or reproducibility) conditions can analogously be extended to the process too. Nevertheless, when doing so, it is important to clearly distinguish between

- the variability due to the manufacturing process (investigated by measuring different manufactured parts)
- the variability due to the measurement process (inspected by considering several repeated measurements per each part)

10.3 Metrology for Micro Injection Molding

Micro nano metrology supports crucial aspects of the micro injection molding process like design, manufacture, quality control, and optimization of the process itself (also, recently, the simulation of micro injection molding processes has become more and more crucial for their optimization–see Section 10.4 [19, 20, 66]).

Length, area, slope, and curvature take part in measurements of distance, width, height, geometry (form), and texture. In the following sub-sections, such measurements are dealt in dimensional metrology (distance, width, height) and surface metrology (geometry, texture).

10.3.1 Dimensional Metrology

In manufacturing, international standards of a series called Geometrical Product Specification (GPS) aim to provide rules considering design, manufacture, and tolerance verification. The classification of all GPS standards is summarized in ISO 14638:2015 [21], which is commonly known as *Matrix model*. The approach is based on a detailed description of geometrical features linked to functional properties. Furthermore, ISO 14253-1:2013 [22] states that the measurement uncertainty is to be accounted for in tolerance verification. In particular, a tolerance is to be verified reducing the *specification zone* (tolerance interval) by the expanded uncertainty of the measurements, resulting in what is called *conformance zone* (see Figure 10.2).

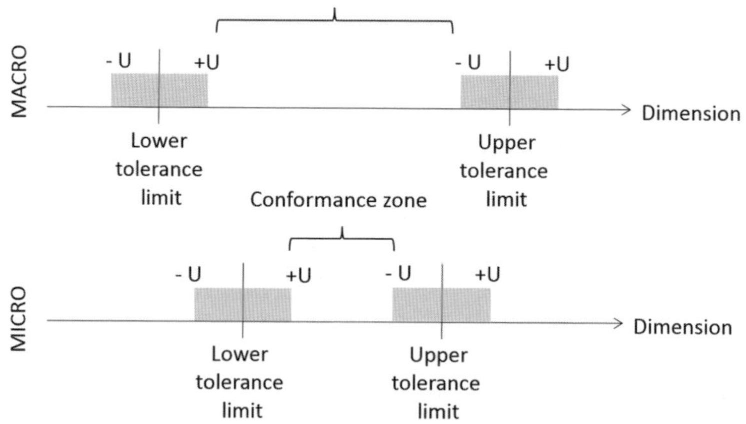

Figure 10.2 Illustration of the relationship between tolerance and measurement uncertainty. In this representation, the measurement uncertainty was kept constant and the tolerance zone decreased [24]

At the micro and nano scale, some problems can be related to the use of GPS standards [5, 23]. The absolute dimensions at the micro and nano scale are small and tolerances should be small too. These facts implicitly require that: (a) the measurement method adopted must be suitable to measure the components, (b) the measurement uncertainty must be sufficiently small to verify the tolerance (see Section 10.5).

Absolute dimensions can be measured according to their definition [5]:

- Distance is defined as the minimum length between two surfaces oriented in the same direction.
- Width is defined as the distance between two opposite surfaces.
- Height is defined as the distance between two surfaces with the same orientation, along the vertical direction.

Software tools are available for such operations [25, 26][2]. In addition, indications for the assessment of the steps height are given in ISO 5436-1:2000 [17]. This ISO standard describes types of material measures with the profile method (see Section 10.3.2). An ISO standard describing material measure types for the areal method is ISO 25178-70:2014 [30], which also includes the profile material measures in [17]. However, no indications are available for the assessment of areal material measures. Hence, the assessment should follow the two-dimensional (2D) specifications. This is not misleading when an artifact is uniform along one lateral direction. Examples are shown in Chapter 4, Section 4.3 (Figure 4.5 and Figure 4.6).

An alternative evaluation of the height of steps from areal measurements (see Section 10.3.2) is the calculation of the frequentist distribution of the pixels' height in a measurement, resulting in as many histograms as height levels in a step. The distance between two successive histograms' dominant values is the height of a step level.

An example of multilevel step is in Figure 10.3, where three levels can be recognized. The frequentist distribution of the pixels in the map (see Section 10.4.1) of the related optical measurement is represented in Figure 10.4. There are three histograms corresponding to the three height levels: the height is in abscissa and the relative distributions of the pixels in ordinate. The graph was produced using SPIP™ [26]. A tool is available in the software for selecting the central values (maximum) of the histograms and evaluating the ranges corresponding to the heights.

In the following, there are some solutions for the dimensional metrology of micro tool inserts and micro injection molded parts (see also Chapter 4).

[2] Software reference tools are available from different National Metrology Institutes (NMIs) worldwide for the validation of developed software [27–29].

Figure 10.3 Three-level step (3D view) acquired by a focus-variation instrument (see Section 10.4.1) with a 5× magnification objective lens

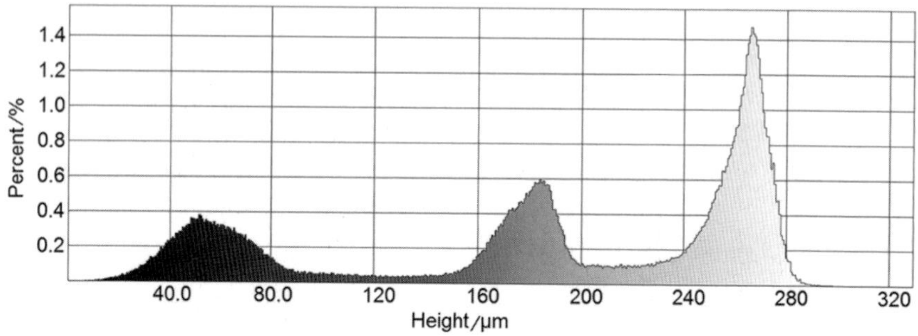

Figure 10.4 Distribution of the pixels in the measurement shown in Figure 10.3

Manufacture of a High Density Optical Disc (HD-DVD) with Surface Nano Features

Quality control of the ultra-high precision manufacture of optical discs was performed by atomic force microscopy (Section 10.4) on both master tool insert (stamper of nickel) and replicated polymer (polycarbonate) discs (an example is shown in Figure 10.5) [31].

The HD-DVD surface is nanostructured with nano pillars on the stamper. The nano pillars, organized in adjacent tracks on a spiral at a certain distance from one another, are replicated in nano pits on the discs. The pits are responsible for data storage.

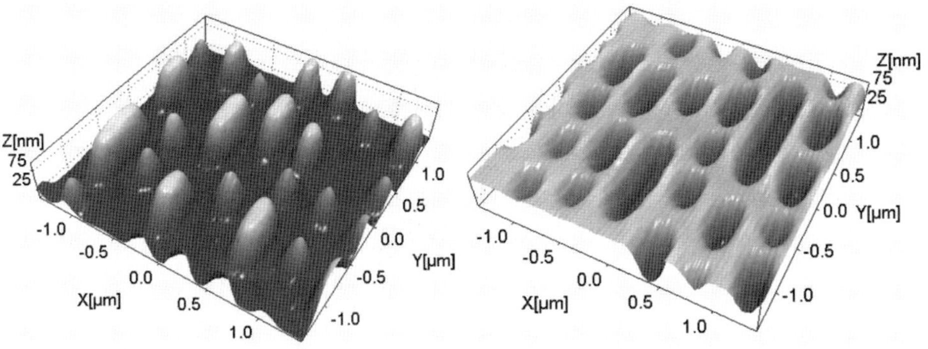

Figure 10.5 Atomic force microscope (see Section 10.4) acquisitions (3D view) of HD-DVD features on a nickel stamper (left) and polymer molded disc (right) (axis magnifications: $X = 1\times$, $Y = 1\times$, $Z = 5\times$) [31]

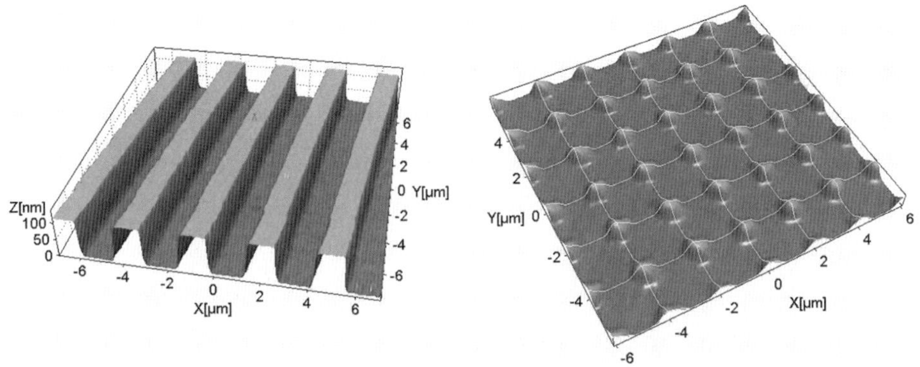

Figure 10.6 Atomic force microscope (see Section 10.4) acquisitions (3D view) of silicon calibrated objects. Left: 100 nm step height (TGZ02) for vertical calibration. Right: 2.121 µm grid (TGT) for *XY* calibration (axis magnifications: $X = 1\times$, $Y = 1\times$, $Z = 20\times$) [31]

The evaluation consisted of the pitch of the tracks, together with the height of the pillars for the stamper and the depth of the pits for the replicated polymer discs. It resulted in a pitch of (82 ± 4) nm for both stamper and discs, a height of (419 ± 10) nm for the pillars, and a depth of (420 ± 14) nm for the pits. The traceability was established using the artifacts shown in Figure 10.6 (see Section 10.5 for further details on uncertainty evaluation).

Some pillars/pits do not have a well-defined level; therefore, in this case, an evaluation using histograms on local features is more adequate. Furthermore, the pitch of the tracks can be measured as the distance between adjacent edges of two adjacent features or, more appropriately, as the distance between the center of mass of two adjacent features (post-processing software can be used for this purpose [25]).

Manufacture of Polymer Microfluidic Systems

The replication process capability of high aspect ratio microfluidic systems was investigated by comparing measurements of a tool insert and replicated components. The proper operations are strictly dependent on the correct dimensions of the micro channels on the device, specifically the two parallel micro channels shown in Figure 10.7 (right). The investigation was conducted using an optical coordinate measuring machine for width measurements and a coherent scanning interferometer for depth measurements (see Section 10.4) [32, 33]. In particular, considering the high aspect ratio, the dimensions were evaluated in different spots along the channels to inspect the uniformity. Such an evaluation can easily be programmed in the software of an optical coordinate measuring machine for the width and can be successfully performed for the depth using the profile method described in ISO 5436-1:2000 [17]. The resulting width was (48.4 ± 3.1) μm for the tool and (48.0 ± 4.0) μm for the parts, whilst the depth was (110.4 ± 1.0) μm for the tool and (110.0 ± 7.0) μm for the parts.

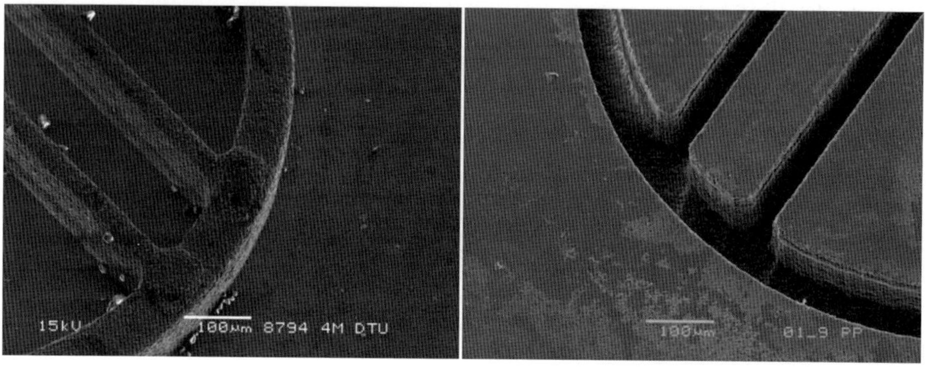

Figure 10.7 Replication of a high aspect ratio micro structure: silicon substrate machined by indirect hybrid tooling (left) and polymer replica (right) [33] (SEM measurements–see Section 10.4)

10.3.2 Surface Metrology

Surfaces are interfaces between a component and its environment, i.e., they are continuous boundaries over certain regions. They can have geometry (or form), determined as the distance of a generic surface from a reference surface, and texture, which indicates a structure of smaller geometries on a component's surface: specifically, such geometries are small in comparison to the dimension of the component itself [5]. Geometries locally contributing to form and texture together, i.e., responsible for a departure of the surface being evaluated from an ideal flat plane, are generically identified as topography.

Traditionally surface topography is characterized using the profile method. Terms and definitions can be found in ISO 4287:1997 [34]. In particular, it states:

- Real surface is the "surface limiting the body and separating it from the surrounding medium".
- Surface profile is the "profile that results from the intersection of the real surface by a specified plane".

According to the profile method, the surface profile thus defined is successively filtered by a low-pass filter[3] with cut-off[4] wavelength λ_s to obtain a primary profile, which is the basis for successive evaluations. The primary profile is further separated into a waviness profile, applying a band-pass filter (λ_f, λ_c) and a roughness profile, applying a high-pass filter λ_c. The longest components discarded by the cut-off wavelength λ_f correspond to the form of the surface under investigation. This process is schematically simplified in Figure 10.8.

Rules for the selection of cut-off wavelengths λ_c are given in ISO 4288:1996 [35]. They are intended for measured values with specified tolerance limits when using a stylus instrument.

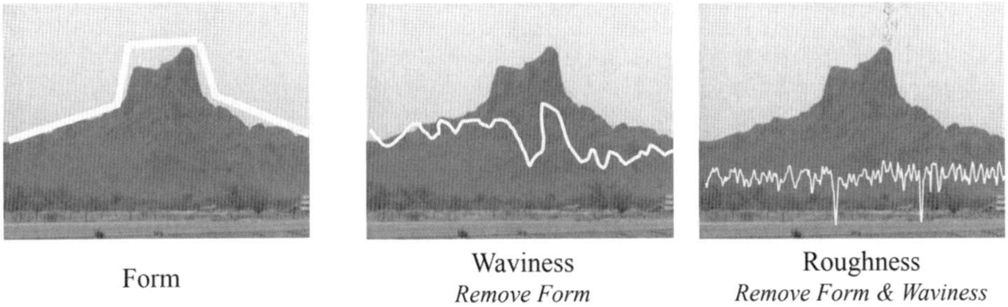

Figure 10.8 Profile method–filtering: form, waviness, and roughness [36]

Roughness parameters are defined on roughness profiles [34], corresponding to the components (passband or transmission band) selected by the cut-off wavelengths (λ_c, λ_s). There are

- 9 amplitude parameters
- 1 spacing parameter (in this regard, see also [37])
- 1 hybrid parameter
- Two functional curves and related parameters

---

[3] This filter is traditionally used to account for the smoothing effect due to the finite radius of the tip of a contact instrument, which is not able to completely penetrate into narrow valleys on a surface. A similar low pass filter can also be used for eliminating high frequency noise in optical areal measurements. Nevertheless, the use of filters is not encouraged here, except for specific investigations where the use of a filter for separating specific harmonic components may be justified by a physical model of a running experiment.

[4] The cut-off defines the wavelength (or the frequency) at which the filter reduces the transmission to 50% of the passband (also called the transmission band).

Corresponding waviness parameters are also defined on waviness profiles. In addition, it should be noted that similar definitions of the roughness parameters are given in ASME B46.1-2009 [36].

Further information can be found in the cited documents. Nevertheless, the profile method is a simplified approach for tolerance verification in manufacturing and is considered here as not adequate for micro injection molding topography characterization.

The sought functional requirements determine the assessment to be done. In a tool insert for micro injection molding applications, in fact, the insert surface is to be completely transferred to the replicated surface. Hence, splitting the surface in form, waviness, and roughness profile is not useful (if not harmful) to describe the replication process. Moreover, surfaces are areal entities in nature; when applying the profile method, there is a significant loss of information.

10.3.2.1 Areal Topography Measurements

More information can be extracted from an areal topographic characterization, allowing for a **functional analysis** of the surfaces too. It is an investigation and **quantification** of surfaces' properties, including the analysis of specific characteristics related to their intended purpose, e. g., aesthetics, the impact of the efficiency of different cutting tools on the produced surfaces, the flow-ability of a microchannel in lab-on-a-chip devices, etc.

Areal parameters are defined in ISO 25178-2:2012 [38]. They are categorized as

- Field parameters, which include S-parameters (averages on a surface) and V-parameters (averages of volume)
- Feature parameters, which involve the selection and investigation of specific features on portions of a definition area (segmentation)

Indeed, areal parameters refer to an average behavior within a definition area. However, it should always be kept in mind that the measurand is typically a dataset of pixels (see Section 10.4.1 and [18]) mapping a surface.

The average behavior requires that the areal acquisitions have the same discretization level or, referring to the Discrete Fourier Transform (DFT) theory, the same bandwidth. This is particularly critical when dealing with heterogeneous measurements[5] [39]. An equal discretization level can be achieved when the areal acquisitions have about the same sampling width[6] or, equivalently, the same field of view and number of pixels. Consequently, it is important to plan heterogeneous measurement sessions with this ultimate purpose.

[5] They can be measurements from different instruments, from different objective lenses of the same instruments, etc.

[6] The definitions of sampling width, field of view, etc. are elsewhere [40].

In this regard, the ISO 25178-2:2012 defines areal parameters over the *scale-limited surface,* i.e., the data set of points filtered using S-filter and F-operation or, equivalently, S-filter and L-filter (see also [41]). Specifically,

- S-filter is a surface filter for removing small lateral components (areal equivalent of a low-pass filter). An S-filtered surface is called primary surface.
- F-operation removes the form from the primary surface (S-F surface).
- L-filter is a surface filter for removing large lateral components (areal equivalent of a high-pass filter) from the primary surface or S-F surface (S-L surface).

Form removal (F-operation) is normally based on fitting of a mathematical surface and subtraction from the surface under investigation. Hence, it can be considered a non-destructive operation because it does not change the texture and, in addition, it can be reversible. Nevertheless, it has been shown elsewhere [42] how a good choice of the nesting index (the equivalent of a cut-off in areal filters) in an S-filter preserves the amplitude of a surface evaluation but locally changes the slope of the texture's micro and nano geometries.

A filtered surface cannot be counted as a measurement anymore. Indeed, it is the result of a calculation, which is normally irreversible and, thus, not suitable to be treated by statistical methods. Hence, at micro and nano scales, filtering may hide or transform micro and nano features on the surface under investigation.

Indeed, using a post-processing software tool (see, e.g., [25, 26]), it is certainly straightforward to equally size the field of view of different areal acquisitions and to re-sample to the same number of pixels, maintaining the bandwidth requirements. Re-sampling means to increase or decrease the number of pixels in the data set of points of the surface map. It might be a destructive operation; however, it does not significantly affect the relative proportions of micro and nano geometries on the surface under evaluation. Therefore, form removal and/or levelling (i.e., subtraction of a least-square plane and/or reference surface) of the acquired surfaces are the recommended processing, provided in addition to the same bandwidth requirements when dealing with heterogeneous areal measurements.

To follow are the definitions of the areal parameters S_a, S_q, S_z, and S_{dq}. S_a and S_q are both amplitude parameters and can be used, e.g., for an investigation of amplitude replication (see Chapter 4, Section 4.3.3) or amplitude variation. They are correlated and express the same quantity in different ways, thus one of the two is normally enough. S_a is easy to understand but S_q has a more significant statistical meaning. After removing a least-square plane from a surface under investigation (an operation that is always required for texture evaluation), S_q is the variance of the pixels' height in the data set of points of the surface map. S_z is also an amplitude parameter, which gives the maximum peak-to-valley extension. S_z is quite sensitive to local disturbances, which can considerably change its value. Hence, care is needed when using this parameter.

S_{dq} is sensitive to the local slope of the surface and is suggested for investigation of the slope replication or slope variation.

Arithmetic Mean Height of the Scale-Limited Surface

The S_a roughness parameter is defined as the arithmetic mean of the absolute of the ordinate values within a definition area A. The corresponding equation is

$$S_a = \frac{1}{A}\iint_A |z(x,y)|\; dx\, dy \tag{10.1}$$

Root Mean Square Height of the Scale-Limited Surface

The S_q roughness parameter is defined as the root mean square value of the ordinate values within a definition area A. The corresponding equation is

$$S_q = \sqrt{\frac{1}{A}\iint_A z^2(x,y)\; dx\, dy} \tag{10.2}$$

Maximum Height of the Scale-Limited Surface

The S_z roughness parameter is defined as the sum of the maximum peak height value and the maximum pit height value within a definition area A.

Root Mean Square Gradient of the Scale-Limited Surface

The S_{dq} roughness parameter is defined as the root mean square of the surface gradient within the definition area A. The corresponding equation is

$$S_{dq} = \sqrt{\frac{1}{A}\iint_A \left[\left(\frac{\partial z(x,y)}{\partial x}\right)^2 + \left(\frac{\partial z(x,y)}{\partial y}\right)^2\right] dx\, dy} \tag{10.3}$$

To follow are some solutions of surface metrology of micro tool inserts and micro injection molded parts.

Manufacture of Polymer Microfluidic Systems

Continuing with the example of microfluidic systems, a complete layout is shown in Figure 10.9. Three reservoir chambers are connected to the micro channels. The surface texture in the reservoir chamber was also investigated.

Figure 10.9 Micro tool (nickel substrate) for microfluidic systems manufactured by a hybrid tooling process chain (μEDM, electroforming, and selective etching) [33]

The surface texture can vary substantially from one spot of the surface to another. In this regard, for a larger evaluation of the surface, a matrix of areal acquisitions, one adjacent to the others, can be performed to extend the field of view of an instrument to a larger one. This technique is called "stitching" and is normally performed automatically by many instruments for topography measurements. Such a technique can be affected by the positioning errors of the stage of the instrument used.

Another solution can be averaging over several acquisitions obtained by sampling the surface with a certain field of view over a constant period in the X and Y directions. In one case, the average roughness of the polymer microfluidic reservoir chambers was up to 33% lower with respect to the tool, which was $S_a = (448 \pm 27)$ nm (see Figure 10.10).

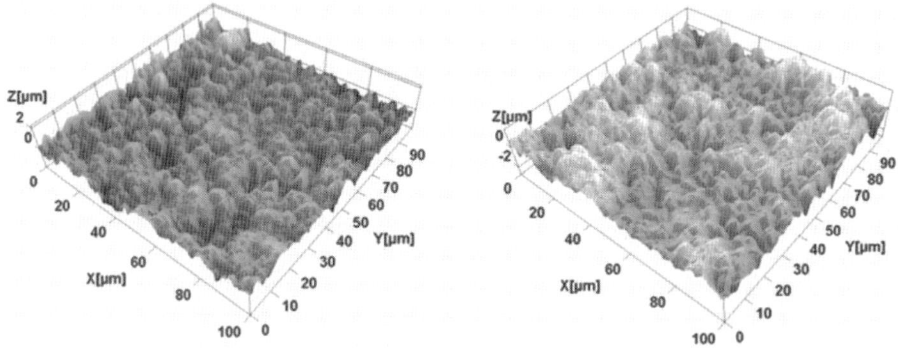

Figure 10.10 Atomic force microscopy scan of the surfaces of reservoir chambers (3D view), in a tool insert for microfluidic systems (left) and in a replicated polymer part (right) [33]

Replication of Sub-Micro Structured Polymer Surfaces

In another example, artifacts were replicated as transfer standards to be used in an international comparison of optical instruments for areal topography measurements [43]. The artifacts are roughness standards manufactured by Rubert & Co Ltd., UK. Two of them, namely 503 and 529, are shown in Figure 10.11, while examples of replicated specimens are shown in Figure 10.12; in the figure, a general profile indicates the type of texture–random and sinusoidal, respectively. Furthermore, Figure 10.13 and Figure 10.14 show examples of atomic force acquisitions for both types of roughness standards and the corresponding replicated surfaces.

The particularity of the solution here is the presence of a mark defining a univocal reference system on the surface of the standards, which was consequently transferred to the surface of the replicated specimens. Taking advantage of the mark, the measurements were acquired in the same spot of each specimen and of the same size. To evaluate the replication, in fact, it was important that the same portions of the surfaces were matched in the comparison between the master geometry and the replicated one. Furthermore, as already explained earlier, the same discretization level was required. The complete solution is in Section 10.5.1.

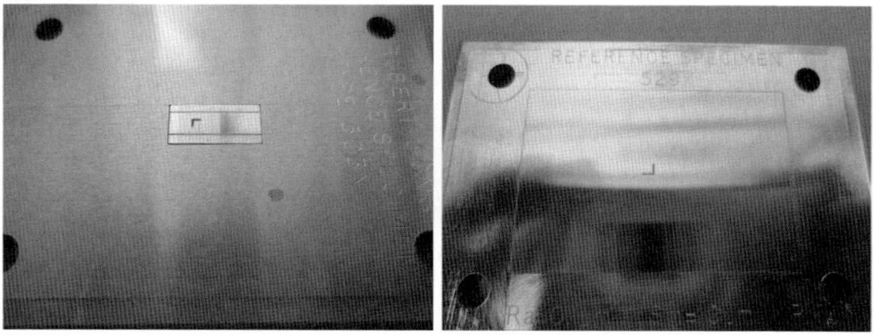

Figure 10.11 Nickel roughness standards 503 (a) and 529 (b) [43, 44]

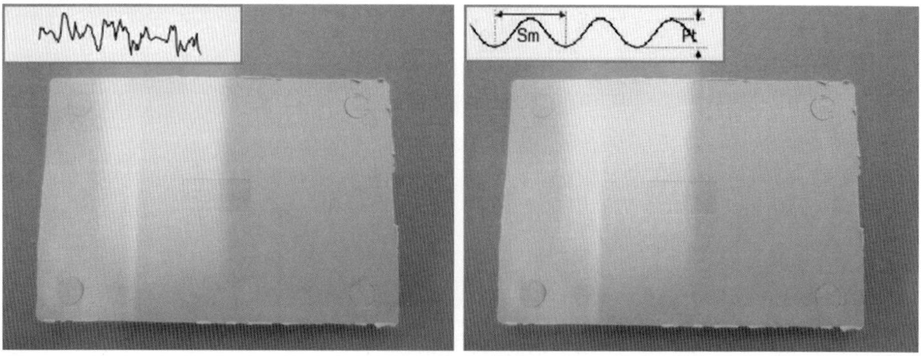

Figure 10.12 Examples of replicated polymer specimens by roughness standards 503 (a) and 529 (b). For both specimens, a general profile is given to indicate the type of surface topography [43, 44]

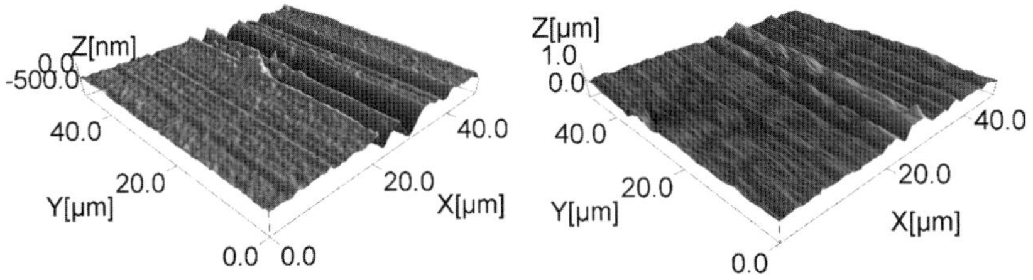

Figure 10.13 Examples of atomic force microscope acquired surfaces (3D view): 503 nickel reference (left) and corresponding polymer replicated (right) surfaces [43, 44]

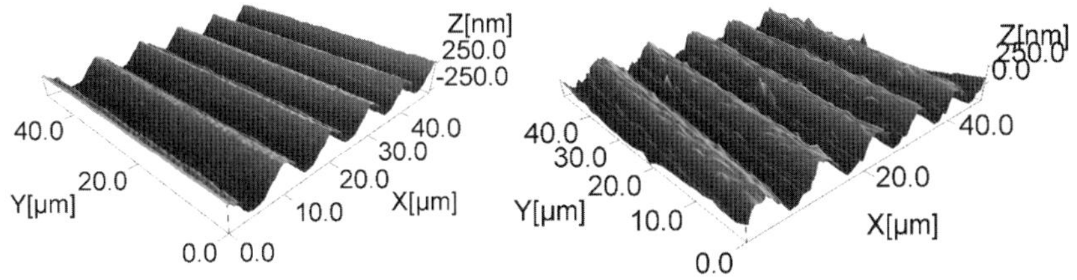

Figure 10.14 Examples of atomic force microscope acquired surfaces (3D view): 529 nickel reference (left) and corresponding polymer replicated (right) surfaces [43, 44]

10.4 Instrumentation for Micro Injection Molded Parts and Micro Injection Molding Tools

The instruments for micro and nano dimensional and geometrical measurements can be grouped according to the following classification [5, 45, 46]:

- Micro topography measurement instruments, which include
  - Contact instruments
  - Scanning probe and particle beam microscopy
  - Optical scanning instruments
  - Areal optical instruments
- Interferometry
- Coordinate metrology
- Computer tomography

The capabilities of the major working principles are summarized in Figure 10.15. The measurement range of each working principle is given as a box, in which the lateral range is a function of the vertical one.

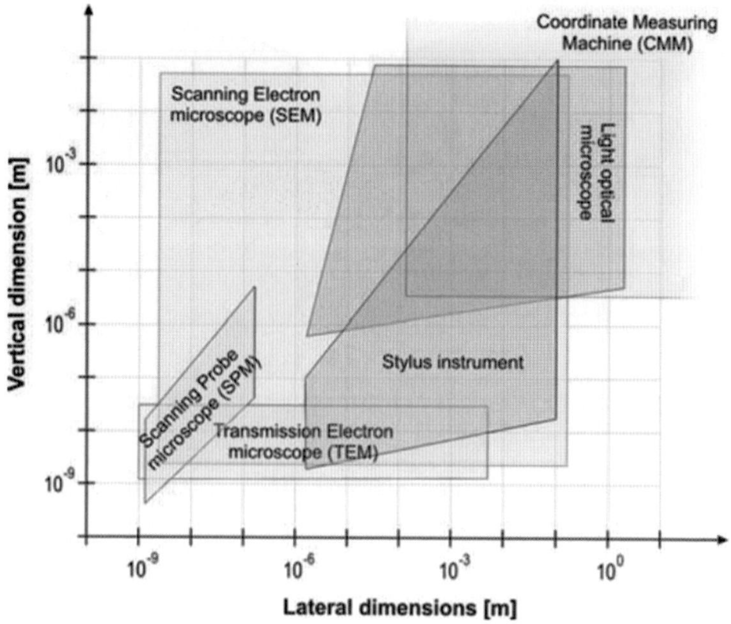

Figure 10.15 Measurement range—lateral range vs. vertical range—of the instruments' main working principles for dimensional and geometrical micro/nano metrology [5]

Contact instruments have a probe (stylus) that is physically in contact with the surface under measurement. The vertical movement of the stylus is converted into an electrical signal by a transducer. The probe is usually a diamond tip with a cone angle of 60° or 90° and a radius in the range (1–10) µm [5, 14]. The main drawback of contact instruments is that, due to their size, some probes do not completely penetrate into the valleys of the surface texture. In addition, the force applied by the probe can produce plastic deformation (scratches) in soft or polished surfaces [47]. Moreover, areal acquisitions may require extensive measuring times [45].

Scanning probe and particle beam microscopy can be suitable for measurements on the nanoscale and are identified in the two sub-categories: Scanning Probe Microscopy (SPM) and electron microscopy, which comprises Scanning Electron Microscopy (SEM) and Transmission Electron Microscopy (TEM).

In *SPM*, a probe, with a radius of a few nanometers and connected to a piezoelectric scanner, scans in close proximity of a surface but without direct contact. Hence, it is considered a non-destructive technique. It is suitable for measurements of several materials like metals, semiconductors, or biological specimens. *Atomic Force Microscopy* (AFM) is the most common among SPM technologies (see Figure 10.16). The probe, a tip some microns long and with a radius often less than 10 nm, is located at one end of a cantilever that has a length normally between 100 µm and 400 µm.

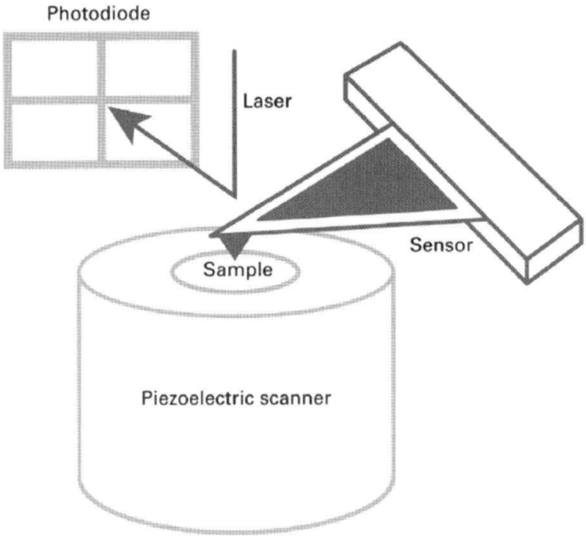

Figure 10.16 Schematic image of an atomic force microscope [45]

During a measurement, the tip is subjected to an equilibrium of forces (elastic force of the cantilever and repulsive forces of electric (intermolecular) nature), with resultant force sensitivity in the range $(10^{-7} - 10^{-12})$ N that allows for vertical deflections of the probe while scanning, detected by means of an interferometric system. The measurement scanning area is typically 100 µm × 100 µm or less, and the vertical range is less than 10 µm. The vertical resolution is about 0.1 nm, while the lateral resolution for most AFMs is usually from 2 nm to 10 nm [5]. AFM measurements are time-consuming though consistent at the nanoscale. Examples of AFM measurements have been shown in Figure 10.5, Figure 10.6, Figure 10.10, Figure 10.13, and Figure 10.14.

In *SEM*, an electron beam scans the surface of the specimen under measurement. When electrons hit the specimen, a characteristic X-ray radiation is emitted which, collected by an X-ray detector, allows for measurements of the physical and chemical properties of the specimen. Typical image magnifications are in the range 100×–100000×, with a resolution down to 2 nm, large field of view, high depth of focus, long working distance, elemental analysis capability, and minimum diffraction effects [48]. However, the dataset extracted is inherently two-dimensional (2D) and no height information is achievable (unless reconstructed from stereo pairs or triplets of SEM images). Hence, SEM is mostly used for qualitative image analysis due to its high-resolution imaging and significant visualization of the specimen under measurement [5]. Other disadvantages are the need for measurements in high vacuum conditions, a relatively low throughput, and an electric potential as a result of the specimen charging, which causes image degradation [49]. An example of an SEM measurement is shown in Figure 10.7. Another example is shown in Figure 10.17, where SEM measurements were used for a quantification of flushes resulting from a micro injection molding process of a structure with three pins [50]. Flushes were in the x-y plane of the SEM measurements.

In *TEM*, an electron beam is operated according to the same basic principle as that of an optical microscope. This allows TEM to spread the electron beam over the whole specimen (different from SEM in which the focus is on only one spot at a time). The electron beam is directed through a series of lenses to the specimen and, after passing through the specimen, it is finally collected at the end of its optical axis. For this reason, a thickness in the range of (50–500) nm is required to limit the chromatic aberration due to energy loss. The resolution is about 0.05 nm with aberration-corrected instruments [14, 45–47].

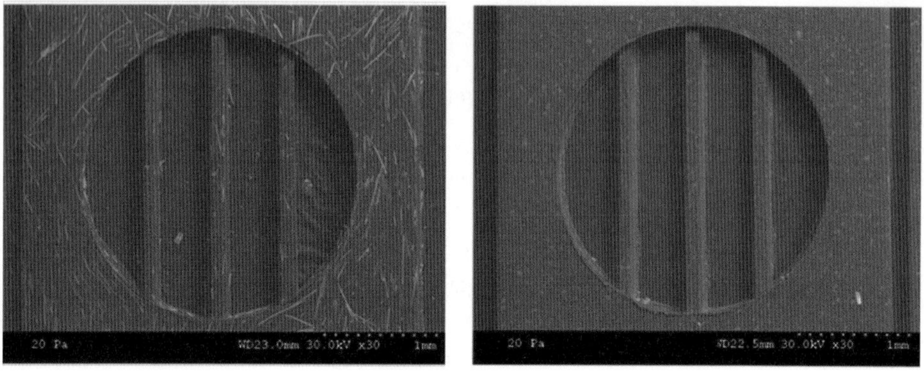

Figure 10.17 Images of three-pin structures manufactured by micro injection molding using a liquid crystal polymer (LCP–left) and polyoxymethylene (POM–right) [50]

Optical instruments perform non-contact measurements, with the advantage of not damaging the surface of the specimen under investigation. Furthermore, they have relatively high resolution in the vertical direction and can acquire relatively large areas in a reasonable time. For such reasons, they are becoming increasingly popular in metrology for micro injection molding and micro nano metrology in general. Section 10.4.1 presents the most common technologies for optical instruments.

Interferometry is based on the analysis of the interference pattern created by the interference of two or more waves. Different interferometric configurations exist. Many of them are variants of a basic setup, which is the Michelson interferometer [45]. The resolution is in the range of (0.1–10) nm, obtained in the direction of the beam propagation, i. e., interferometers are 1D instruments. The measurements range depends on the coherence length of the source (300 nm for white light, centimeters for spectral lamps and (1–10) m for commercially available laser interferometers) [5]. Interferometers are used for the calibration of 1D scales, as links to the primary standards for laser interferometer systems, and have applications in optical microscopy.

A *coordinate measuring machine* (CMM) is commonly an instrument with a probe that measures single points on a specimen surface or scans a selected surface in order to continuously collect data as the probe is dragged across the part. The acquired points are then fitted in geometric entities, normally by software equipped with the CMM. The probe in contact with the surface is usually a synthetic ruby ball and typical diameter dimensions are in the range of (0.5–10) mm [45].

In recent years, smaller CMMs have been developed to enable 3D measurements of nanometer features, typically with ranges from tens of millimeters down to tens of nanometers in the *x*, *y*, and *z* directions. They are generically called miniature-CMMs or micro-CMMs (μCMMs). Different μCMMs are described in [45].

Most μCMMs usually have standard probe tips with a diameter of 0.3 mm, which is too large to measure structures like MEMS or micro holes. Therefore, smaller probes have been developed. Silicon flexures, meshes or membranes are used to suspend the probe shaft in order to reduce the probing force.

In further developments, probes take optical measurements from illuminated glass fibers to reduce the surface damage caused by probe interactions [51]. In Figure 10.18, the stage of a multisensor CMM (Werth Video Check HA [52]) hosts multi-scale micro powder injection molded components, with a feature of size spanning from tenths of a micrometer to ten millimeters. The feature of size can be a cylindrical or a spherical surface, a set of two opposite parallel surfaces, or a cone or a wedge, which have a linear or an angular dimension (ISO 14405-1:2010, ASME Y14.5-2009). The instrument is equipped with several sensors, including an imaging video sensor, a laser sensor, a touch trigger and dynamic scanning probes, a fiber probe, Computer Tomography (CT), etc.

In *micro Computer Tomography* (μCT), a micro-focused X-ray source illuminates a specimen during image acquisition. The specimen is rotated stepwise of 180° and images are recorded at each position. Resolution down to 0.15 μm can be reached.

Recently, CT systems have been developed to replace classical CMMs or to be integrated into multisensory CMMs. Due to CT instruments' complexity, their uncertainty has not completely been quantified, yet. Nevertheless, some measurement uncertainty contributions have been assessed [53, 54] and more complete approaches to the uncertainty evaluation for CT systems can be found in recent work [55, 56], also dealing with μCT [57].

These technologies give precise quantitative information on the whole structure of a body, without destroying it, within a short period of time [58]. This is what has made CT very interesting for testing and inspecting manufactured components like engine blocks, gear-boxes, and injection nozzles [59]. Hence, their main applications are for non-destructive testing and for dimensional measurements [60].

In the following, an example of the dimensional measurements of the same quantities, using a tactile CMM (TCMM), an optical CMM (OCMM), and a μCT, highlights the pros and cons of such technologies.

Figure 10.18 Detail of the stage of a multisensor measuring machine (Werth Video Check HA [52]) and multiscale micro powder injection molded parts. The largest dimension for each component is about 8 mm

Toggle for Hearing Aid Applications

Toggles for hearing aid applications (see Figure 10.19), manufactured by micro injection molding using a liquid crystal polymer (LCP), were measured by different technologies in order to assess the possibility of performing in-line quality control by a μCT [61, 62].

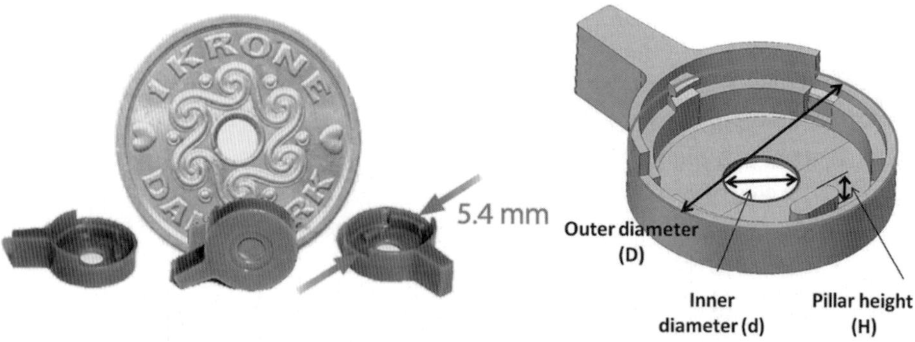

Figure 10.19 Toggles for hearing aid applications (left) and dimensions measured in the investigation (right) [61, 62]

Height, inner and outer diameters, and their concentricity were initially measured by a TCMM (see Figure 10.19 (right) and Figure 10.20). The uncertainty was evaluated according to ISO 15530 [63] (see Section 10.5). Except for few cases, the tolerance limits are verified for most (see Figure 10.21 to Figure 10.24).

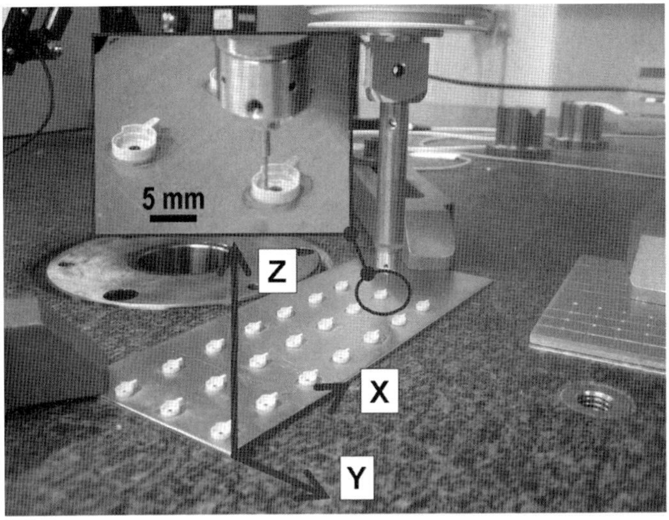

Figure 10.20 Measurement setup on a tactile CMM (TCMM) [61, 62]

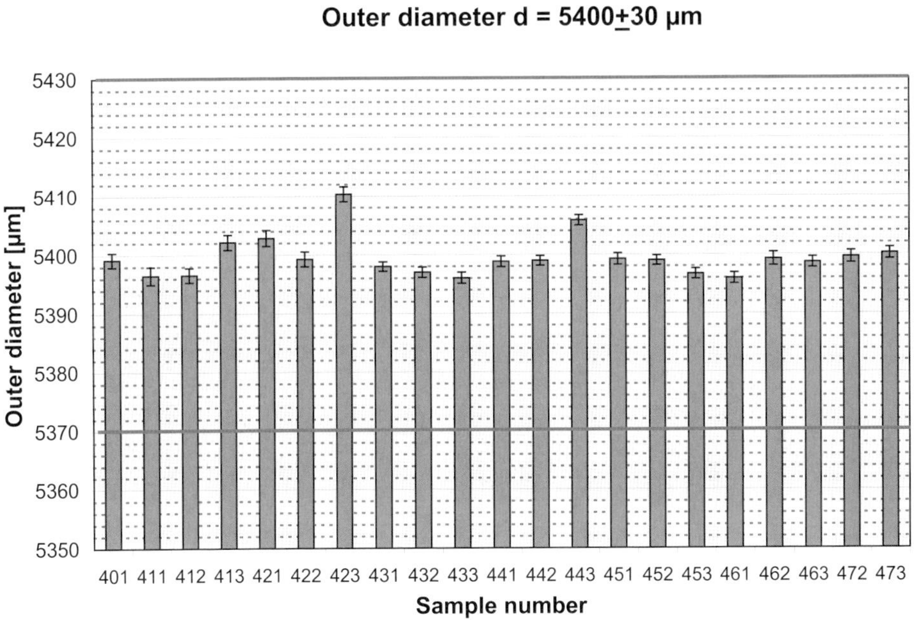

Figure 10.21 TCMM results of the toggles' outer diameter $D = (5400 \pm 30)$ µm (the measured parts are indicated by numbers). The thick horizontal lines indicate the tolerance limits [61, 62]

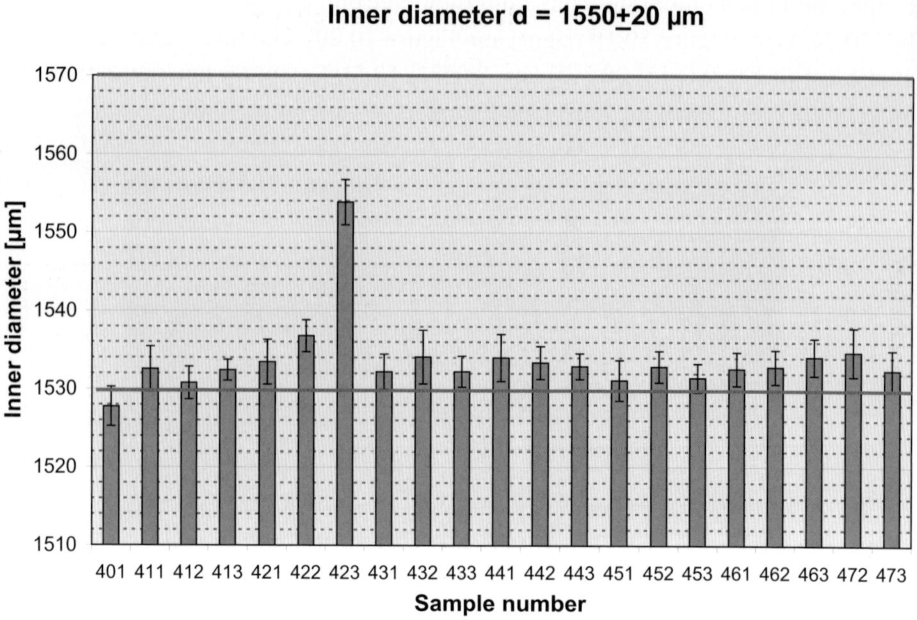

Figure 10.22 TCMM results of the toggles' inner diameter d = (1550 ± 20) µm (the measured parts are indicated by numbers). The thick horizontal lines indicate the tolerance limits [61, 62]

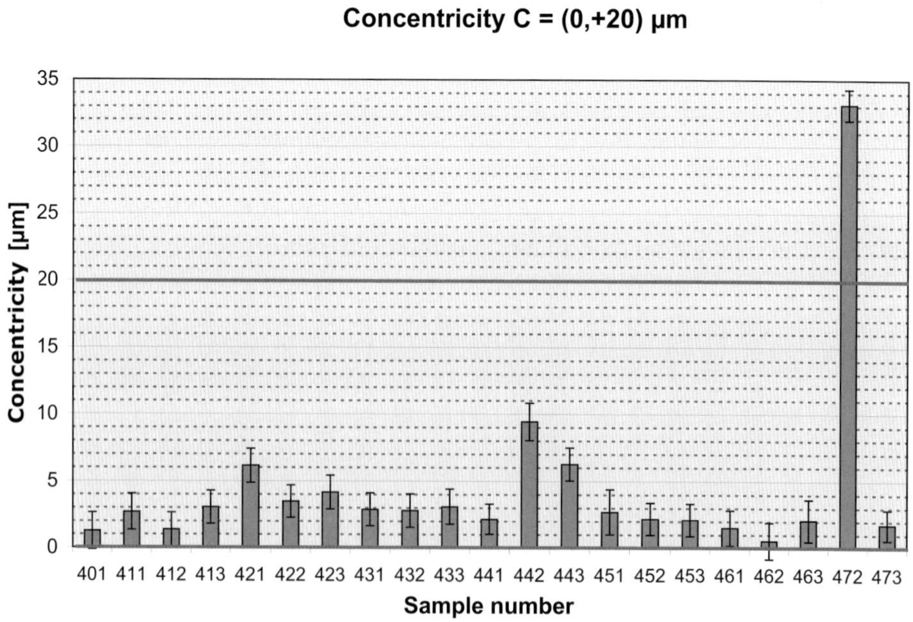

Figure 10.23 TCMM results of the toggles' inner and outer diameters' concentricity C = 0, 20 µm (the measured parts are indicated by numbers). The thick horizontal lines indicate the tolerance limits [61, 62]

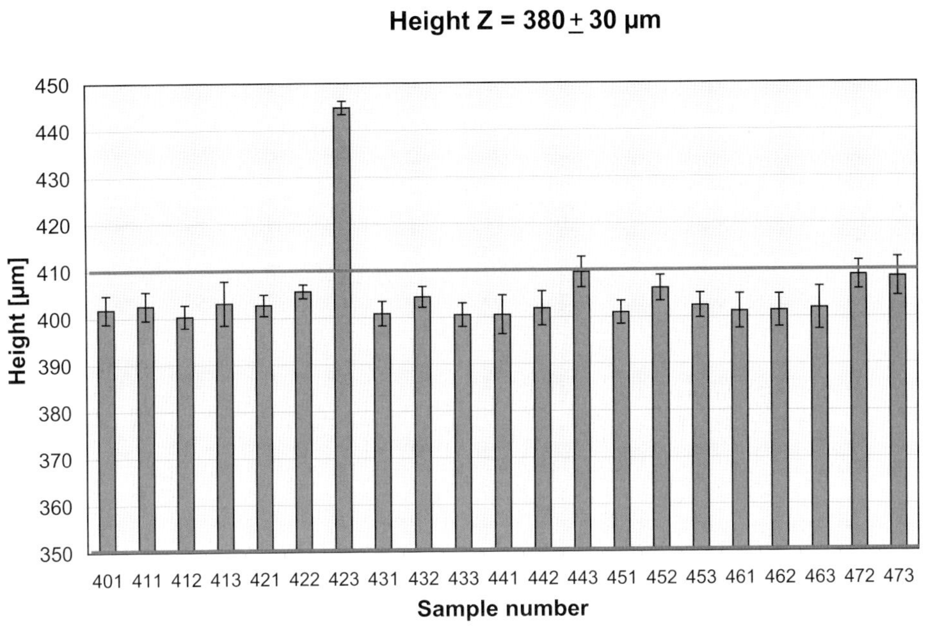

Figure 10.24 TCMM results of the toggles' height $H = (380 \pm 30)$ µm (the measured parts are indicated by numbers). The thick horizontal lines indicate the tolerance limits [61, 62]

TCMM measurements were successively compared with OCMM measurements of the same quantities. The results are shown in Figure 10.25 to Figure 10.28. The large uncertainty evaluated for these measurements counteracted the verification of the tolerance limits. A different calculation of the systematic components in the uncertainty assessment (i.e., the compensation, indicated as "comp" in the figures) obtained a more adequate uncertainty. Nevertheless, a certain ambiguity in the tolerances' verification remained, indicating that the OCMM is not completely suitable, above all for smaller dimensions (inner diameters).

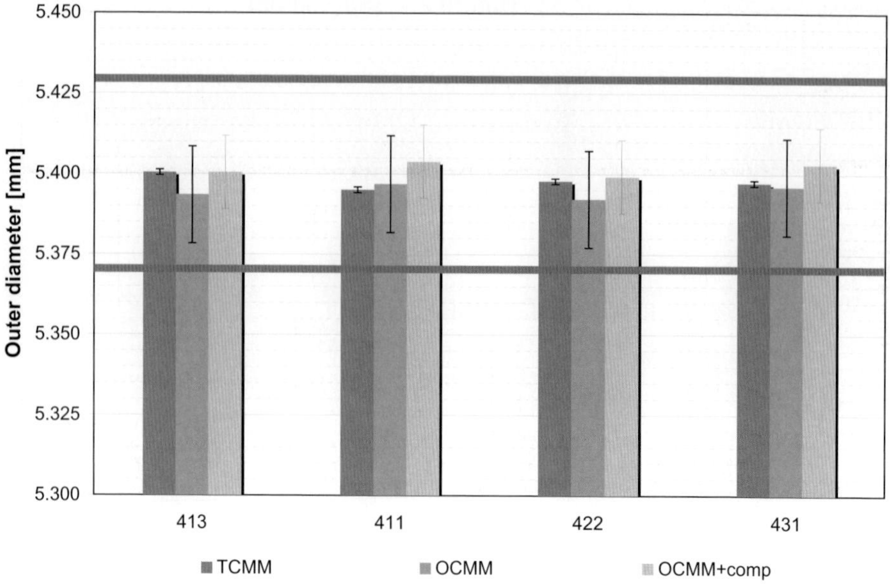

Figure 10.25 Comparison between tactile and optical CMMs' results of the toggles' outer diameter. The thick horizontal lines indicate the tolerance limits [61, 62]

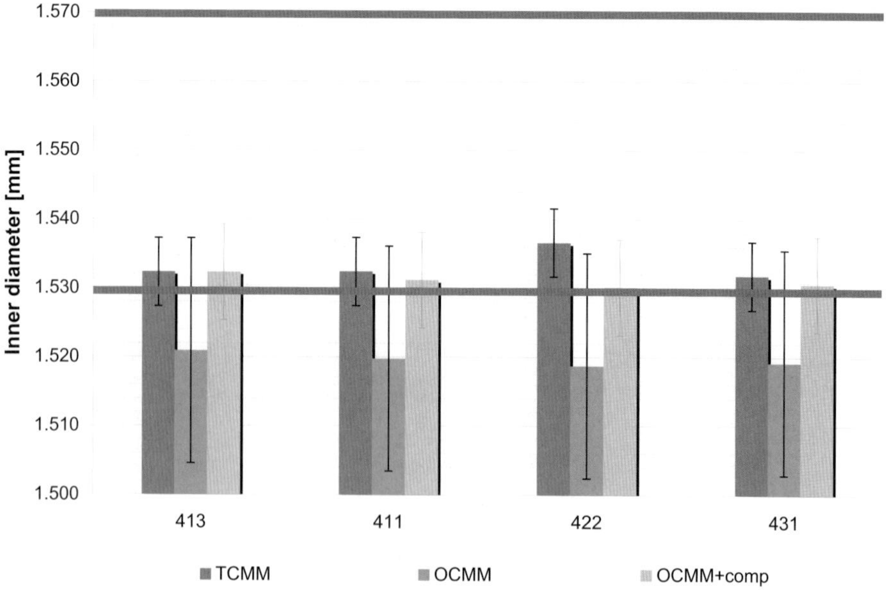

Figure 10.26 Comparison between tactile and optical CMMs' results of the toggles' inner diameter. The thick horizontal lines indicate the tolerance limits [61, 62]

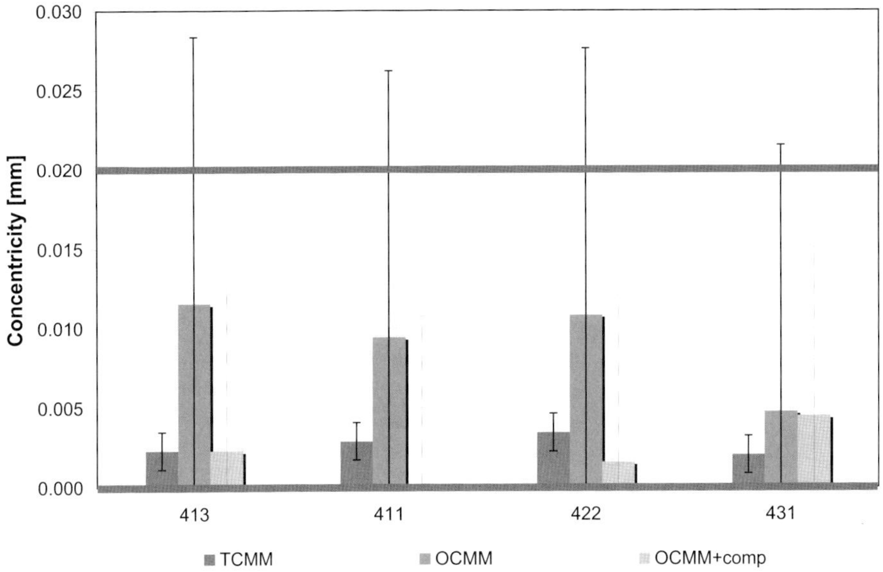

Figure 10.27 Comparison between tactile and optical CMMs' results of the toggles' inner and outer diameters' concentricity. The thick horizontal lines indicate the tolerance limits [61, 62]

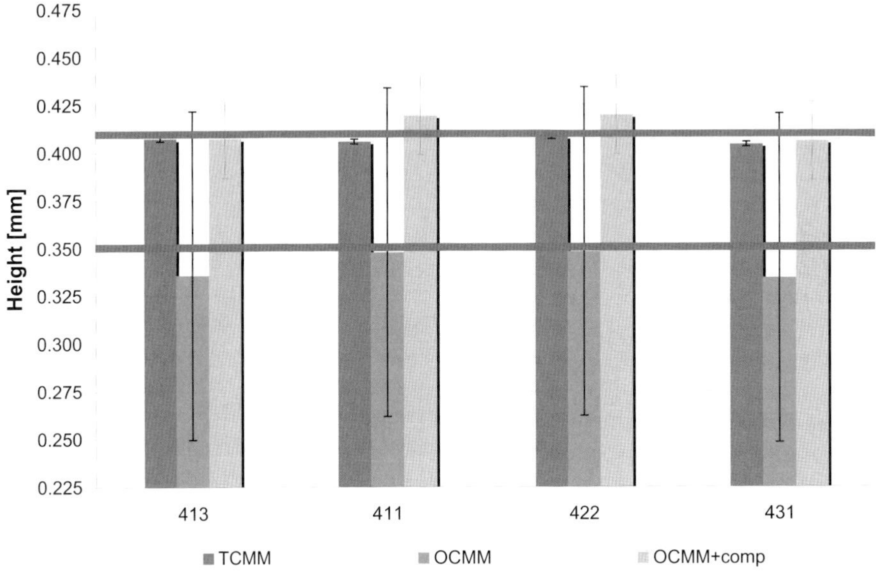

Figure 10.28 Comparison between tactile and optical CMMs' results of the toggles' height. The thick horizontal lines indicate the tolerance limits [61, 62]

The analysis of the μCT results gave a trend similar to those from OCMM, including the uncertainty, which was evaluated by modifying the method of ISO 15530 to adapt it for CT measurements. In general, deviations of CT measurements with respect to the reference dimensions were always smaller than 1%.

Despite these encouraging results, CT measurements are still complex, as shown by the flow of operations in Figure 10.29 and by the error sources to be considered (see Figure 10.30). In addition, Figure 10.31 shows typical defects noticed on the surfaces of the CT-scanned specimens. Hence, further improvements are still expected for this technology.

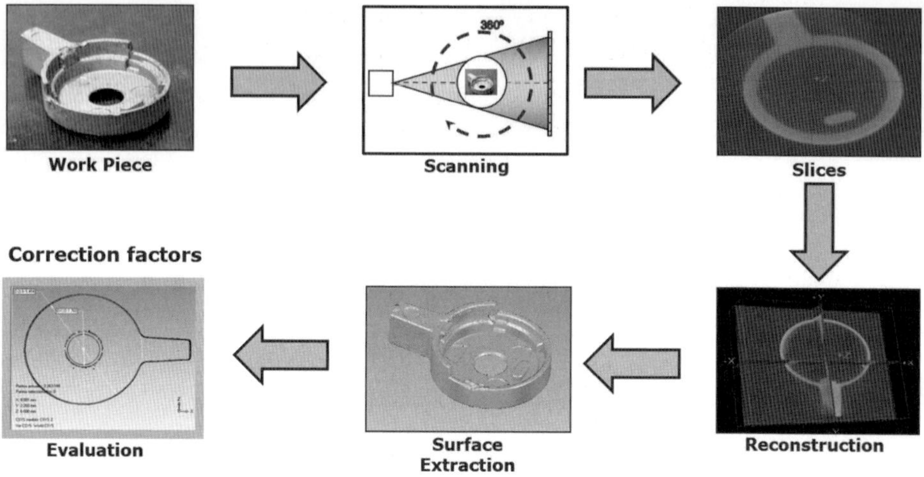

Figure 10.29 Flow of operations for single CT measurements [61, 62]

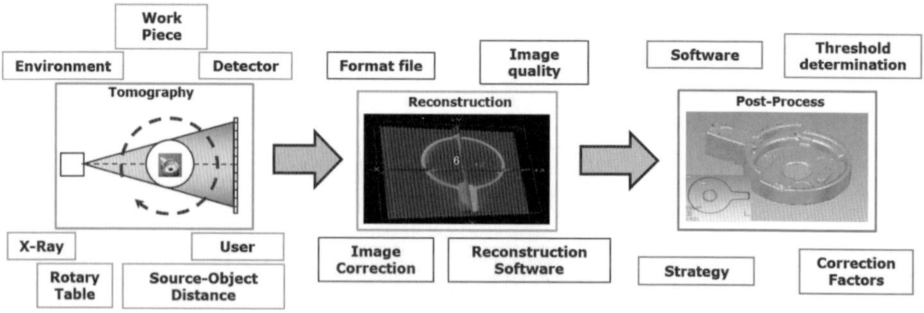

Figure 10.30 Error sources considered in the investigation [61, 62]

On the other hand, CT instruments have unique and extremely useful capabilities, like providing morphological information such as suctions (i.e., valleys) on the parts' surfaces and cavities inside the moldings. In addition, the collection of complete point clouds is independent from any material properties or form errors, in contrast to other non-contact measuring techniques such as OCMMs.

The following example points out such important capabilities of CT instruments.

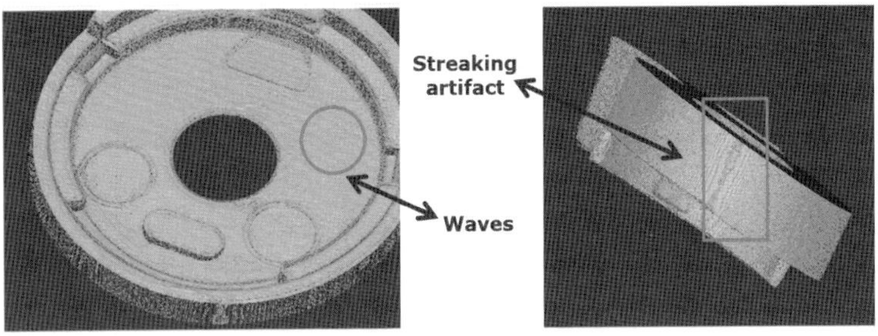

Figure 10.31 Typical defects on the measurements of toggles from the μCT used in the investigation

Manufacturing of a Tool Insert for Micro Injection Molding by Selective Laser Melting (SLM)

Taking advantage of a SLM process, one of the 3D printing technologies, a tool insert for micro injection molding was manufactured with internal channels for conformal cooling [64]. The design of the mold is shown in Figure 10.32, while the manufactured mold insert with an inlet and outlet of channels is shown in Figure 10.33 (left). On the right of the same figure, instead, CT measurements visualize the cooling channels.

Since the insert was a monolithic component, CT measurements were the only non-destructive method for analyzing the continuity of the channels and for measuring their diameter.

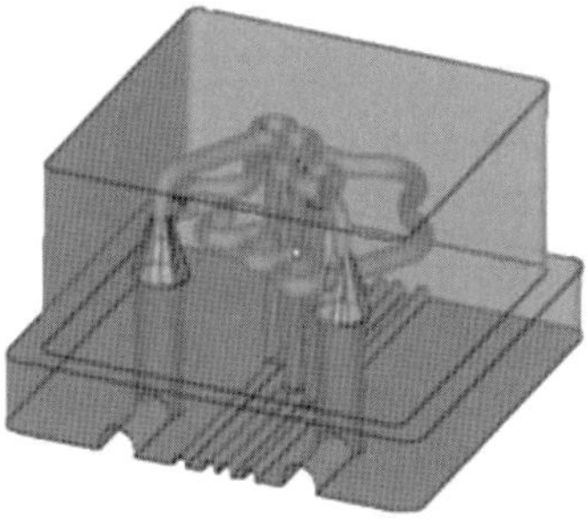

Figure 10.32 Design of a micro injection molding tool insert with cooling channels [64]

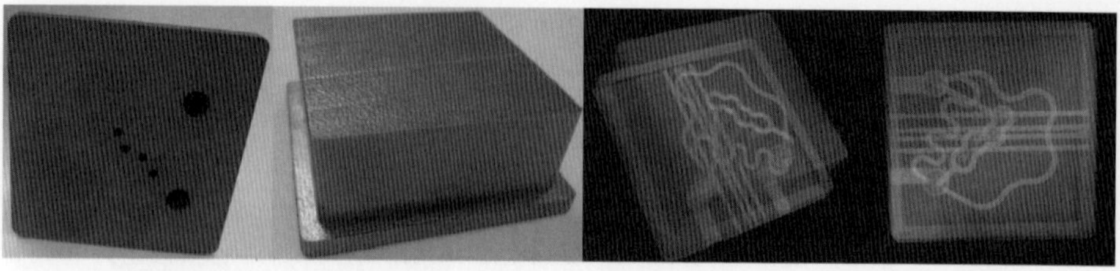

Figure 10.33 Left: tool insert for micro injection molding with cooling channels. Inlet and outlet are visible in the picture. Right: CT measurement of the tool insert. The complete channels are visible [64]

Further research on measurements from different instruments compared with CT ones can be found elsewhere [65]. Micro gears were measured by a TCMM, tactile-optical CMM (TOCMM optical glass fiber probe), focus-variation microscope (see Section 10.4.1), and industrial CT [65]. The comparison among measurements from different instruments showed that the deviations with CT measurements could be kept below a few micrometers.

Moreover, simulation in micro injection molding is increasingly being used for optimization of the process. However, due to the complexity of the physical phenomena involved, reliable physical models may be based on measurements sessions from different instruments [19]. The choice of the measurement instrument is of central importance and may lead to disagreement among different studies when the measurands are critical, like an inner diameter of a microinjection-molded component (low manufacturing repeatability and/or small dimension, almost out of the range of CMMs) [20]. In this regard, it is worth mentioning a recent study [66] in which the data from CT measurements were merged and, hence, enhanced with data from confocal microscopy. The investigation was intended for quality control but also to support the simulation of a micro injection molding process.

Finally, a review of CT systems that surveys CTs' market, capabilities, technologies, and examples from the manufacturing, electrical, electronic, and food industries can be found in [67].

10.4.1 Optical Instruments

Whereas all the instruments briefly discussed here can potentially be used for micro injection molding applications, optical instruments are assuming increasing importance in geometrical and dimensional assessments of micro molded components, which more and more often have structured and freeform surfaces. Although such instruments have limitations, they can easily overcome many difficulties connected with the use of other technologies (e.g., contact instruments). Furthermore, many manufacturers are optimizing and automating optical instruments for in-line use, i. e., for direct use in the production-line environment, allowing for a fast and convenient quality assurance of parts.

This sub-section deals with the most common technologies for optical instruments available nowadays. They are divided into scanning (intended as lateral scanning, i. e., along a direction parallel to the surface under measurement) and areal optical instruments [11, 45, 46]: confocal instruments in the first class group and focus-variation instruments, phase-shifting interferometry, and coherence scanning interferometry in the other group. However, both are areal topography measuring instruments, i. e., instruments that can produce a three-dimensional (3D) map of surfaces. Strictly speaking, the measurand ("quantity intended to be measured" [12]) is a 3D map of the height of the surface under measurement with respect to its least-squares fit plane (pixels) [18]. A set of metrological characteristics (MCs) have been defined (measurement noise, residual flatness, amplification, linearity, squareness of the axes, resolution), which identify features common to all the areal topography measuring instruments, with respect to the acquired 3D data sets of points and, concurrently, define possible sources of variation of the output [68–71].

As already noticed, such instruments have relatively high resolution in the vertical direction and can acquire relatively large areas in a reasonable time. On the other hand, dimensional measurements can be limited by the optical diffraction of light. Furthermore, when an objective lens is used, the optical resolution is restricted by the wavelength of the light used, but also by the numerical aperture A_N of the optical system, which determines the theoretical maximum slope on the surface that can be measured (Eq. 10.4).

In a medium with refractive index n, the half-angle of the maximum one accepted into the objective is the angle α, given by

$$A_N = n \sin \alpha \quad (10.4)$$

Hence the optical resolution of the objective, i. e., the minimum distance between two lateral features on a surface that can be measured, is given by

$$r = \xi \frac{\lambda}{A_N} \quad (10.5)$$

where λ is the wavelength of the incident radiation and ξ is a constant depending on the criterion used for the evaluation of the resolution. The most common ones are the Rayleigh criterion (for non-coherent systems) and the Sparrow criterion (for coherent systems) [45].

In *confocal microscopes*, a monochromatic light source is normally used together with at least two mechanical filters (see Figure 10.34), namely two pinholes, placed along the optical path: one after the light source and one before the detector. The pinholes discard the light that is not in focus, increasing the lateral optical resolution. The instruments scan vertically within a certain range. Hence, the height is determined by fitting the data corresponding to a maximum intensity [72].

The optical resolution can be further increased by the pinholes, limiting the field of view to an area smaller than the limit of Eq. 10.5. The vertical measurements range, instead, is limited by the working distance of the objective and, thus, by A_N. Increased working distance, governed by the microscope objectives, decreases the A_N, resulting in reduced lateral and axial resolution.

Advantages of this method are the possibility of using objectives with high A_N, high lateral resolution, and low influence of the surface color. Conversely, drawbacks are a longer measuring time with respect to non-scanning instruments, due to light beam scanning, and a short working distance of the objective that raises the risk of collision on the surface under measurement.

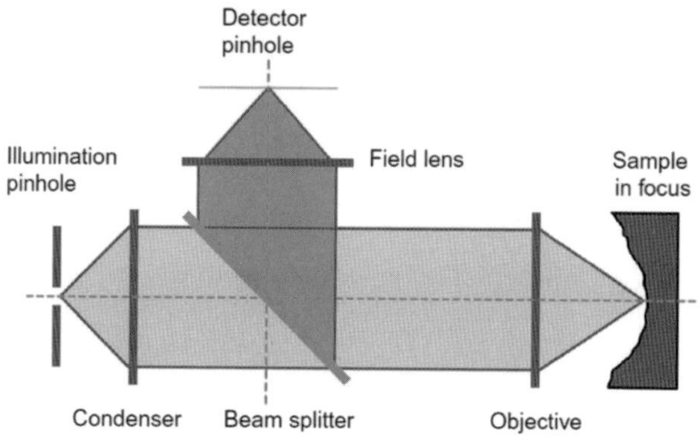

Figure 10.34 Confocal setup with object in focus [45]

A non-scanning version of a confocal microscope *(chromatic confocal microscope)* adopts non-monochromatic light and, consequently, chromatic objective lenses with reduced numerical aperture and an increased working distance. A different focus distance is due to different wavelengths. The advantage is that there is no vertical scanning. The vertical range is limited by the difference between the focal distance of the shortest optical wavelength and the longest one, resulting in a faster acquisition with respect to the scanning version. However, it has a lower vertical resolution and can be influenced by the color of the surface [11].

The *focus-variation* working principle combines a small depth of focus with vertical scanning by the objective lens, sweeping the focus along the vertical features of a surface and continuously capturing data. Such variation of the focus provides topographical and color information together. In addition, the maximum detectable slope is not dependent on the numerical aperture of the objective. Different illumination sources and the focus-variation system allow for measurements of slope angle exceeding 80°. The vertical resolution is dependent on the chosen objective but independent from the scan height and it can reach 10 nm. Nevertheless, the focus-variation working principle requires that the focus varies sufficiently during the vertical scanning in order to detect and reconstruct surfaces and structures. Hence, transparent specimens or components with low local roughness cannot be measured [73].

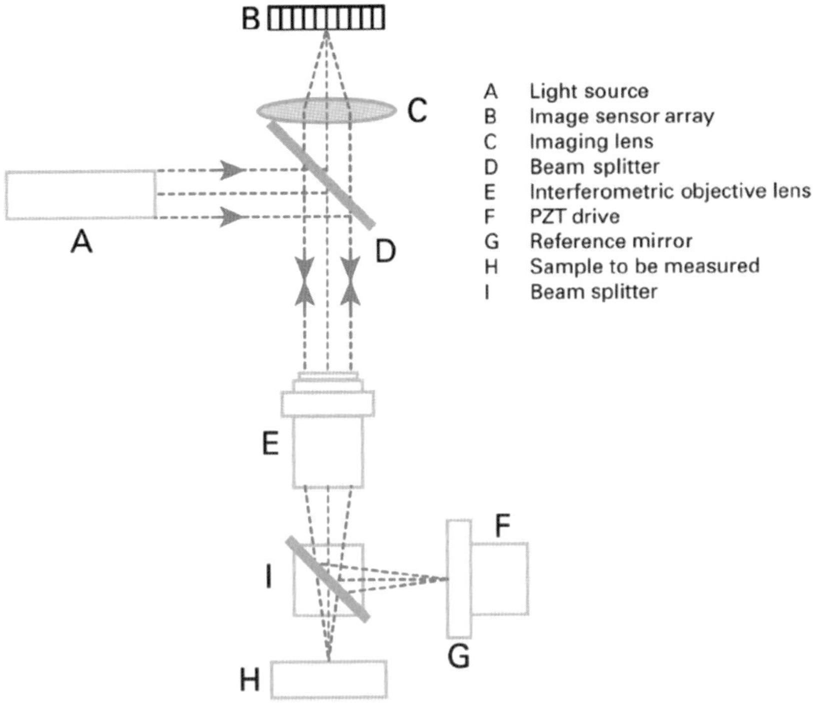

Figure 10.35 Schema of a phase-shifting interferometer [45]

Phase-Shifting Interferometry (PSI) involves the analysis of interference fringes between incident light from the specimen under measurement and reference light. It consists of an interferometer integrated into a microscope (see Figure 10.35) [45]. Within the interferometer, a beam-splitter directs one beam to a reference path and a second beam to the specimen surface under measurement. The two reflected beams return to the beam-splitter where they recombine and form, on the imaging sensor array (camera), an image of the measured surface with a superimposed interference pattern.

PSI instruments can have two different configurations–Mirau and Linnik interferometers–depending on the microscope objective (see Figure 10.36). Mirau is more compact and needs less adjustment than Linnik. However, in the Linnik configuration, the central area of the objective is not blocked, and no space is needed underneath the objective for attaching an extra mirror and beam splitter. Therefore, with the Linnik objective, magnifications and resolutions are the highest among standard optical microscope objectives. PSI instruments can have sub-nanometer resolution and repeatability. Nonetheless, it is very difficult to determine their accuracy, which is greatly dependent on the surface under measurement.

Many PSI instruments usually require adjacent points on a surface with a height difference less then $\lambda/4$. The range of PSI is limited to one fringe, or approximately half the central wavelength of the light source. Therefore, PSI instruments are usually used for measuring approximately flat surfaces (as a rule of thumb, surfaces with an R_a or S_a less than $\lambda/10$) [45].

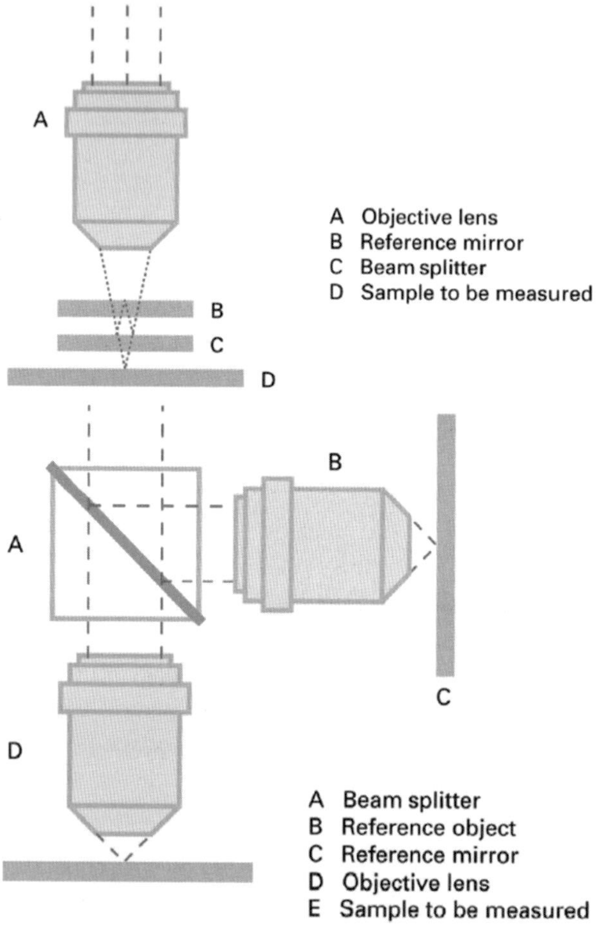

Figure 10.36 The different configurations of PSI [45]. Top: schematic diagram of a Mirau objective. Bottom: schematic diagram of a Linnik objective

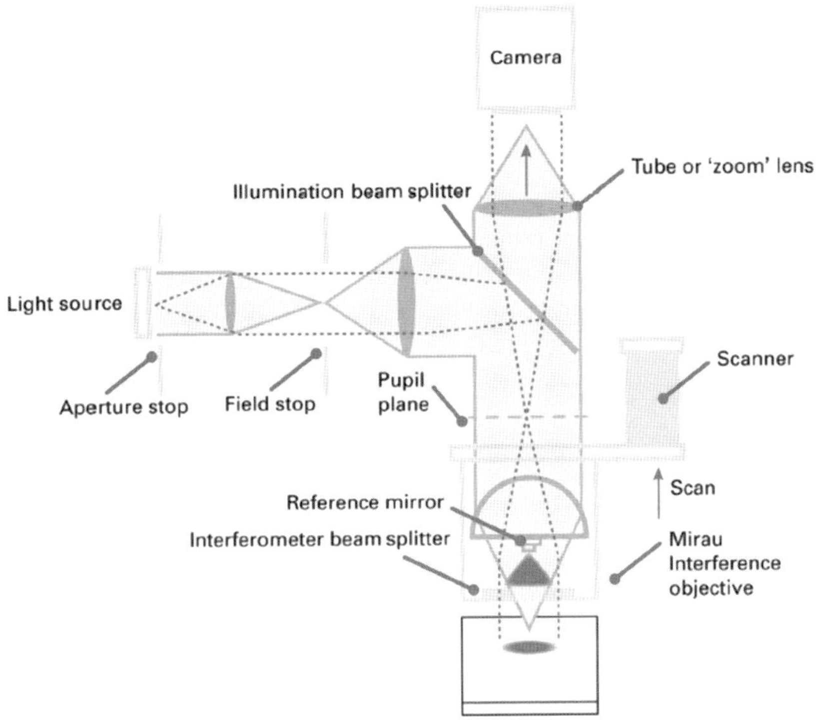

Figure 10.37 Schema of a coherence scanning interferometer [45]

Coherence Scanning Interferometry (CSI) is commonly known as scanning white light interferometry (SWLI) and is based on interferometric techniques too. Due to the low coherence of the source (non-monochromatic), to observe interference, the length of the optical path to the specimen and the reference must be almost identical. The detector measures the intensity of the interference fringes as the optical path is varied in the vertical direction and finds the interference maximum (see Figure 10.37). The advantage of this system is a complete 3D areal surface dataset produced by a single vertical scan using a Charge-Coupled Device (CCD) imaging sensor. CSI instruments can have sub-nanometer vertical resolution and repeatability but, as with all instruments that scan vertically, such resolution is somehow limited by the vertical scanning system. In addition, objective lenses have small numerical apertures.

In the following, qualitative examples of measurements performed using different optical technologies are presented. The same pseudo-random roughness metal artifact (AIR-B70) was measured in the same (central) spot on the surface (see Figure 10.38). AIR-B70 belongs to a set of artifacts, for the areal calibration of optical instruments, denominated "Bento Box" and produced at the National Physical Laboratory (NPL) in Teddington, UK [74]. Measurements were performed using the same 50× magnification objective, by imaging confocal (Figure 10.39), focus-variation (Figure 10.40), PSI (Figure 10.41), and CSI (Figure 10.42) instruments. Differences of field of view and sampling distance [40] among the instruments are indicated in each figure's caption.

Figure 10.38 Pseudo-random roughness metal artifact of the "Bento Box" areal calibration set (National Physical Laboratory (NPL), Teddington, UK [74])

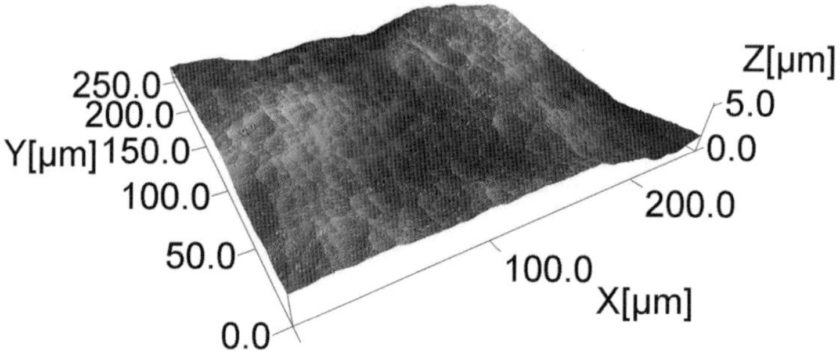

Figure 10.39 Qualitative example of imaging laser scanning confocal microscope acquisition (3D view) of an irregular artifact, using a 50× magnification objective with field of view 258 µm × 258 µm and minimum sampling distance of 0.06 µm

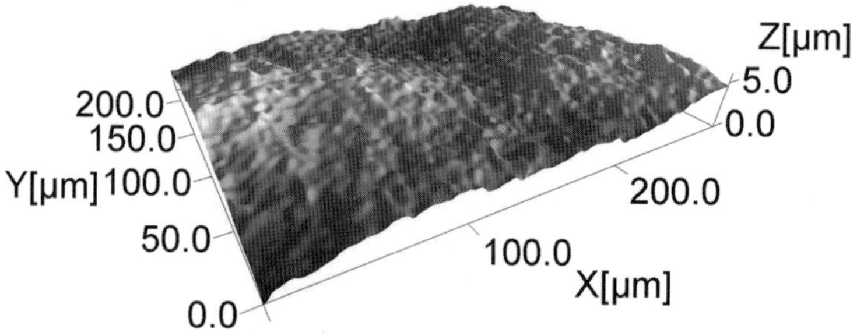

Figure 10.40 Qualitative example of focus-variation microscope acquisition (3D view) of an irregular artifact, using a 50× magnification objective with field of view 288 µm × 218 µm and minimum sampling distance of 0.18 µm

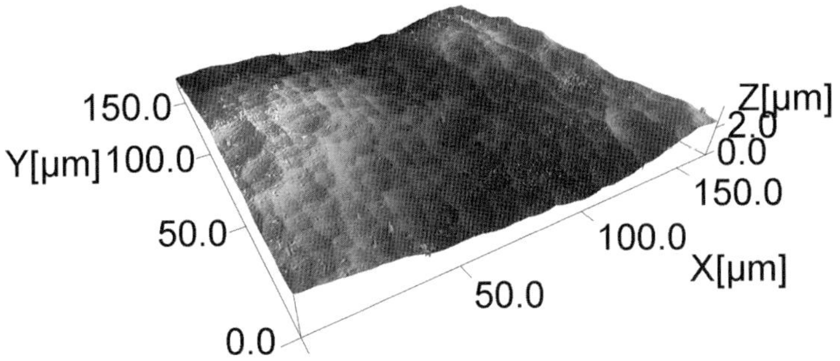

Figure 10.41 Qualitative example of PSI microscope acquisition (3D view) of an irregular artifact, using a 50× magnification objective with field of view 168 µm × 168 µm and minimum sampling distance of 0.16 µm

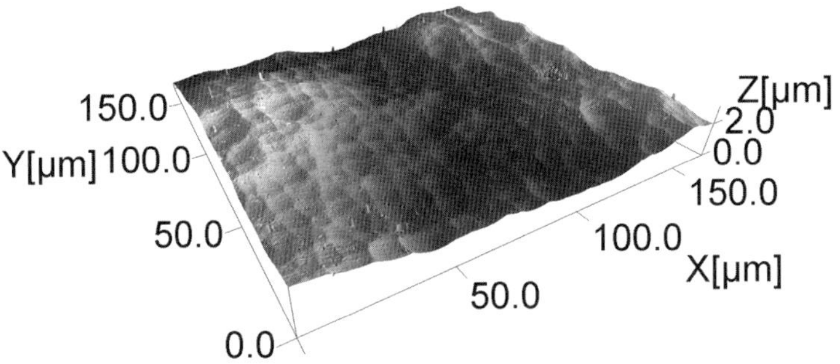

Figure 10.42 Qualitative example of CSI microscope acquisition (3D view) of an irregular artifact, using a 50× magnification objective with field of view 168 µm × 168 µm and minimum sampling distance of 0.16 µm

10.5 Uncertainty of Dimensional and Surface Topography Measurements of Micro Molded Parts and Micro Tools

The evaluation of the *measurement uncertainty* is fundamental in metrology and must express the results of a measurement session, "characterizing the dispersion of the quantity values being attributed to a measurand, based on the information used" [12] associated with a number of repeated measurements.

The official document dealing with the assessment of the measurement uncertainty is the *Guide to the expression of Uncertainty in Measurement*, commonly shortened to GUM [75]. It is based on the well-known frequentist approach and is, usually, considered a well-founded and exhaustive procedure.

According to the GUM, the ideal method for evaluating and expressing the uncertainty of a measurement result should be

- *Universal*: the method should be applicable to all kinds of measurements and to all types of input data used in measurements.

The actual quantity used to express uncertainty should be

- *Internally consistent*: it should be directly derivable from the components that contribute to it, as well as independent of how these components are grouped and of the decomposition of the components into subcomponents.
- *Transferable*: it should be possible to use directly the uncertainty evaluated for one result as a component in evaluating the uncertainty of another measurement in which the first result is used.

The method involves the identification of all sources of influence contributing to the uncertainty, which have to be estimated and combined in order to evaluate the final uncertainty. The contributors are related to a model equation

$$Y = f\left(X_1,\cdots,X_n\right) \tag{10.6}$$

where Y is the measurand and X_i are the input quantities.

The contributors of each source of uncertainty in Eq. 10.6 and other possible influence factors not in the model equation are evaluated according to two approaches:

1. Contributors estimated by statistical analysis of the observations (usually from repeated readings) are *Type A* evaluation.

   The mean value $\overline{x}$ of a number n of measurement results x_i is an estimate of the true value of the mean μ of the distribution

$$\overline{x} = \frac{1}{n}\sum_{i=1}^{n} x_i \tag{10.7}$$

   The experimental standard deviation of the distribution based on n repeated measurements u_x is an estimate of the standard deviation of the distribution σ

$$u_x = \sqrt{\frac{\sum_{i=1}^{n}\left(\overline{x}-x_i\right)^2}{n-1}} \tag{10.8}$$

   The standard deviation of the mean value $u_{\overline{x}}$ is equal to the standard deviation of the distribution divided by the square root of the number of measurements n

$$u_{\bar{x}} = \sqrt{\frac{\sum_{i=1}^{n}\left(\bar{x}-x_i\right)^2}{n(n-1)}} = \frac{u_x}{\sqrt{n}} \tag{10.9}$$

The experimental standard deviation in Eq. 10.8 is used to characterize the variability of the observed values x_i (single observation of n repeated measurements).

The experimental standard deviation of the mean in Eq. 10.9 is used in the uncertainty budget when the measurement result is obtained using the mean of m independent repeated observations $X_{i,k}$.

Contributors calculated from assigned probability distributions are *Type B* evaluation. The probability distribution can be identified from past measurement experience, calibration certificates, manufacturer's specifications, calculations, published information, and from common sense.

In Figure 10.43, some common distributions are given together with the corresponding uncertainty contributor.

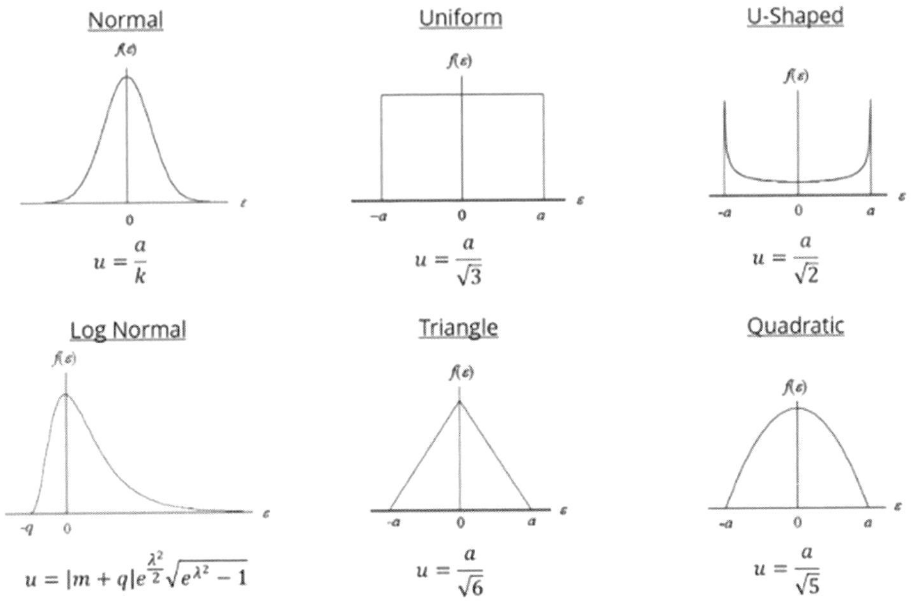

Figure 10.43 Experimental distributions for the Type B uncertainty evaluation

The combined standard uncertainty associated with the result is the positive square root of the combined variance, i. e., the variance obtained by combination of the all variances of the considered uncertainty contributors, using the law of the propagation of the uncertainty.

The combined variance for uncorrelated quantities is

$$u_c^2(y)=\sum_{i=1}^{n}\left(\frac{\partial f}{\partial x_i}\right)^2 u^2(x_i) \tag{10.10}$$

while the combined variance for correlated quantities is

$$u_c^2(y)=\sum_{i=1}^{n}\left(\frac{\partial f}{\partial x_i}\right)^2 u^2(x_i)+2\sum_{i=1}^{n-1}\sum_{j=i+1}^{n}\frac{\partial f}{\partial x_i}\frac{\partial f}{\partial x_j}u(x_i,x_j) \tag{10.11}$$

where $u(x_i,x_j)$ is the covariance with respect to the variables x_i and x_j.

Eventually, an expanded uncertainty U is evaluated extending the combined standard uncertainty of the measurement result to an interval that encompasses a large fraction p of the probability derived interpreting the experimental distribution as probability density function distribution, where p is the coverage probability or level of confidence of the interval. Hence, U is obtained multiplying the combined standard uncertainty u_c by a coverage factor k, corresponding to the desired level of confidence.

The estimate y of the measurand Y and its expanded uncertainty U are expressed as

$$y \pm U = y \pm k \cdot u_c \tag{10.12}$$

The procedure described in the GUM allows for the correction of the systematic behavior, which is performed through Eq. 10.4 [75]. Even so, such an equation is not always easy to determine and several repeated measurements (>15) are necessary per each measurand to correct for the systematics. For this reason, the GUM approach is considered complex and time consuming: it is, in fact, a method more suitable for calibration laboratories and National Metrology Institutes (NMIs) [16].

The GUM approach is beyond the purposes of this chapter. Nevertheless, Eq. 10.6–10.12 have general validity and can be used with any other procedure for the evaluation of the uncertainty and its propagation (see, e.g., Chapter 4, Section 4.3.3, Eq. 4.3). Furthermore, a comprehensive procedure for micro and nano geometrical and dimensional measurements of micro injection molded parts and micro injection molding tool inserts, based on the frequentist approach and in full agreement with the GUM, is proposed and validated elsewhere [42, 76–78].

In manufacturing, the *Procedure for Uncertainty Management*, known as PUMA [79], is a simplified approach to the GUM. It can be summarized with the following characteristics:

- "Approximated uncertainty".
- Upper bound strategy (overestimation).
- Intended for industrial users.

As explained in Section 10.1, the PUMA method is not considered suitable for micro injection molding applications, where an overestimation of the evaluated uncertainty can have a severe impact on tolerance verification.

Different general approaches for the uncertainty evaluation in micro and nano manufacturing are listed, e.g., in [14] and [80]. They are based, for the most part, on the "profile method", which, again, is deemed outdated here considering the trend toward large area acquisitions using optical instruments. Indeed, rules and procedures for the uncertainty evaluation of areal acquisitions are at an early stage. Besides the frequentist approach in micro and nano manufacturing already mentioned [42], a review of the methods available for areal topography measurement instruments is in [73].

Some indications are given in the recent ISO 25178 series (International Organization for Standardization) but the suggested contributors are mostly related to the instruments' metrological characteristics [40] and there are no suggestions about the measurand.

Other works propose a correction of the systematic behavior in an uncertainty evaluation achieved by the use of metrological characteristics [81] and uncertainty evaluation by the Monte Carlo method [82] (see also [83]).

Attempts to use the GUM approach, instead, can be found in [84], limited to the propagation of the uncertainty of correlated filtered profile roughness measurements, and in [85]. In this last work, the Type A uncertainty was directly considered for the measurand, i.e., the height (z values) of each pixel of the surface texture measurements (3D map of pixels), arriving at final equations that were difficult to manage because of the excessive amount of data.

Eventually, the approach known as *substitution method* [63] was conceived specifically for CMMs. The following applies to the method:

- A number of calibrated measurement standards.
- Repeated measurements on calibrated measurement standards, in the same conditions of the actual measurands (alignment, handling, etc.), for estimating the influence of the instrument.
- Repeated measurements on the actual measurands.
- Calculation of uncertainty by a given formula.

Details of the method are elsewhere [63]. However, a similar method is proposed in Section 10.5.1, which was inspired by such an approach. It is proposed here for its ease of use, and because specific knowledge is not required. The method is suggested for dimensional and topographic measurements performed by areal topography measurement instruments (measurements performed by μCMM can be processed according to [63]).

10.5.1 Assessment of the Uncertainty for Micro Injection Molding Applications

Following [63], two groups of uncertainty contributors are evaluated. One group is assessed measuring calibrated artifacts, in the same conditions as those of the actual measurands (alignment, handling, dimensions, etc.). These contributors estimate the influence of the instrument on the measurement process. The other group refers to the components under measurement. The complete uncertainty model is summarized in Figure 10.44. Each uncertainty contributor is explained in the following.

1. Contributors evaluated on calibrated artifacts:

   a) u_n. Background noise is an estimate of the disturbances introduced during the measurement, possibly due to the instrument (objective aberration, internal noise, etc.) or to the surrounding environment. It is evaluated here by measuring the surface of an optical flat or silicon carbide flat and averaging the S_z areal roughness parameter over at least fifteen repeated measurements (≥15) (this contributor is evaluated using S_z in both cases of dimensional and topography measurements). The uncertainty is evaluated from the average S_z, maximum height of the scale-limited surface, considered to be uniformly distributed (see Figure 10.43 and Figure 10.44). A more accurate assessment would be the evaluation of the metrological characteristics *measurement noise* and *residual flatness*, as they are defined in [40]. The procedure for their evaluation can be found elsewhere [66].

   b) u_p. The instrument repeatability (called measuring procedure in Figure 10.44 because it may be dependent on the operator) is assessed on calibrated standards and on the measurands of interest, according to the components under measurement. For example, if a dimension of 100 μm is to be measured on a certain component, this contributor is evaluated on a calibrated artifact with a nominal dimension very close to 100 μm (better if it is the same). Hence, the standard deviation of the mean (see Eq. 10.9) is calculated over at least fifteen repeated measurements per measurand (≥15).

   c) u_{cal}. This is the combined uncertainty stated in the calibration certificate of all the calibrated standards used (e.g., related to u_n and u_p). It is the expanded uncertainty U divided by the coverage factor (see Eq. 10.12). If the coverage factor is not indicated in the calibration certificate, then the choice is $k = 2$.

d) u_b. This contributor takes into account the effect of temperature on the standard. The formula used is shown in Figure 10.44. It considers the relative temperature deviation with respect to the reference temperature of 20 °C (393.15 K) multiplied by the average measurand relative to *the* calibrated standards and multiplied again by the uncertainty of the coefficient of thermal expansion (CTE) u_α. A straightforward way for estimating u_α is to run a search in books, manuals, on the web, etc., and to choose the maximum and minimum values found. In fact, it is likely to found slightly different values since the CTE of a certain material is not easy to determine univocally. Hence, the CTE uncertainty is calculated by the semi amplitude of the maximum variation considered to be uniformly distributed (see Figure 10.43).

2. Contributors evaluated on specimens under measurement:

   a) u_{wt}, similar to u_b, takes into account the effect of temperature on the part under measurement. It is evaluated by the same formula of u_b, except for the average measurand that is relative to the component under measurement and for u_α, which is also related to the material of the component under measurement.

   b) u_{wp} takes into account the variability due to the manufacturing process calculating the uncertainty contributor as the semi amplitude of the maximum variation in the measurements, considered to be uniformly distributed (see Figure 10.44). According to what has already been specified in Section 10.2.1.1, the measurements are different produced components chosen from different batches of production. However, in the case of tool inserts for micro injection molding, their number is normally reduced, if not unique. Hence, the variability in that case can be obtained by several repeated measurements for each measurand (at least fifteen measurements (≥15)).

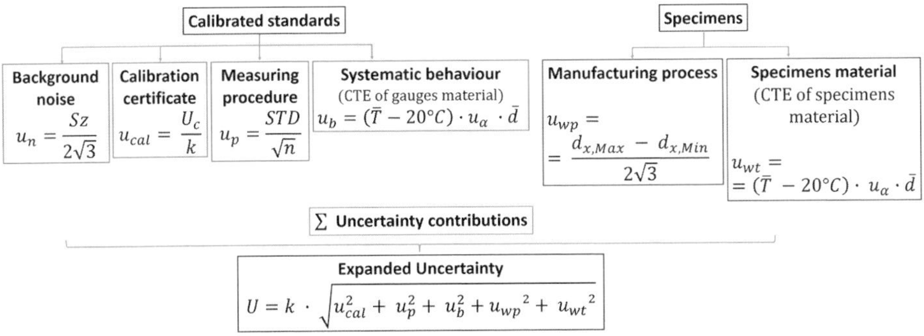

Figure 10.44 Uncertainty model for assessment in micro injection molding

Eventually, the combined uncertainty is calculated summing in quadrature all the contributors, according to the formula in Figure 10.44. The expanded uncertainty is calculated by a coverage factor $k = 2$, which approximately corresponds to a confidence level of 95% (see Eq. 10.12).

Considering again the example of the replication of sub-micro structured polymer surfaces in Section 10.3.2.1 (Figure 10.11 to Figure 10.14), the results and a complete uncertainty budget are given in Table 10.1.

Table 10.1 Averages, Expanded Uncertainties, and Uncertainty Contributors (See Figure 10.44) of the Unfiltered AFM Measurements

		Nickel masters		Polymer replicas	
		503/nm	529/nm	503/nm	529/nm
Average	S_a	83	117	64	108
Background noise	u_n	0.3	0.3	0.3	0.3
Calibration certificate	u_{cal}	0.7	0.7	0.7	0.7
Procedure	u_p	1.0	1.0	1.0	1.0
Uncertainty due to systematics	u_b	0.0	0.0	0.0	0.0
Workpiece–manufacturing	u_{wp}	0.7	0.2	1.4	1.4
Workpiece–material	u_{wt}	0.0	0.0	0.0	0.0
Expanded uncertainty	U	3	3	4	4

References

[1] Heckele, M., Schomburg, W. K., *Review on micro molding of thermoplastic polymers. J. Micromech. Microeng.* (2004) 14, pp. R1–14

[2] Hansen, H. N., Hocken, R. J., Tosello, G., *Replication of micro and nano surface geometries. CIRP Ann. - Manuf. Technol.* (2011) 60, pp. 695–714

[3] Tosello, G., Hansen, H. N., Marinello, F., Gasparin, S., *Replication and dimensional quality control of industrial nanoscale surfaces using calibrated AFM measurements and SEM image processing. CIRP Ann. - Manuf. Technol.* (2010) 59(1), pp. 563–568

[4] Tosello, G., Marinello, F., Hansen, H. N., *Characterisation and analysis of microchannels and submicrometre surface roughness of injection moulded microfluidic systems using optical metrology. Plast., Rubber Compos.: Macromol. Eng.* (2012) 41(1), pp. 29–39

[5] Hansen, H. N., Carneiro, K., Haitjema, H., De Chiffre, L., *Dimensional Micro and Nano Metrology. CIRP Ann. - Manuf. Technol.* (2006) 55(2), pp. 721–743

[6] ISO 14253-2:2011 Geometrical product specification (GPS)—Inspection by measurement of workpieces and measuring equipment – Part 2: Guidance for the estimation of uncertainty in GPS measurement, in calibration of measuring equipment and in product verification. ISO 14253 part 2 (Geneva: International Organization for Standardization)

[7] Tosello, G., Hansen, H. N., Calaon, M., Gasparin, S., *Challenges in high accuracy surface replication for micro optics and micro fluidics manufacture. Int. J. Precis. Technol.* (2014) 4, pp. 122–144

[8] Kunzmann, H., Pfeifer, T., Schmitt, R., Schwenke, H., Weckenmann, A., *Productive Metrology - Adding Value to Manufacture. CIRP Ann. - Manuf. Technol.* (2005) 54, 155-168

[9] Weckenmann, A., Rinnagl, M., *Acceptance of processes: do we need decision rules? Precis. Eng.* (2000) 24, pp. 264-269

[10] Leach, R., *Characterisation of Areal Surface Texture*, Leach, R. (Ed.) (2013) Springer Heidelberg, Berlin

[11] Leach, R., *Optical Measurement of Surface Topography*, Leach, R. (Ed.) (2011) Springer Heidelberg, Berlin

[12] JCGM 200:2012 International vocabulary of metrology–Basic and general concepts and associated terms (VIM), Joint Committee for Guides in Metrology (Sèvres: Bureau International des Poids et Mesures, BIPM)

[13] www.bipm.org/en/measurement-units

[14] De Chiffre, L., *Geometrical metrology and machine testing* (2011) Technical University of Denmark, Kgs. Lyngby

[15] Gasparin, S., *Verification of Tolerance Chains in Micro Manufacturing* (2012) Technical University of Denmark, Denmark, Doctoral Thesis

[16] www.bipm.org/en/worldwide-metrology

[17] ISO 5436-1:2000 Geometrical product specification (GPS)–Surface texture: Profile method; Measurement standards - Part 1: Material measures. ISO 5436 part 1 (Geneva: International Organization for Standardization)

[18] Evans, C. J., Davies, A. D., *Certification, self-calibration and uncertainty in optical surface testing. Int. J. Precis. Technol.* (2013) 3(4), pp. 388-402

[19] Zhao, J., Mayes, R. H., Chen, G., Xie, H. Chan, P. S., *Effects of process parameters on the micro molding process. Polym. Eng. Sci.* (2003) 43, pp. 1542-1554

[20] Aufiero, R., *The effect of process conditions on part quality control in microinjection molding. Annual technical conference ANTEC* (2005) Boston, MA, pp. 3721-3725

[21] ISO 14638:2015 Geometrical product specification (GPS)–Matrix model. ISO 14638 (Geneva: International Organization for Standardization)

[22] ISO 14253-1:2013 Geometrical product specification (GPS)–Inspection by measurement of workpieces and measuring equipment - Part 1: Decision rules for proving conformity or nonconformity with specifications. ISO 14253 part 1 (Geneva: International Organization for Standardization)

[23] Weckenmann, A., Wiedenhoefer, T., *The use of the GPS-Standard in Nanometrology. Proc. of 5th euspen international conference* (2005), pp. 169-172

[24] Tosello, G., Hansen, H. N., Gasparin, S., *Applications of dimensional micro metrology to the product and process quality control in manufacturing of precision polymer micro components. CIRP Ann. - Manuf. Technol.* (2009) 58, pp. 467-472

[25] MountainsMap® Premium–www.digitalsurf.com

[26] SPIP™–www.imagemet.com

[27] NIST reference software–https://physics.nist.gov/VSC/jsp/About.jsp

[28] PTB reference software–www.ptb.de/cms/en/ptb/fachabteilungen/abt5/fb-51/ag-515/referenzsoftware515.html

[29] NPL reference software–www.npl.co.uk/measurement-services/dimensional/surface-topography-and-nano-metrology-measurements

[30] ISO 25178-70:2014 Geometrical product specification (GPS)–Surface texture: Areal - Part 70: Material measures. ISO 25178 part 70 (Geneva: International Organization for Standardization)

[31] Tosello, G., Hansen, H. N., Marinello, F., Gasparin, S., *Replication and dimensional quality control of industrial nanoscale surfaces using calibrated AFM measurements and SEM image processing. CIRP Ann. - Manuf. Technol.* (2010) 59(1), pp. 563-568

[32] Tosello, G., Marinello, F., Hansen, H. N., *Multiscale optical metrology for the quality control of polymer microfluidic systems. 9th euspen International Conference* (2009) San Sebastian, Spain

[33] Tosello, G., Bissacco, G., Tang, P.T., Hansen, H. N., Nielsen, P.C., *High aspect ratio micro tool manufacturing for polymer replication using μEDM of silicon, selective etching and electroforming. Microsyst. Technol.* (2008) 14, pp. 1757–1764

[34] ISO 4287:1997 Geometrical Product Specifications (GPS)—Surface texture: Profile method – Terms, definitions and surface texture parameters. ISO 4287 (Geneva: International Organization for Standardization)

[35] ISO 4288:1996 Geometrical Product Specifications (GPS)—Surface texture: Profile method – Rules and procedures for the assessment of surface texture. ISO 4288 (Geneva: International Organization for Standardization)

[36] ASME B46.1-2009 Introduction to Surface Texture Measurement and Analysis (New York: The American Society of Mechanical Engineers)

[37] Leach, R.K., Harris, P.M., *Ambiguities in the definition of spacing parameters for surface-texture characterization. Meas. Sci. Technol.* (2002) 13, pp. 1924–1930

[38] ISO 25178-2:2012 Geometrical product specification (GPS)—Surface texture: Areal – Part 2: Terms, definitions and surface texture parameters. ISO 25178 part 2 (Geneva: International Organization for Standardization)

[39] Leach, R., Haitjema, H., *Bandwidth characteristics and comparisons of surface texture measuring instruments. Meas. Sci. Technol.* (2010) 21, pp. 1–9

[40] ISO/DIS 25178-600: 2016 Geometrical product specification (GPS)—Surface texture: Areal – Part 600: Nominal characteristics of non-contact (variable focus) instruments. ISO DIS 25178 part 600 (Geneva: International Organization for Standardization)

[41] ISO 25178-3: 2012 Geometrical product specification (GPS)—Surface texture: Areal – Part 3: Specification operators. ISO 25178 part 3 (Geneva: International Organization for Standardization)

[42] Quagliotti, D., *Multi Scale Micro and Nano Metrology for Advanced Precision Moulding Technologies* (2017) Technical University of Denmark, Denmark, Doctoral Thesis

[43] Tosello, G., Haitjema, H., Leach, R.K., Quagliotti, D., Gasparin, S., Hansen, H. N., *An international comparison of surface texture parameters on polymer artefacts using optical instruments. CIRP Ann. – Manuf. Technol.* (2016) 65, pp. 529–532

[44] Quagliotti, D., Tosello, G., Hansen, H. N., *Replication assessment of nanostructured polymer surfaces using atomic force microscopy: Uncertainty evaluation in the surface replication fidelity assessment of moulded specimens at the 100 nm scale. VDI-Berichte – Messunsicherheit praxisgerecht bestimmen – Prüfprozesse in der industrielle Praxis* (2015) 2269, pp. 49–55

[45] Leach, R., *Fundamental principle of engineering nanometrology* (2014), Elsevier

[46] ISO 25178-6: 2012 Geometrical product specification (GPS)—Surface texture: Areal – Part 6: Classification of methods for measuring surface texture. ISO 25178 part 6 (Geneva: International Organization for Standardization)

[47] Whitehouse, D., *Handbook of Surface and Nanometrology* (2011) Second Edition, CRC Press

[48] Rai-Choudhury, P., *Handbook of microlithography, micromachining, and microfabrication* (1997) volume I, SPIE Optical Engineering Press

[49] Goodhew, J.P., Humphrey, J., Beanland, R., *Electron microscopy and analysis* (2001) Third Edition, CRC Press

[50] Mani, M.R., Surace, R., Ferreira, P., Segal, J., Fassi, I., Ratchev, S., *Process Parameter Effects on Dimensional Accuracy of Micro-Injection Moulded Part. J. Micro Nano-Manuf.* (2013) 1, p. 31003

[51] Schwenke, H., Wäldele, F., Weiskirch, C., Kunzmann, H., *Opto-tactile Sensor for 2D and 3D Measurement of Small Structures on Coordinate Measuring Machines. CIRP Ann. – Manuf. Technol.* (2001) 50, pp. 361–364

[52] http://www.werth.de/en/unser-angebot/products/multisensor-cmms/

[53] Carmignato, S., *Traceability of dimensional measurements in computed tomography. Proceedings of 8th A.I.Te.M. Conference* (2007)

[54] Weckenmann, A., Kramer, P., *Predetermination of measurement uncertainty in the application of computed tomography. 11th CIRP Int. Conf. Computer Aided Tolerancing* (2009)

[55] Kruth, J.P., Bartscher, M., Carmignato, S., Schmitt, R., De Chiffre, L., Weckenmann, A., *Computed tomography for dimensional metrology. CIRP Ann. - Manuf. Technol.* (2011) 60, pp. 821–842

[56] Stolfi, A., *Integrated Quality Control of Precision Assemblies using Computed Tomography* (2017) Technical University of Denmark, Denmark, Doctoral Thesis

[57] Jiménez, R., Torralba, M., Yagüe-Fabra, J., Ontiveros, S., Tosello, G., *Experimental Approach for the Uncertainty Assessment of 3D Complex Geometry Dimensional Measurements Using Computed Tomography at the mm and Sub-mm Scales. Sensors* (2017) 17, p. 1137

[58] Kak, A.C., Slaney, M., *Principles of computerized tomographic imaging* (1988), IEEE press, New York

[59] Bartscher, M., Hilpert, U., Goebbels, J., Weidemann, G., *Enhancement and Proof of Accuracy of Industrial Computed Tomography (CT) Measurements. CIRP Ann. - Manuf. Technol.* (2007) 56, pp. 495–498

[60] Schwenke, H., Neuschaefer-Rube, U., Pfeifer, T., Kunzmann, H., *Optical Methods for Dimensional Metrology in Production Engineering. CIRP Ann. - Manuf. Technol.* (2002) 51, pp. 685–699

[61] Yagüe, J.A., Tosello, G., Carmignato, S., Ontiveros, S., Jiménez, R., Gasparin, S., Hansen, H.N., Pierobon, A., *Measurement of micro moulded parts by computed tomography and comparison to optical and tactile techniques. Proc. of the 11th International Conference of the European Society for Precision Engineering and Nanotechnology (euspen)* (2011) 1, pp. 76–79

[62] Ontiveros, S., Yagüe-Fabra, J., Jiménez, R., Tosello, G., Gasparin, S., Pierobon, A., Carmignato, S., Hansen, H.N., *Dimensional measurement of micro-moulded parts by computed tomography. Meas. Sci. Technol.* (2012) 23, p. 125401

[63] ISO 15530-3:2011 Geometrical product specifications (GPS)—Coordinate measuring machines (CMM): Technique for determining the uncertainty of measurement – Part 3: Use of calibrated workpieces or measurement standards. ISO 15530 part 3 (Geneva: International Organization for Standardization)

[64] Qian, J., Kempen, K., Welkenhuyzen, F., Vanderauwera, W., Kuijpers, A., Wang, J., Yang, F., Kruth, J.P., Reynaerts, D., *Manufacturing of μIM Mould Inserts with AMed Cooling Channels. Proceedings of the 4M/ICOMM2015 Conference* (2015) Research Publishing Services, Singapore, pp. 558–561

[65] Neuschaefer-Rube, U., Bartscher, M., Neukamm, M., Neugebauer, M., Härtig, F., Goebbels, J., Ehrig, K., Staude, A., *Measurement of micro gears: comparison of optical, tactile-optical, and CT measurements*, Beraldin, J.A., Cheok, G.S., McCarthy, M.B., Neuschaefer-Rube, U., Baskurt, A.M., McDowall, I.E., Dolinsky, M. (Ed.) (2011) 7864:78640H, IS&T/SPIE Electronic Imaging, 2011, San Francisco Airport, California, United States

[66] Rathore, J.S., Carmignato, S., Lucchetta, G., Improving CT quality control of a micro injection molded part by multisensor data fusion. *6th Conference on Industrial Computed Tomography (iCT 2016)* (2016) Wels, Austria, pp. 4–10

[67] De Chiffre, L., Carmignato, S., Kruth, J.P., Schmitt, R., Weckenmann, A., *Industrial applications of computed tomography. CIRP Ann. - Manuf. Technol.* (2014) 63, pp. 655–677

[68] Leach, R.K., Giusca, C.L., Haitjema, H., Evans, C., Jiang, X., *Calibration and verification of areal surface texture measuring instruments. CIRP Ann. - Manuf. Technol.* (2015) 64, pp. 797–813

[69] Giusca, C.L., Leach, R.K., Helary, F., Gutauskas, T., Nimishakavi, L., *Calibration of the scales of areal surface topography-measuring instruments: part 1. Measurement noise and residual flatness. Meas. Sci. Technol.* (2012) 23, p. 35008

[70] Giusca, C. L., Leach, R. K., Helary, F., *Calibration of the scales of areal surface topography measuring instruments: part 2. Amplification, linearity and squareness. Meas. Sci. Technol.* (2012) 23, p. 65005

[71] Giusca, C. L., Leach, R. K., *Calibration of the scales of areal surface topography measuring instruments: part 3. Resolution. Meas. Sci. Technol.* (2013) 24, p. 105010

[72] Wilson, T., *Theory and practice of scanning optical microscopy* (1984), Academic Press

[73] Danzl, R., Helmli, F., Rubert, P., Prantl, M., *Optical roughness measurements on specially designed roughness standards. Proceedings of SPIE,* Duparré, A., Geyl, R. (Ed.) (2008) 7102, pp. 71020M

[74] www.npl.co.uk/nanometrology

[75] JCGM 100:2008 Evaluation of measurement data–Guide to the expression of uncertainty in measurement, Joint Committee for Guides in Metrology (Sèvres: Bureau International des Poids et Mesures, BIPM)

[76] Quagliotti, D., Tosello, G., Islam, A. M., Hansen, H. N., Zeidler, H., Martin, A., Schubert, A., Brandao, C., Riemer, O., *A method for dimensional and surface optical measurements uncertainty assessment on micro structured surfaces manufactured by Jet-ECM. 4M 2015 Proc. of the International Conference on Micro-Manufacturing Milan* (2015) Italy, 31 March–2 April

[77] Quagliotti, D., Tosello, G., Islam, A. M., Hansen, H. N., Zeidler, H., Martin, A., Schubert, A., Brandao, C., Riemer, O., *Optical micro-metrology of structured surfaces micro-machined by jet-ECM. Proc. of the 15th international conference of the european society for precision engineering and nanotechnology (euspen)* Leach, R. (Ed.) (2015) Leuven, Belgium, pp. 167–168

[78] Quagliotti, D., Baruffi, F., Tosello, G., Gasparin, S., Annoni, M., Parenti, P., Sobiecki, R., Hansen, H. N., *Correction of systematic behaviour in topographical surface analysis. Proc. of the 4M/IWMF2016 Conference,* Tosello, G., Hansen, H. N., Kornel, E., Dimov, S. (Ed.) (2016) Copenhagen, DK, 13–15 September, pp. 277–280

[79] ISO 14253-2:2011 Geometrical product specification (GPS)–Inspection by measurement of workpieces and measuring equipment – Part 2: Guidance for the estimation of uncertainty in GPS measurement, in calibration of measuring equipment and in product verification. ISO 14253 part 2 (Geneva: International Organization for Standardization)

[80] www.npl.co.uk/publications/guides/dimensional-good-practice-guides

[81] MacAulay, G. D., Giusca, C. L., *Assessment of uncertainty in structured surfaces using metrological characteristics. CIRP Ann. – Manuf. Technol.* (2016) 65, pp. 533–536

[82] Leach, R. K., Giusca, C. L., Naoi, K., *Development and characterization of a new instrument for the traceable measurement of areal surface texture. Meas. Sci. Technol.* (2009) 20, p. 125102

[83] JCGM 101:2008 Evaluation of measurement data–Supplement 1 to the "Guide to the expression of uncertainty in measurement"–Propagation of distributions using a Monte Carlo method, Joint Committee for Guides in Metrology (Sèvres: Bureau International des Poids et Mesures, BIPM)

[84] Krystek, M., *Measurement uncertainty propagation in the case of filtering in roughness measurement. Meas. Sci. Technol.* (2001) 12, pp. 63–67

[85] Haitjema, H., *Uncertainty in measurement of surface topography. Surf. Topogr.: Metrol. Prop.* (2015) 3, p. 35004

11 Additive Manufacturing for Micro Tooling and Micro Part Rapid Prototyping

Ali Davoudinejad, David Bue Pedersen, Guido Tosello

This chapter highlights some general additive manufacturing (AM) methods, applications, and technologies, and relates these to tooling for micro injection molding and micro plastic parts. AM is represented by a family of processes that fabricate the workpiece directly from 3D model data by means of a layer-based consolidation, in contrast to conventional subtractive and forming manufacturing methods. AM processes have been developed in a variety of technologies; however, only specific methods of such processes are appropriate for the fabrication of functional and 3D micro components. The chapter will show how different AM methods are used for tooling, by linking these processes to mass production. It will present micro injection molding hard and soft tooling applications and AM for direct micro component manufacturing.

11.1 Additive Manufacturing Process Technologies and Materials

The capability to manufacture a wide variety of components with a time- and cost-effective method is the main competitive objective of the industrial market. Rapid prototyping (RP) is an essential technique in product design and development, and it has been used in many fields (medical, product development, engineering etc.). RP is a dominant communication tool that bridges together design, marketing, process planning, and manufacturing. Originally, the field began as a means to produce prototype mechanical parts rapidly and, hence, was called "rapid prototyping". Rapid prototyping was utilized successfully for assembly process planning. These technologies give the designer the ability to evaluate the outline of the product, validate the fitness of part assembly, and test the functionality of the product model. Most commercial rapid prototyping processes are based on layer-by-layer additive manufacturing of horizontal cross sections from process materials in order to complete the component manufacture. More recently, RP has been utilized for much more than just making prototypes, and the term "additive manufacturing", or AM, was used to highlight the additive nature of the processes and their growing potential for applications in direct component manufacturing. AM processes with the fundamental ability to fabricate three-dimensional parts with high aspect ratios are thus becoming standard methods in part production and rapid tooling applications [1–4].

Though AM processes fabricate parts by adding material in contrast to, e.g., machining that removes material for part production, the emphasis in AM is on the capability of the processes to fabricate complex geometric shapes, both internal and external to the outer surface of a component. AM processes have a lot in common, mainly in terms of general procedure.

The process starts with a 3D model input to the system, which is translated into the additive manufacturing file format (AMF) or the de facto standard file format standard tessellation language (STL). The converted file is processed by an AM machine-specific job generator that can be compared to the post processor used for NC machining. Here, the software slices the solid body layer by layer, and a sequence of processing instructions is sent to the AM equipment where the job is finally executed. Depending on the AM technology, post-processing could be required. Currently, most AM processes work in a layerwise fashion and encompass many materials. The surface finish is affected by the layer thickness and varies according to the surface inclination; the orientation in the workspace can also influence the functionality of certain surfaces. In 2012, the International Committee F42 on Additive Manufacturing Technologies categorized the AM processes into seven areas as follows [5, 6]:

- Material extrusion: an additive manufacturing process in which material is selectively dispensed through a nozzle or orifice (e.g., fused deposition modeling, FDM).
- Material jetting: an additive manufacturing process in which droplets of build material are selectively deposited. Example materials include photopolymers and wax (e.g., Polyjet).
- Binder jetting: an additive manufacturing process in which a liquid bonding agent is selectively deposited to join powder materials (e.g., some 3D printers by powder and binder).
- Sheet lamination: an additive manufacturing process in which sheets of material are bonded to form an object (e.g., sheet forming).
- Vat photopolymerization: an additive manufacturing process in which a liquid photopolymer in a vat is selectively cured by light-activated polymerization (e.g., stereolithography, SLA).

- Powder bed fusion: an additive manufacturing process in which thermal energy selectively fuses regions of a powder bed (e.g., selective laser sintering, SLS).
- Directed energy deposition: an additive manufacturing process in which focused thermal energy is used to fuse materials by melting as they are being deposited. "Focused thermal energy" means that an energy source (e.g., laser, electron beam, or plasma arc) is focused to melt the materials being deposited.

There are three hierarchy levels for AM standards as shown in Figure 11.1. General top-level AM standards: general concepts, common requirements, generally applicable. Category AM standards: specific to materials or process category. Specialized AM standards: specific to material, process, or application [6].

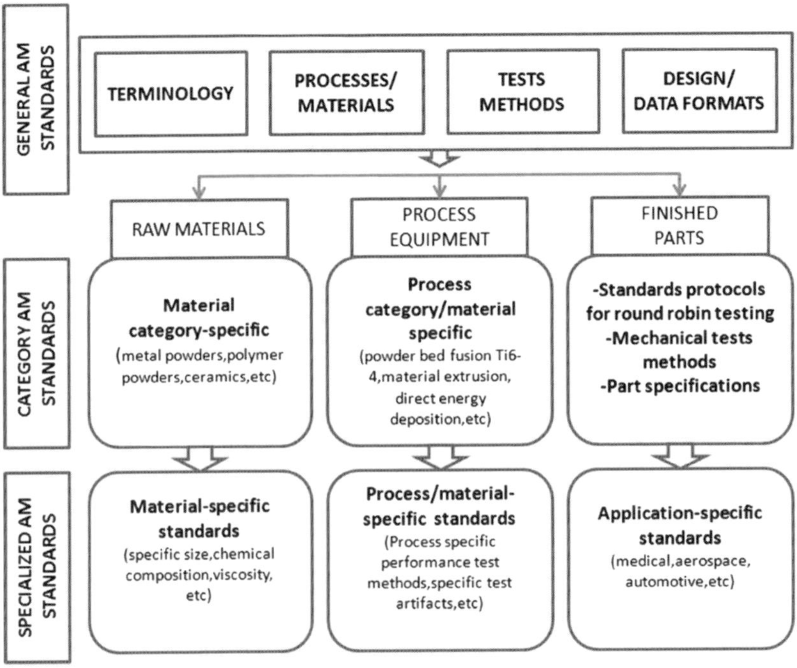

Figure 11.1 Structure of AM standards agreed by ASTM-ISO [6]

Another potential of AM processes is the ability to produce components with a variety of materials that are difficult to process in conventional machining process such as hard metals, ceramics, composites, etc. During the past few decades, remarkable progress has been achieved in developing a range of materials for additive manufacturing processes [7]. In addition to the two main material categories of plastics and metal, a variety of filled and composite materials are also available, as well as polymers (thermoplastics, thermosets, photopolymers), carbon, glass, metal, and wood composite polymers, ceramics and ceramic-metal hybrids. Different additive manufacturing techniques can produce a variety of parts in these material ranges [7]. Broad ranges of biocompatible materials are also processed by AM machine tools and can be used to fabricate medical devices [8]. Ceramics with superior material strength and thermal and wear resistance are capable of fulfilling demands in various medical applications [9]. Ceramics display difficulties in machining and casting; however, AM enables a big leap in geometrical flexibility [10].

The selection of appropriate AM processes depends on different factors such as the ease of manufacturing, speed, fabrication cost, part design, material, process limitations, surface finish, and dimensional accuracy of the part. The dimensional accuracy ranges of the various AM processes for polymer and metal materials are presented in Figure 11.2, which shows the lowest and highest values expressed by the machine manufacturers [11]. Additionally, processes such as injection molding (IM), investment casting, electro-discharge machining (EDM), and milling are presented to provide better comparison. The dimensional accuracy values are assumed to represent the best case of build parameters and working conditions of the machine used.

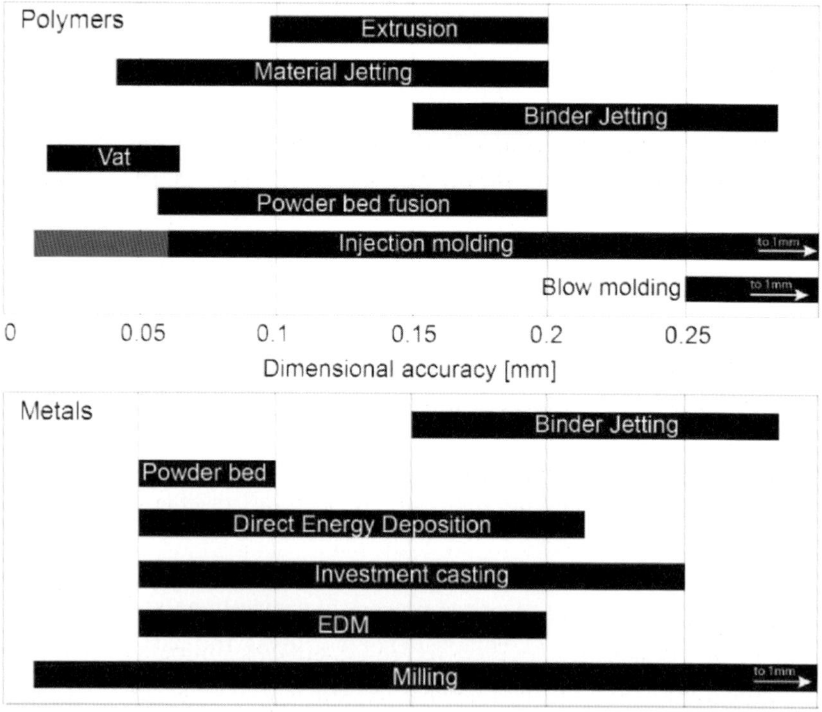

Figure 11.2 Comparison of AM and CM processes in respect of dimensional accuracy [11]

11.1.1 Additive Manufacturing Methods for Polymer Materials

Polymer materials with low melting points or those in a liquid state at room temperature are widely used in the AM industry because of their low weight, low cost, and processing flexibility. Despite the geometric complexity of the manufactured polymer products, lack of mechanical strength and functionality is a big challenge for their wide applications. To obtain desired mechanical and functional properties, different materials are combined to overcome some of these problems [12]. Different AM methods have been employed for printing polymer composite materials; some well-recognized techniques to fabricate polymer parts are as follows:

- Fused Deposition Modeling (FDM®) or fused filament modeling (FFM): this is a common method used for thermoplastic materials with low melting points. It works by extrusion of thin (sub-1-mm) thermoplastic filaments of molten thermoplastic (normally from a feedstock of a larger filament or powder) through a heated nozzle onto a room-temperature or heated build platform and the part quality is affected by printing parameters [13, 14]. There are very few reports on the utilization of extrusion-based systems for fabrication of true complex 3D micro parts. Some limitations for applying extrusion-based systems with material melting in micro fabrication areas still exist. However, FDM was used for micro periodic structures such as tissue engineering (TE) scaffolds with micro-sized rods/pores in poly(e-caprolactone) (PCL), polypropylene/tricalcium phosphate (PP/TCP), polycaprolactone/hydroxylapatite (PCL/HA), PCL/TCP, PP/TCP, and poly(lactide-coglycolide) (PLGA) with a resolution of 250 μm [15–17].
- Powder bed additive manufacturing and inkjet head 3D printing: this method is based on powder processing. Powders are primarily spread on the build platform and are selectively joined into a patterned layer by depositing a liquid binder through an inkjet print head, which is able to move in the X and Y directions. After a chosen 2D pattern is formed, the platform lowers, and the next layer of powder is spread. This process is repeated to fabricate the final part and at the end of process, the unbounded powder should be removed. The quality of the final parts is affected by powder size, binder viscosity, the interaction between the binder and the powder, and the binder deposition speed [18, 19]. Figure 11.3 shows sample micro parts with 400 μm blade thicknesses, which have been fabricated via different commercially available inkjet-based systems, including Eden 260 V, ProJet HD 3000plus, as well as Solidscape's T76 3D printer, which is a commercial piezoelectric DOD micro AM system for printing polymer inks [20].
- Stereolithography (SL): this method uses photopolymers that can be cured by UV laser. The UV laser is controlled in a way to shoot in the resin reservoir, and the photocurable resin will polymerize into a 2D patterned layer. After each layer is cured, the platform lowers, and another layer of uncured resin is ready to be patterned. The process takes place at room temperature and does not require a controlled atmosphere [21].

- Selective Laser Sintering (SLS®): in this method, a high-temperature laser is applied to sinter layers of fine powders selectively. After each layer is produced, a roller spreads a fresh layer of powder on the bed and the process repeats until the part is completed. The layer information of the part is used to control a laser scanner/controller that directs to reproduce the mathematical layer by selectively sintering or fusing polymer powder. Typical layer thicknesses are controlled to lie in the 75–125 μm range [22]. SLS® is a similar technique to 3D printing; however, rather than using a liquid binder, in SLS®, a laser beam with a controlled path scans the powders to sinter them by heating [23]. This method is capable of producing micro components from ceramics with an aspect ratio of up to 12, a resolution of less than 30 μm, and a minimum surface roughness of 1.5 μm [22, 24].
- Laser Beam Melting (LBM): this method was first used for building prototypes or parts in a small series of polymers in 1986 [25] by melting preheated polymer powders layerwise by selective energy deposition [26].
- 3D plotting/direct-write: this method works based on extruding a viscous material from a pressurized syringe to create 3D shapes of materials that are added layer by layer. Curing reactions can be performed by dispensing two reactive components using mixing nozzles or can be induced either by heat or UV light [27].
- Vat photopolymerization: an AM process in which liquid photopolymer in a vat is selectively cured by light activated polymerization. For example, Direct Light Processing (DLP) uses projector light to solidify a liquid photopolymer. By changing the light pattern and vertical position of the workpiece, the favored geometry is built up layer by layer [28].
- Liquid Deposition Modeling (LDM): in this method, 3D micro structures with arbitrary shapes are produced via the deposition of a new conductive nanocomposite from liquid dispersion by means of a low-cost commercial benchtop 3D printer equipped with a syringe dispenser [29].
- Fiber Encapsulation Additive Manufacturing (FEAM): this method enables fibers and extrudable matrix material to be printed simultaneously in a single, potentially low cost, machine [30].
- Direct Laser Writing (DLW): multiphoton polymerization (MPP) is a three-dimensional (3D) printing technology that allows the construction of readily assembled structures with sub-100-nm resolution [31]. Two-Photon Polymerization (2PP) has become a significant process for polymers with a high resolution. 2PP is the reaction of photopolymerization induced by two-photon absorption (2PA), enabling the creation of 3D structures with sub-micron features. The efficiency depends highly on the square of the intensity of the laser light so an ultra-fast laser is used instead of a conventional UV laser [32].

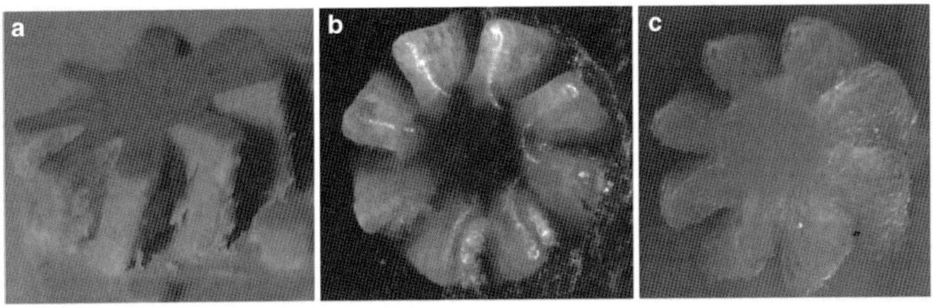

Figure 11.3 Sample micro parts with 400 µm blade thickness produced using different inkjet printing systems: (a) Solidscape's T76, (b) 3D System's ProJet HD 3000plus, and (c) Objet's Eden 260 V 3D printer [20]

11.1.2 Additive Manufacturing Methods for Metal Materials

It has become possible to manufacture dense parts with certain AM processes and for a number of metallic materials, such as steel, aluminum, and titanium. Metal AM technology is divided into two main categories: powder bed fusion (PBF) and directed energy deposition (DED). PBF contains the processes in which focused energy (electron beam or laser beam) is used to selectively melt or sinter a layer of a powder bed. DED includes all processes in which focused energy generates a melt pool into which feedstock is deposited. This process can use a laser, arc, or e-beam as a heat source [33]. Various metal AM processes are currently available, depending on the heat source, such as electron beam, laser, or arc, and on how the raw material is supplied, as discussed in the following:

- Laser Beam Melting (LBM): this process is also known as Selective Laser Melting (SLM). LBM is a powder-bed-based process in which metal powder is spread in thin layers with a typical layer thickness of 20–100 µm. The spot sizes of the laser beam in the focal plane are between 50 µm and 180 µm depending on the manufacturing system used [34, 35]. Stainless steel was produced by SLM with a particle size below 50 µm and the micro structure was investigated by optical microscopy [36].
- Electron Beam Melting (EBM): similar to the LBM method, a powder bed is created, and metal powder is fed from a hopper and distributed by a rake across a build plate. Commonly, the powder layer thickness amounts to 50–200 µm [34, 35].
- Laser Metal Deposition (LMD): this procedure is based on wire- or powder-feeding systems. Generally, powder is used, which is fused to the workpiece or substrate in a melting pool and, therefore, forms a strong metallurgical bond. Currently, the minimum thickness of cladded tracks is in the range of 30 µm [37].

- Selective Laser Sintering (SLS®): the aforementioned SLS® technique is used for generating micro metal components with a structural resolution of less than 30 µm. The grain size of the powder must be in the sub-micrometer range to reach the required resolution. Powder materials such as tungsten (300 nm grain size), aluminum (3 µm grain size), copper (10 µm grain size), and silver (2 µm grain size) were used; moreover, metal sintering was performed with mixtures of copper and tungsten [38].
- Direct Metal Laser Sintering (DMLS®): in this method, laser sintering is carried out using a metal powder so that metal parts are produced directly in the building process. This method enabled the production of complex parts with high accuracy and good surface quality and opened up the way to the commercial use of DMLS® for rapid tooling [39].
- Three-dimensional printing (3D printing): this method has demonstrated the capability of producing parts of a variety of materials. It is highly powder-dependent and powders with smaller sizes are required for thinner layers, finer details (micro scale features), and lower surface roughness. 3D printing was used for printing the TiNiHf shape memory alloy (SMA) with 35 µm printing layer thickness to evaluate the effects of printing layer thickness and binder saturation level on the integrity and dimensional accuracy of the 3D mesh structures [40].
- Inkjet printing process: this is based on inkjet technology and includes arrangement and layer-by-layer deposition of a liquid material in droplet form. The material often turns into a solid after the deposition process via cooling, chemical changes, or solvent evaporation. Two main techniques are utilized for droplet creation: drop-on-demand (DOD) and continuous inkjet (CIJ). Generally, CIJ systems operate with fluids of lower viscosity than DOD and at a higher drop velocity and, so, the DOD technique has more potential for 3D micro fabrication [41]. DOD printing of metal inks needs special configurations to control the jetting process for achieving well-defined droplets. For instance, the jetting process should be performed in a low-oxygen environment to avoid the oxidation of the jet and the micro droplets [20].
- Laser chemical vapor deposition (LCVD): in this process, a laser beam is employed to convert gaseous reactants into thin solid layers in a selective manner. 3D aluminum oxide (Al_2O_3) rods of 3 to 20 µm diameters were fabricated with growth rates of up to 80 µm/s. Two orthogonal argon laser beams with 0.2–20 mW power and a spot diameter of 3 µm were used instead of a single laser beam [42]. The growth rate was improved on the same material up to 100 µm/s to fabricate 3D photonic micro parts [43].
- Laser Induced Forward Transfer (LIFT): this is a 3D direct-write method suitable for precision printing of various structural and functional materials, like DNA for sensor applications or nanoparticle inks for electrical applications [44]. The common LIFT process is based on an ablation process, which results in a

deposition of various types of materials. Depending on laser fluence, the donor material is melted and may be vaporized. Due to the melting process and the arising pressure, the donor material is propelled toward the receiver. The volume, size, and quality of the deposited feature depend on the processing parameters. Using the LIFT technique, droplets as small as 300 nm in diameter have been achieved [45].

- Electrochemical Fabrication (EFAB): this is a hybrid process for volume manufacturing of micro devices with features as small as 20 µm and tolerances of ±2 µm. The EFAB process is based on multilayer electrodeposition and planarization of at least two metals: one structural material and one sacrificial material [46]. Figure 11.4 shows different medical devices monolithically fabricated without assembly by EFAB.

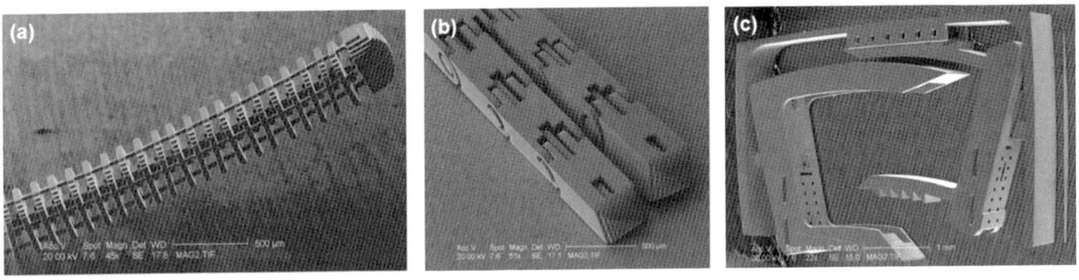

Figure 11.4 (a) A four-direction steerable guidewire tip that is about 0.43 mm in diameter, (b) a SplitClip™ device intended for minimally invasive tissue approximation, and (c) a pair of 0.35 mm two-direction fingers suitable for steerable guidewires [47]

11.2 Additive Manufacturing Technologies for Micro Tooling

Toward increasing the effectiveness of AM in the production cycle, the generation of tooling for large-volume production processes aimed at tools that maintain their consistency over a long term is a critical aspect for AM tooling processes [4]. It bridges the gap between prototype tools and series tools. This method has become more popular mainly due to worldwide markets and the consequent increase in the numbers of competitors facing individual manufacturers; it is becoming more important for manufacturers to be first on the market with their products. With AM for tooling, it is possible to reduce product development time [48]. The main advantages of this method for manufacturing tools are lead time, cost, and, very importantly, tool behavior that is similar to the conventional ones produced for high-volume replication [49].

Rapid tooling is classified into four categories: in terms of procedures, direct and indirect, and based on materials, referred to as soft and hard tooling (metal or polymer tooling) [50, 51]. Figure 11.5 shows some examples of processes and materials with the user preference classification (not all technologies in the table are applied to the micro scale) [50].

- Indirect tooling process: incorporates pattern-based methods where the tool is cast from a rapid prototype that depicts the part to be molded. For instance, room temperature vulcanized (RTV) rubber molding. Normally, the pattern applied to produce this type of indirect rapid tooling is made using an AM machine [52].
- Direct tooling process: is a method where the mold, inserts, or other components are made directly by an AM process. These processes are still developing as new technologies, materials, and system features are introduced. The main principle advantage of direct tooling is its time- and cost-saving aspects [52].
- Soft tooling: generally, tooling for shorter manufacturing runs is often known as soft tooling as these tools are commonly made from materials such as silicon rubber, epoxy resins, etc., with low-melting-point alloys, or aluminum, which are easier to work with than tooling steels [49].
- Hard tooling: for longer manufacturing runs, hard tooling is applied, which is usually made of hard tooling steel [49].

Material	Direct technologies	Indirect technologies
Polymer	Bridge Tooling, CuPA-SLS (3D-Systems) SLS/SLA soft shells	Silicone rubber pattern RTV Swift™ Tooling (SWIFT™ Tech.)
Metal	DMLS™ (EOS) Rapid Steel 2 LaserForm (3D-Systems) 3D Printing (ProMetal™)	KelTool™ (3D-Systems) Cast tools Metal spraying (HEK) Metal deposition

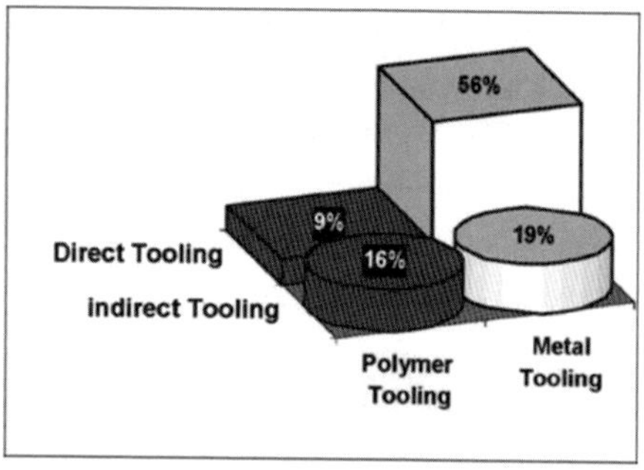

Figure 11.5 Process chain selections for mold manufacturing, rapid tooling classification, and user preferences [50]

11.2.1 AM Technologies for Micro Injection Molding Hard Tooling Applications

Micro hard tooling technologies can be characterized by the materials used to form the core and cavity, level of precision, strength and durability, and speed of production of the mold. The fabrication of AISI 420 stainless steel using SLM for the application of a plastic injection mold was investigated using a fiber laser beam of 100 µm and an average particle size of 20 µm. The densification mechanism was initially provided and the evolution of the micro structure, phase, and mechanical properties with the SLM parameters were evaluated. The micro structure, phase, and hardness results reveal that SLM is a potential method for the application of a plastic injection mold [53]. The DMLS method developed by EOS GmbH was applied for rapid tooling processing. Accuracy, wear resistance, and mechanical properties are critical when choosing the rapid tooling mold as the production-grade tooling. The material used for this study was a mixture of nickel, bronze, and copper phosphide. The average grain size of the powder was 30 µm and the dimensional accuracy, surface roughness, impact toughness, hardness, and strength were evaluated [54]. The injection molding tools made by SLM GmbH with H13 Tool Steel material were evaluated to investigate technological parameters that influence the accuracy of the SLM process and the accuracy of injection molding tools. Finite element analyses were carried out to investigate the significant effects of laser power, scanning speed, and powder bed temperature to find out the optimum closing of the active elements of the molds [55].

11.2.2 AM Technologies for Micro Injection Molding Soft Tooling Applications

The use of AM technologies promises to produce parts with complex geometries and on a small scale. As mentioned before, AM offers numerous benefits, mainly in the fabrication of composites with high accuracy, in a cost-effective way, and with customized geometry. The capabilities of soft tooling by micro stereolithography (µSL) on a micro injection molding system was investigated. The micro stereolithography photopolymer tooling was evaluated based on the different layers' thicknesses and manufacturing resolution. However, the resolution of the 100 µm × 100 µm channel features was relatively poor and they had an improper layer thickness, which resulted in an undercut as shown in Figure 11.6. The mold was tested by three thermoplastic materials, namely acrylonitrile butadiene styrene (ABS), polycarbonate (PC), and polypropylene (PP). The polymer is transported into the runner and cavity via a metering unit and plunger using a combination of fast servo drives with switchover times of 2.5 ms at an injection speed of 800 mm/s. To optimize the process, condition monitoring was applied on the pressure inside the mold cavity due to the injected polymer and temperature during the filling stage of the process, which have a critical effect on tool life. Experiments designed with four factors and two levels were selected to plan the experiments, which reveal dissimilar material combinations of settings for PP compared to ABS and PC, for which the best results were obtained [56].

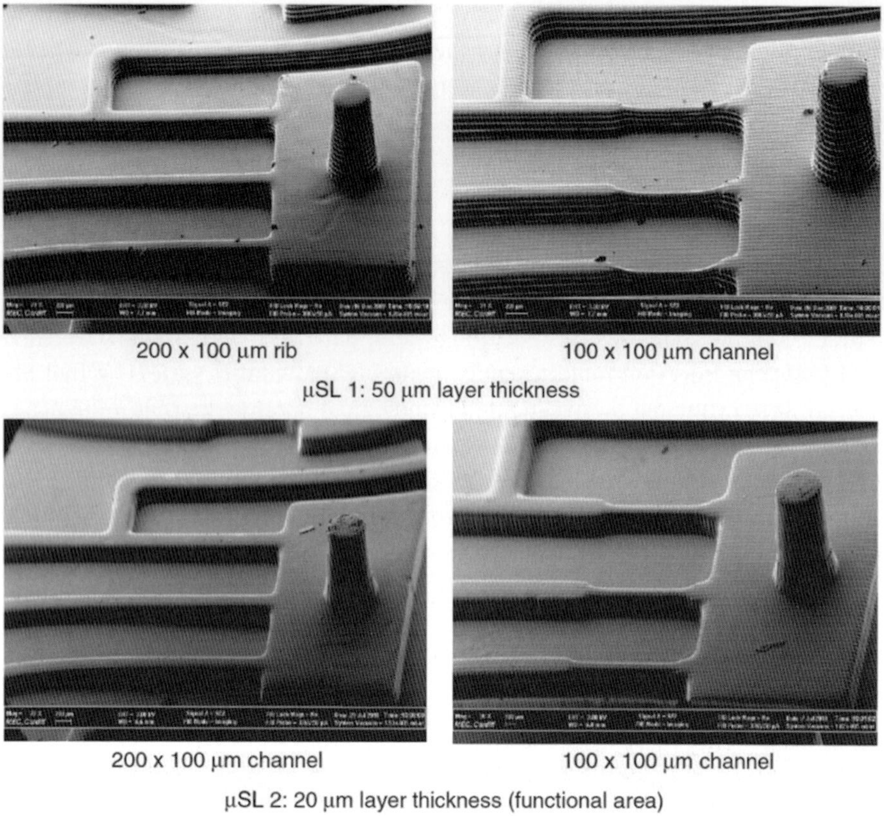

Figure 11.6 Micro stereolithography microfluidics features [56]

Micro stereolithography was used to fabricate soft inserts for producing injection molding polymer prototypes of a lens micro device. The micro mold inserts were built in three different orientations–X, Y, and Z–with a constant 20 µm layer thickness. The capabilities of the micro stereolithography inserts were evaluated by an experimental study with controlling the filling performance of the melt temperature of the mold cavity. A material commonly used in injection molding–polystyrene–was selected for the experiments. The micro stereolithography molds in different orientations were not deviated more than 5 per cent from the target dimensions of the test part. The same micro injection molding settings (injection velocity 100 mm/s and holding pressure 0.2 mm/s) were used for polystyrene parts. The inserts' capabilities were evaluated by visual inspection and major cracks parallel to the build orientations were observed in the inserts built in the X and Y directions after molding only one part. These cracks led to a breakaway of the mold inserts together with the polystyrene parts during demolding [57].

An investigation was carried out to evaluate the capabilities of the soft tooling process chain. The soft tooling insert was manufactured in a high temperature photopolymer by DLP. Two inserts were fabricated to form the cavity. An EnvisionTEC perfactory® 3 MML with an enhanced resolution module (ERM) was used in order to produce complex micro parts. Injection molding tests were operated by an Arburg (370 A 600 – 70) with a micro unit. The screw had a diameter of 8 mm and the clamping force of the machine was 100 kN. Polyethylene (PE Purell 1840) was chosen as the material for injection molding mainly due to the low processing temperatures. IM parts were measured randomly in each batch on the top and back areas. The analysis results reveal the replication process was stable during 100 cycles, considering the uncertainty for the production. In terms of the injection speed, a lower speed is recommended in order to get a high replication degree regarding the height and diameter of the pillars. The pillars far from the gate are lower, but the diameters of the pillars are not different when they are far or close to the gate; the pillars are also less uniform regarding both the height and the diameter replication. When the mold temperature is lower, the influence from the position relative to the gate is much less. Figure 11.7(a) shows the injection molding inserts produced by DLP and the product. Figure 11.7(b) and (c) show scanning electron microscopy images of the parts [58, 59].

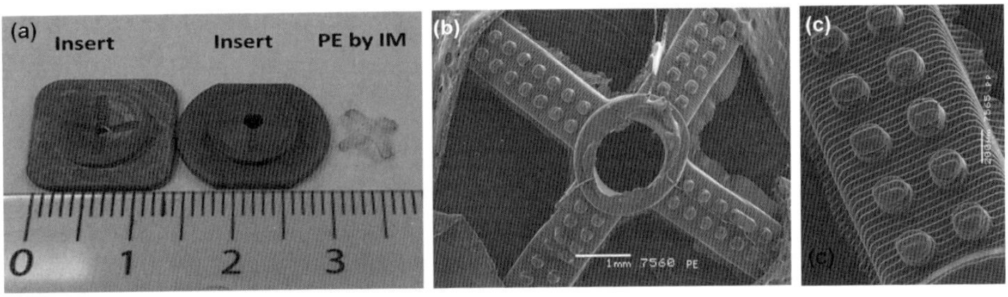

Figure 11.7 (a) The polymer replica is produced by injection molding using the two inserts, which forms the mold cavity, (b) the layers are replicated onto the polymer surface by an injection molding process, and (c) the magnification of the part [58]

A case study was investigated to introduce a new approach to modeling manufacturing process chains to support process selection, concurrent engineering, and design for manufacturing in the early phases of conceptual and detailed design. This approach was used to consider the role of additive manufacturing technologies in process chains for the small-scale production of a redesigned part with micro scale features. The test parts were produced using five process chains: machining, DLP AM with polymers and ceramics, DLP AM plus machining, and DLP AM plus machining to create an insert that is used for injection molding. Three sets of parts were produced using DLP AM. The first two sets of parts were produced using RCP30 and HTM140 on an EnvisionTEC Perfactory Mini Multi Lens machine with an 85 mm lens. The third set of DLP parts was produced using LithaLox HP 500 on a Lithoz CeraFab 7500 machine. These methods were used due to the high resolution of printing. In the fourth process chain, raw blocks of workpieces material were produced using the EnvisionTEC for subsequent machining. In the final process chain, mold inserts were printed in HTM140 and LithaLox HP 500. This was done to investigate DLP's potential to quickly produce low-cost tooling for injection molding. The insert geometry was a direct inversion of the part geometry. Figure 11.8 shows the results from different processes. It was observed that each process chain offered different benefits and imposed different limitations. Milling and DLP AM processes with low viscosity materials were viable alternatives to injection molding with a steel insert for the production of a part with high aspect ratio micro scale features. High viscosity DLP AM materials were unable to adequately produce the micro geometries or sharp corners due to the post-curing of liquid residue. Injection molding with DLP AM inserts was shown to be promising but requires further development and a different type of process optimization than traditional injection molding [60].

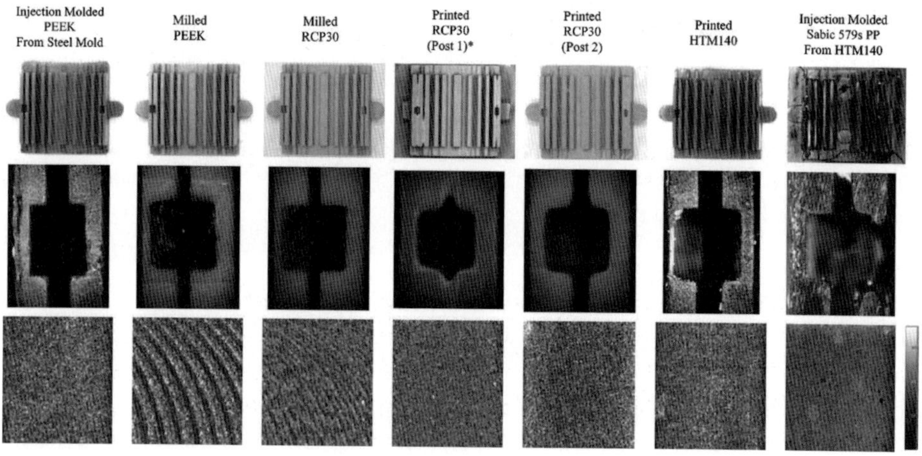

Figure 11.8 Test parts produced in a case study; row 1: top view of samples, row 2: large pocket imaged, and row 3: surface roughness [60]

The DLP method has been used to manufacture injection molding (IM) inserts reinforced with oriented short carbon fibers. The effect of the fibers on the lifetime and surface deterioration of the IM inserts was investigated. A photopolymer resin with 5 wt % short carbon fiber content equipped with an average diameter of 7.2 µm and an average length of 100 µm was used. Post processing was applied at the back and sides of the inserts by milling to reduce warpage of the inserts and ensure the tolerances. The inserts were used on a steel mold block mounted on an IM machine injecting low-density polyethylene (LD-PE) at 210 bar of maximum injection pressure during a 3 s filling time followed by a 10 s packing time and 10 s cooling phase in order for the insert to cool down to a temperature of 36 °C. Insert dimensions of 20 × 20 × 2.7 mm^3 resulted in a total cycle time for the molding of 23 s. Surface investigations of the parts produced showed cracks in the insert in the micrometer dimensional range after about 300 shots, leading to a change in roughness of the manufactured part. Figure 11.9 shows fibers in the original IM insert and cracks in the insert after 500 and 2100 shots, respectively. It was observed that the lifetime of the insert in fiber-reinforced IM AM inserts could be extended compared to plain IM AM inserts. The surface wear in terms of mean surface roughness was negligible when inspecting the surface without crack-like features [69, 70].

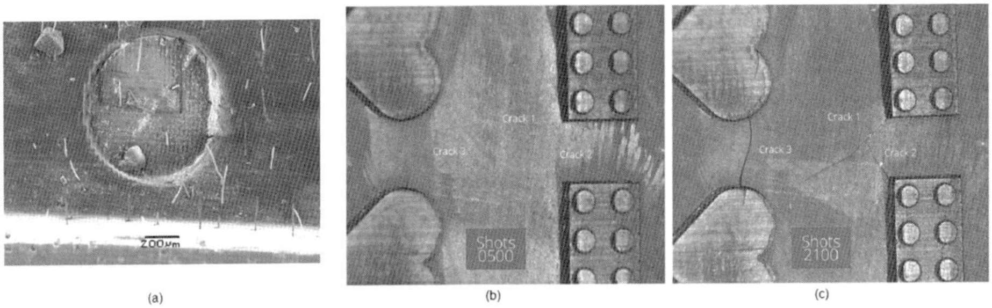

Figure 11.9 (a) Fibers standing out of the produced layers, (b) cracks in the insert after shot 500, and (c) cracks in the insert after shot 2100 [69]

11.2.3 Indirect Methods for Micro Tooling Production

This method is called indirect due to the use of a pattern obtained by a proper AM process as a model for mold and die making. This tooling process produces molds by forming them from a pattern. The mold will be available for production runs of up to several hundred parts made on the same material as, or on a material very similar to, that of the final production part [61]. Indirect tooling mainly includes a die and insert for injection molding, diecasting, vacuum casting, metal forming operations, and sacrificial patterns or shells for investment casting. Several processes have been developed, such as metal deposition tools using an RP model with a good surface. This process technology enables possibilities for micro applications with a thickness of up to 30 µm of cladded tracks [62]. The epoxy tools, which include the pattern (master model), are created by an AM process. They are then embedded in a parting line block to create the parting line of the mold. Epoxy tools are mostly used for molds for prototype injection plastics, molds for castings, compression molds, and reaction injection molds [63]. With ceramic tools, the master model can be cast with any plaster ceramics to create a tool cavity. Ceramic tools can be applied for plastics processing, metal forming, and metal casting [64]. For example, the 3D Keltool process is based on a metal sintering process that converts RP master patterns into production tool inserts with high quality definition and surface finish. The material properties allow inserts produced using this process to withstand more than 1000000 molding cycles [65]. This process requires a master pattern, typically an ACES SLA model, which can be used to develop a silicone mold that will then be used to produce the Keltool mold. An example of the Keltool process is shown in Figure 11.10.

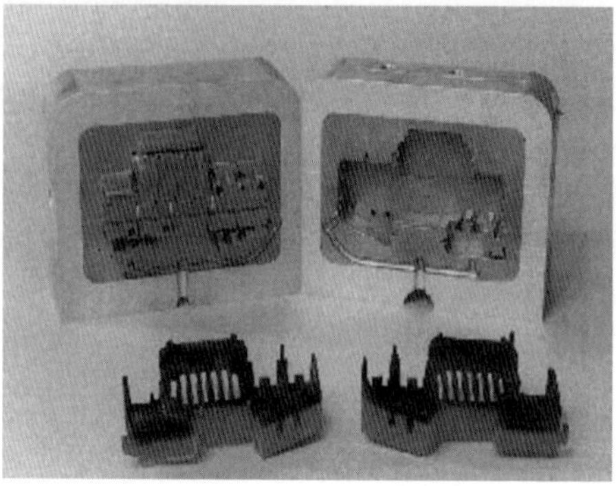

Figure 11.10 Plastic injection molding core and cavity set [66]

A study on the production of micro-textured micro devices was aimed at rapidly connecting the complex geometries attainable by additive technologies with mass production procedures. DLP was used to manufacture a mold insert of a micro-textured fractal surface-based cell culture platform or tissue engineering scaffold and a micro system for studying cell motility and addressing the effect of surface texture on cell migration. In an evaporation process, master and substrate were coated with a 7-nm-thick chromium layer and 50-nm-thick layer of gold. These metallic layers support a precise metal deposition along the micro structures on top of the master. This process leads to a stiff and homogeneous metal block, which can withstand the forces applied in the injection molding process. PMMA and PC are two interesting thermoplastic materials for medical applications, which were used to produce injection molded parts. An injection speed of 33 mm/s, a packing pressure of 1050 bar and a maximum injection pressure of 1300 bar were applied for both components and more than 200 parts were produced. The manufacturing process chain appears very promising in terms of speed, and led to the possibility of obtaining final low-cost replicas with a high accuracy of the final parts and micrometric details that are perfectly replicated. Figure 11.11 shows an electroplated mold insert for fractal cell culture platforms and a culture micro system with the micro-injected parts [67].

Figure 11.11 Electroplated mold inserts and micro-injected parts [67]

The fabrication of micro-patterned ceramics and ceramic micro parts is highly demanding with respect to the precision and resolution of the molding process. Stereolithography (SLA), MultiJet Modelling (MJM), and Rapid Micro Product Development (RMPD) as rapid prototyping processes have been investigated for the manufacture of micro ceramic parts. SLA was used produce the epoxide master model with a positioning accuracy of 10 µm, and a solid-state laser with a spot diameter of 100 µm was applied to expose a liquid monomer on a construction platform in a layer-by-layer scanning mode. MJM was used in order to reduce the partly cured resin in the narrow gaps of the master model. Additionally, the master model for higher resolution was made of acrylates using the RMPD technique. Then, the polymer master models were molded with liquid silicone rubber, which can be easily demolded after curing. Low-pressure injection molding was used at temperatures of about 80–100 °C and at pressures below 0.7 MPa. In these conditions, each silicone mold usually allows more than 100 ceramic parts to be produced. The rapid prototyping technique should be carefully selected to meet the requirements of the ceramic part to be molded. A comparison of the molding is shown in Figure 11.12 [68].

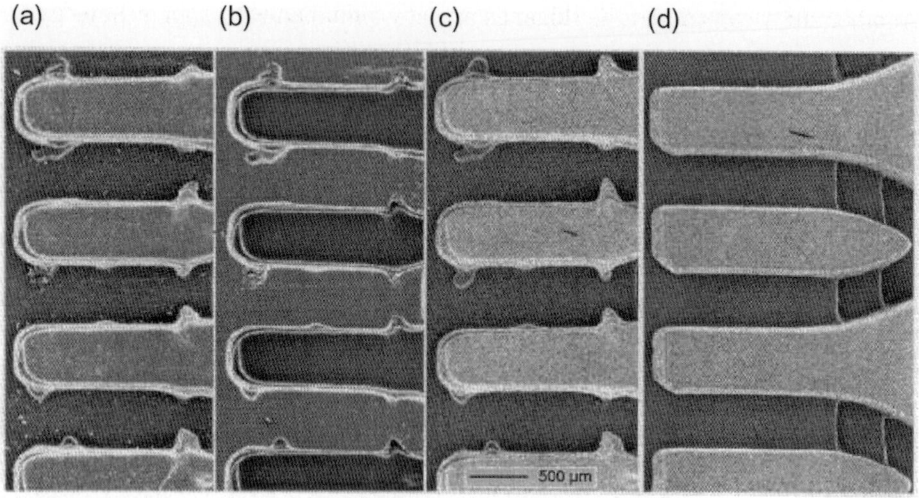

Figure 11.12 Comparison of the different molding steps: (a) polymer master model, (b) mirrored image of the silicone mold, and (c) sintered alumina part. The defects in the master model are affected by incorrect laser control in stereolithography and are replicated throughout the process chain into the ceramic part. (d) A ceramic part molded from a defect-free master model [68]

11.3 Additive Manufacturing for the Direct Manufacturing of Micro Products

The capabilities of different AM processes in the direct high-precision production of micro components were presented in a number of studies. One study investigated a new technology, called Höganäs, for a faster and less expensive production of investment casting shells for micro component manufacturing. The Direct Shell Process (DSP) was compared against the Lost Wax Process (LWP). The Thermojet process was used to produce a pattern in a wax-like material and Höganäs Direct Shell 3D printing technology was performed on a four-axis 'c50' demonstrator to produce ceramic shells. Three different alloys were used in the experiments to cast the test parts: aluminum (LM25), zinc (ZA12), and 316 S16 stainless steel (SS). The comparison results revealed that the surface quality of the DSP casting components was higher than that produced employing LWP due to the lower reactivity with the zirconia molds. The lead time of the DSP was about 21% shorter than that of the conventional block mold casting process, which makes Höganäs technology a viable alternative for the rapid prototyping and short-run manufacture of metal parts [71].

Another study considered the Höganäs process combined with spin technology for manufacturing a four-sided pyramid with light diodes on the top and driving electronics on the sides. 3D tubes were filled with a silver paste and sintered afterward. In order to overcome the porosity problem in the ceramic material, the ceramic parts must be infiltrated with a filling material; epoxy-infiltrated ceramics were used in this study. Additionally, an alternative process based on electroless copper plating was used to achieve plating inside the tubes; the plating solution was pumped through the 3D tubes. Figure 11.13 shows the ceramic inertial measurement unit (IMU) platform, the silver paste that infiltrates the porous ceramics and causes the tubes to short circuit, and the copper-plated tubes [72].

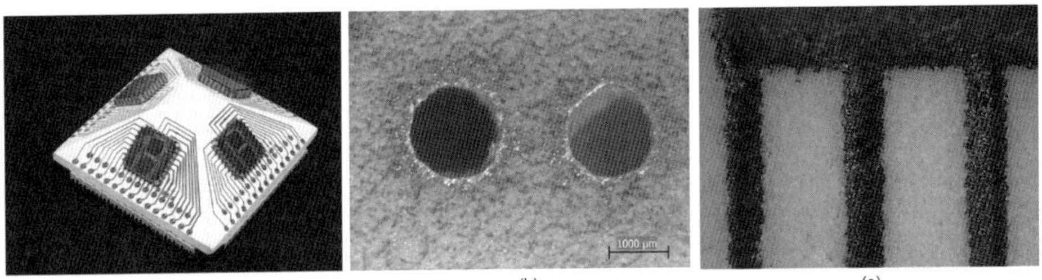

(a) (b) (c)

Figure 11.13 (a) A ceramic IMU platform with gyro components, (b) a short circuit between silver-filled 3D tubes, and (c) copper-plated tubes in epoxy-infiltrated ceramics [72]

A study carried out with the 2PP technique demonstrated the fabrication of structures with a high aspect ratio with sub-micrometer process resolution. The laser power was precisely adjusted by a combination of a $\lambda/2$ wave-plate and polarizing beam splitter cube. The photopolymerized structure was generated layer-by-layer with an x–y galvanometric mirror scanner. The laser beam at a power of 33 mW was scanned at a speed of 4 mm/s. The z-axis was moved stepwise by 0.5 μm to build 3D structures layer by layer. The SEM images in Figure 11.12(c) and (d) show 3D structures; the smallest lateral feature sizes in these structures are 771 nm and 1.2 μm, respectively. Figure 11.14(a) and (b) show schematic and fabricated configurations of various types of tube structures with straight, funnel, and hourglass shapes. 3D structures with a height of 7 mm are demonstrated, which could hardly be built with the conventional 2PP set-up due to refractive index mismatch and laser beam disturbances [73]. 2PP was involved in a variety of medical device component fabrications including photoreactive acrylate-based polymer micro needles, artificial tissues, microfluidic devices, pumps, sensors, and valves [74].

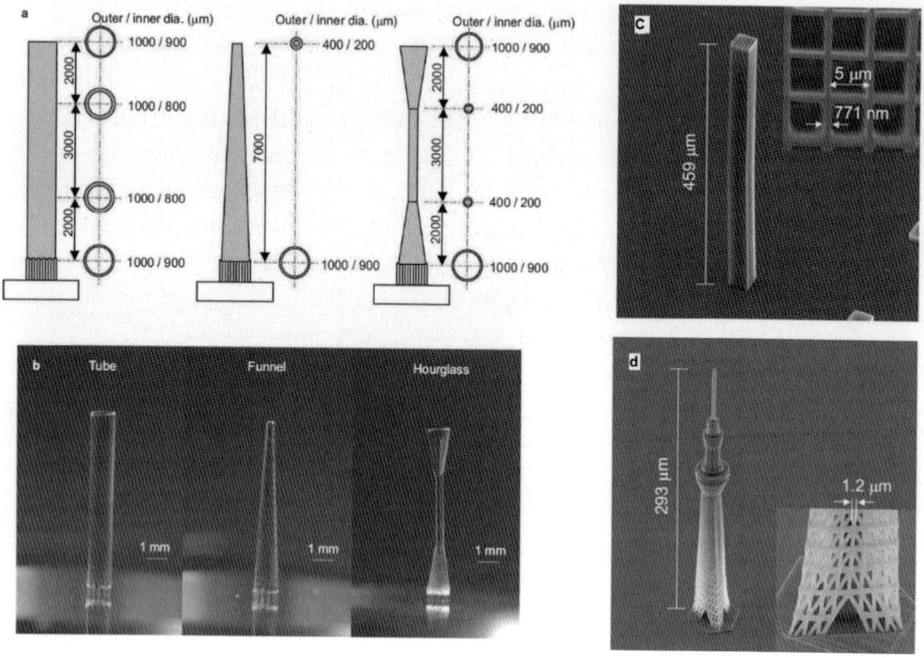

Figure 11.14 (a) Schematic and (b) fabricated configurations of various types of tube structures

Another study with the 2PP method was carried out to evaluate the manufacture of 3D micro polymer parts. The laser power, printing strategy, and printed feature sizes ranging from 8 µm to 200 µm were assessed. The laser power was selected as a variable parameter in order to find out the effect of various powers on printing size and strategy. Six different powers were selected, from 0.6 mW to 1.6 mW, for each set of structures. The smaller samples were printed better with less distortion at different laser powers. However, the bigger samples were burned out on the surface with high laser powers of 1.6–1.4 mW. The measurement results in the X and Y directions of the samples reveal, in terms of feature size, that a laser power of 1.2 mW was a suitable value for printing, but, in terms of the burned-out surfaces of the features, a laser power of 1.0 mW was appropriate for all sample sizes. Figure 11.15 shows SEM pictures of the printed structures of the test part at six different powers from 0.6–1.6 mW in 0.2 mW increments [75].

A study was carried out to investigate the iterative design of a test part to evaluate the resolution of AM processes with voxel sizes at the micro scale. Sample test parts were produced using RCP30 on an EnvisionTEC Perfactory® Mini Multi Lens 3D Printer with standard settings and an 85 mm lens. The test part features were measured using an Alicona InfiniteFocus G4 microscope with different magnifications. Different test part designs were developed as a general benchmark and included features with many shapes and sizes in order to have a reliable comparison for the resolution of the system [76].

Figure 11.15 Features fabricated by 2PP [75]

The minimum feature size and obtainable tolerances of micro AM processes were determined by the smallest printable element, i.e., the resolution of the AM machine. A study analyzed the accuracy of the printable features with a vat photopolymerization method on a test part that was designed with different sizes and aspect ratios in order to evaluate the additively manufactured parts. In order to have an initial optimal range of parameter values, a sensitivity analysis was carried out prior to the final experimental plan. The printing parameters selected for the evaluation were exposure time, light intensity, and layer thickness. The results showed the importance of different process parameters in micro AM processes. Figure 11.16 shows the results of printing with different printing parameter settings [77].

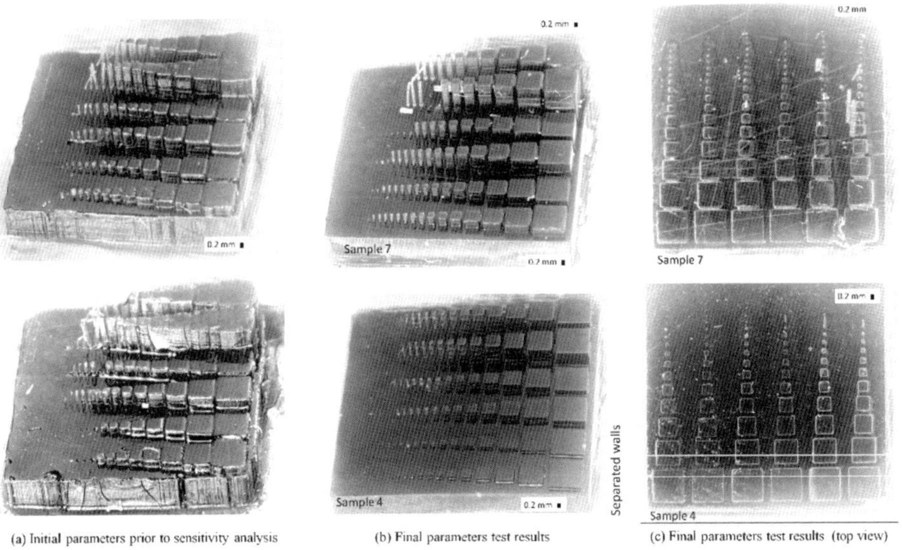

Figure 11.16 Test part sample prints: (a) results with initial parameters, (b) results after sensitivity analysis with final plan, and (c) top view [78]

References

[1] Ahmad, A., Darmoul, S., Ameen, W., Abidi, M.H., Al-Ahmari, A.M., Rapid Prototyping for Assembly Training and Validation. *IFAC-PapersOnLine* (2015) 48(3), pp. 412–417

[2] Yan, X., Gu, P., A review of rapid prototyping technologies and systems. *Comput. Aided Des.* (1996) 28(4), pp. 307–318

[3] ASTM International, F2792-12a – Standard Terminology for Additive Manufacturing Technologies (2013)

[4] Onuh, S.O., Yusuf, Y.Y., Rapid prototyping technology: applications and benefits for rapid product development. *J. Intell. Manuf.* (1999) 10(3), pp. 301–311

[5] Gibson, I., Rosen, D.W., Stucker, B., *Additive Manufacturing Technologies: Rapid Prototyping to Direct Digital Manufacturing* (2009) 54

[6] Monzón, M.D., Ortega, Z., Martínez, A., Ortega, F., Standardization in additive manufacturing: activities carried out by international organizations and projects. *Int. J. Adv. Manuf. Technol.* (2014) 76(5–8), pp. 1111–1121

[7] Kruth, J.-P., Leu, M.C., Nakagawa, T., Progress in Additive Manufacturing and Rapid Prototyping. *CIRP Ann. - Manuf. Technol.* (1998) 47(2), pp. 525–540

[8] Munsch, M., 15 – Laser additive manufacturing of customized prosthetics and implants for biomedical applications, *Laser Additive Manufacturing* (2017), pp. 399–420

[9] Hagedorn, Y., 6 – Laser additive manufacturing of ceramic components: Materials, processes, and mechanisms, *Laser Additive Manufacturing* (2017), pp. 163–180

[10] Eckel, Z.C., Zhou, C., Martin, J.H., Jacobsen, A.J., Carter, W.B., Schaedler, T.A., Additive manufacturing of polymer-derived ceramics. *Science* (2016) 351(6268), pp. 58–62

[11] Angelo, G.D., A Platform for Multi Material Additive Manufacturing, Modular Tooling and Process Verification, December (2015)

[12] Wang, X., Jiang, M., Zhou, Z., Gou, J., Hui, D., 3D printing of polymer matrix composites: A review and prospective. *Composites, Part B* (2017) 110, pp. 442–458

[13] Sood, A.K., Ohdar, R.K., Mahapatra, S.S., Parametric appraisal of mechanical property of fused deposition modelling processed parts. *Mater. Des.* (2010) 31(1), pp. 287–295

[14] Leigh, S.J., Bradley, R.J., Purssell, C.P., Billson, D.R., Hutchins, D.A., A Simple, Low-Cost Conductive Composite Material for 3D Printing of Electronic Sensors. *PLoS One* (2012) 7(11), pp. 1–6

[15] Yen, H.-J., Tseng, C.-S., Hsu, S., Tsai, C.-L., Evaluation of chondrocyte growth in the highly porous scaffolds made by fused deposition manufacturing (FDM) filled with type II collagen. *Biomed. Microdevices* (2009) 11(3), pp. 615–624

[16] Kalita, S.J., Bose, S., Hosick, H.L., Bandyopadhyay, A., Development of controlled porosity polymer-ceramic composite scaffolds via fused deposition modeling. *Mater. Sci. Eng., C* (2003) 23(5), pp. 611–620

[17] Zein, I., Hutmacher, D.W., Tan, K.C., Teoh, S.H., Fused deposition modeling of novel scaffold architectures for tissue engineering applications. *Biomaterials* (2002) 23(4), pp. 1169–1185

[18] Utela, B., Storti, D., Anderson, R., Ganter, M., A review of process development steps for new material systems in three dimensional printing (3DP). *J. Manuf. Process.* (2008) 10(2), pp. 96–104

[19] Sachs, E.M., Haggerty, J.S., Cima, M.J., Williams, P.A., Three-dimensional printing techniques. *Google Patents* (1994)

[20] Vaezi, M., Seitz, H., Yang, S., A review on 3D micro-additive manufacturing technologies. *Int. J. Adv. Manuf. Technol.* (2013) 67(5), pp. 1721–1754

[21] Melchels, F.P.W., Feijen, J., Grijpma, D.W., A review on stereolithography and its applications in biomedical engineering. *Biomaterials* (2010) 31(24), pp. 6121–6130

[22] Subramanian, K., Vail, N.K., Barlow, J.W., Marcus, H., Selective laser sintering of alumina with polymer binders. *Rapid Prototyping J.* (1995) 1(2), pp. 24–35

[23] Gu, D.D., Meiners, W., Wissenbach, K., Poprawe, R., Laser additive manufacturing of metallic components: materials, processes and mechanisms. *Int. Mater. Rev.* (2012) 57(3), pp. 133–164

[24] Ebert, R., Regenfuss, P., Klötzer, S., Hartwig, L., Exner, H., Process assembly for µm-scale SLS, reaction sintering, and CVD, in *Fourth International Symposium on Laser Precision Microfabrication* (2003) pp. 183–188

[25] Deckard, C.R., Beaman, J.J., Darrah, J.F., Method for selective laser sintering with layerwise cross-scanning. *Google Patents* (1992)

[26] Laumer, T., Stichel, T., Riedlbauer, D., Amend, P., Mergheim, J., Schmidt, M., Realization of multi-material polymer parts by simultaneous laser beam melting. *J. Laser Micro/Nanoeng.* (2015) 10(2), pp. 140–147

[27] Billiet, T., Vandenhaute, M., Schelfhout, J., Van Vlierberghe, S., Dubruel, P., A review of trends and limitations in hydrogel-rapid prototyping for tissue engineering. *Biomaterials* (2012) 33(26), pp. 6020–6041

[28] Cooperstein, I., Layani, M., Magdassi, S., 3D printing of porous structures by UV-curable O/W emulsion for fabrication of conductive objects. *J. Mater. Chem. C* (2015) 3(9), pp. 2040–2044

[29] Postiglione, G., Natale, G., Griffini, G., Levi, M., Turri, S., Conductive 3D microstructures by direct 3D printing of polymer/carbon nanotube nanocomposites via liquid deposition modeling. Composites Part A: Applied Science and Manufacturing (2015) 76, pp. 110–114

[30] Saari, M., Cox, B., Richer, E., Krueger, P.S., Cohen, A.L., Fiber Encapsulation Additive Manufacturing: An Enabling Technology for 3D Printing of Electromechanical Devices and Robotic Components. *3D Print. Addit. Manuf.* (2015) 2(1), pp. 32–39

[31] Selimis, A., Mironov, V., Farsari, M., Microelectronic Engineering Direct laser writing : Principles and materials for scaffold 3D printing. *Microelectron. Eng.* (2015) 132, pp. 83–89

[32] Huis in 't Veld, B., Overmeyer, L., Schmidt, M., Wegener, K., Malshe, A., Bartolo, P., Micro additive manufacturing using ultra short laser pulses. *CIRP Ann. - Manuf. Technol.* (2015) 64(2), pp. 701–724

[33] Sames, W.J., List, F.A., Pannala, S., Dehoff, R.R., Babu, S.S., The metallurgy and processing science of metal additive manufacturing. *Int. Mater. Rev.* (2016) 6608, pp. 1–46

[34] Herzog, D., Seyda, V., Wycisk, E., Emmelmann, C., Additive manufacturing of metals. *Acta Mater.* (2016) 117, pp. 371–392

[35] Murr, L.E., Gaytan, S.M., Ramirez, D.A, Martinez, E., Hernandez, J., Amato, K.N., Shindo, P.W., Medina, F.R., Wicker, R.B., Metal Fabrication by Additive Manufacturing Using Laser and Electron Beam Melting Technologies. *J. Mater. Sci. Technol.* (2012) 28(1), pp. 1–14

[36] Facchini, L., Vicente, N., Lonardelli, I., Magalini, E., Robotti, P., Alberto, M., Metastable austenite in 17-4 precipitation-hardening stainless steel produced by selective laser melting. *Adv. Eng. Mater.* (2010) 12(3), pp. 184–188

[37] Kaierle, S., Barroi, A., Noelke, C., Hermsdorf, J., Overmeyer, L., Haferkamp, H., Review on Laser Deposition Welding: From Micro to Macro. *Phys. Procedia* (2012) 39, pp. 336–345

[38] Exner, H., Regenfuss, P., Hartwig, L., Klötzer, S., Ebert, R., Selective laser micro sintering with a novel process, in *Fourth International Symposium on Laser Precision Microfabrication* (2003), pp. 145–151

[39] Shellabear, M., Nyrhilä, O., DMLS – Development History and State of the Art. *LANE 2004* (2004) pp. 1–12

[40] Lu, K., Reynolds, W.T., 3DP process for fine mesh structure printing. *Powder Technol.* (2008) 187(1), pp. 11–18

[41] Hon, K.K.B., Li, L., Hutchings, I.M., Direct writing technology—Advances and developments. *CIRP Ann. - Manuf. Technol.* (2008) 57(2), pp. 601–620

[42] Lehmann, O., Stuke, M., Three-dimensional laser direct writing of electrically conducting and isolating microstructures. *Mater. Lett.* (1994) 21(2), pp. 131–136

[43] Stuke, M., Mueller, K., Mueller, T., Hagedorn, R., Jaeger, M., Fuhr, G., Laser-direct-write creation of three-dimensional OREST microcages for contact-free trapping, handling and transfer of small polarizable neutral objects in solution. *Appl. Phys. A: Mater. Sci. Process.* (2005) 81(5), pp. 915–922

[44] Baum, M., Kim, H., Alexeev, I., Piqué, A., Schmidt, M., Generation of transparent conductive electrodes by laser consolidation of LIFT printed ITO nanoparticle layers. *Appl. Phys. A* (2013) 111(3), pp. 799–805

[45] Banks, D.P., Grivas, C., Mills, J.D., Eason, R.W., Zergioti, I., Nanodroplets deposited in microarrays by femtosecond Ti:sapphire laser-induced forward transfer. *Appl. Phys. Lett.* (2006) 89(19), pp. 1–4

[46] Gad-el-Hak, M., *The MEMS Handbook* (2001), CRC press, Washington, D.C., p. 1368

[47] Cohen, A.L., EFAB® Technology:Unlocking the Potential of Miniaturized Medical Devices, Microfabrica (2008), pp. 1–31

[48] Denton, K.R., The economics of rapid tooling and rapid prototyping, in proceeding of *the Sixth International Conference on Rapid Prototyping* (1995), The Dayton Convention and Exhibition Center, Dayton, Ohio, U.S.A., pp. 179–188

[49] Chua, C.K., Hong, K.H., Ho, S.L., Rapid tooling technology. Part 1. A comparative study. *Int. J. Adv. Manuf. Technol.* (1999) 15(8), pp. 604–608

[50] Levy, G.N., Schindel, R., Kruth, J.P., Rapid Manufacturing and Rapid Tooling with Layer Manufacturing (LM) Technologies, State of the Art and Future Perspectives. *CIRP Ann. - Manuf. Technol.* (2003) 52(2), pp. 589–609

[51] Gebhardt, A., Hötter, J.-S., Additive manufacturing : 3D printing for prototyping and manufacturing (2016), Hanser Publishers, Munich

[52] Nee, J.G., Fundamentals of tool design, Society of Manufacturing Engineers SME, Michigan, USA (2010), p. 464

[53] Zhao, X., et al., Fabrication and Characterization of AISI 420 Stainless Steel Using Selective Laser Melting. *Mater. Manuf. Process.* (2015) 6914, pp. 37–41

[54] Khaing, M., Fuh, J.Y., Lu, L., Direct metal laser sintering for rapid tooling: processing and characterisation of EOS parts. *J. Mater. Process. Technol.* (2001) 113(1), pp. 269–272

[55] Păcurar, R., Păcurar, A., Bâlc, N., Research on the Accuracy of Injection Molding Tools Made by H13 Material Using the Selective Laser Melting Technology. *Recent Adv. Eng. Mech. Struct. Urban Plan.* (2013) pp. 81–86

[56] Griffiths, C.A., Dimov, S.S., Hirshy, H., Scholz, S., Fischer, S., Spitzbart, M., Prototype tooling for producing small series of polymer microparts. *Proc. Inst. Mech. Eng., Part B* (2011) 225(12), pp. 2189–2205

[57] Griffiths, C.A., Dimov, S.S., Fischer, S., Spitzbart, M., Lacan, F., Micro-stereolithography tools for small-batch manufacture of polymer micro-parts. *Proc. Inst. Mech. Eng., Part B* (2012) 226(4), pp. 708–721

[58] Zhang, Y., Pedersen, D.B., Gøtje, A.S., Mischkot, M., Tosello, G., A Soft Tooling process chain employing Additive Manufacturing for injection molding of a 3D component with micro pillars. *J. Manuf. Process.* (2017) 27, pp. 138–144

[59] Mischkot, M., Zhang, Y., Gøtje, A.S., Pedersen, D.B., Tosello, G., Hansen, H.N., Influence of Injection-Molding Process Parameters on Part Replication of Microstructures with Additively-Manufactured Soft Tooling Inserts. WCMNM World Congress on Micro and Nano Manufacturing (2017)

[60] Thompson, M.K., Stolfi, A., Mischkot, M., Process chain modeling and selection in an additive manufacturing context. *CIRP J. Manuf. Sci. Technol.* (2016) 12, pp. 25–34

[61] Pham, D.T., Dimov, S.S., *Rapid Manufacturing: The Technologies and Applications of Rapid Prototyping and Rapid Tooling* (2001) Springer-Verlag, London

[62] Kaierle, S., Barroi, A., Noelke, C., Hermsdorf, J., Overmeyer, L., Haferkamp, H., Review on Laser Deposition Welding: From Micro to Macro. *Phys. Procedia* (2012) 39, pp. 336–345

[63] Wohlers, T., Caffrey, T., 3D Printing and Additive Manufacturing State of the Industry Annual Worldwide Progress Report, Report 2015

[64] Dickens, P.M. et al., Conversion of RP models to investment castings. *Rapid Prototyp. J.* (1995) 1(4), pp. 4–11

[65] Pham, D.T., Dimov, S.S., Rapid prototyping and rapid tooling - the key enablers for rapid manufacturing. *Proc. Inst. Mech. Eng., Part C* (2003) 217(1), pp. 1–23

[66] Equbal, A., Kumar, A., Rapid Tooling: A Major Shift In Tooling Practice. *J. Manuf. Ind. Eng.* (2015) 14(3–4), pp. 1–9

[67] Lantada, A.D., Piotter, V., Plewa, K., Barié, N., Guttmann, M., Wissmann, M., Toward mass production of microtextured microdevices: linking rapid prototyping with microinjection molding. *Int. J. Adv. Manuf. Technol.* (2014) 76(5–8), pp. 1011–1020

[68] Knitter, R., Bauer, W., Ceramic microfabrication by rapid prototyping process chains. *Sadhana* (2003) 28(1–2), pp. 307–318

[69] Hofstatter, T., Mischkot, M., Pedersen, D.B., Tosello, G., Hansen, H.N., Evolution of Surface Texture and Cracks During Injection Molding of Fiber-Reinforced Additively-Manufactured, Injection Molding Inserts. Proceedings of the American Society for Precision Engineering ASPE, 2016

[70] Mischkot, M., Hansen, H.N., Pedersen, D.B., Additive manufacturing for the production of inserts for micro injection molding. *Euspen's 15th Int. Conf.* (2015) June, pp. 2–3

[71] Charmeux, J., Minev, R., Dimov, S., Minev, E., Su, S., Harrysson, U., Capability Study of the Fcubic Direct Shell Process for Casting Micro Components. **4M/ICOMM** (2009) C, pp. 1–4

[72] Johander, P., Haasl, S., Persson, K., Harrysson, U., Layer Manufacturing as a Generic Tool for Microsystem Integration. *Proc. Third Int. Conf. Multi-Material Micro Manuf.* (2007), pp. 3–5

[73] Obata, K., El-Tamer, A., Koch, L., Hinze, U., Chichkov, B.N., High-aspect 3D two-photon polymerization structuring with widened objective working range (WOW-2PP). *Light: Sci. Appl.* (2013) 2, p. e116

[74] Narayan, R.J., Doraiswamy, A., Chrisey, D.B., Chichkov, B.N., Medical prototyping using two photon polymerization. *Mater. Today* (2010)13(12), pp. 42–48

[75] Davoudinejad, A., Malureanu, R., Palima, D., Tosello, G., Two-Photon Polymerization lithography for three-dimensional micro polymer parts manufacturing evaluation. *International Conference of the European Society for Precision Engineering and Nanotechnology* (2017) 1, pp. 2–4

[76] Thompson, M.K., Mischkot, M., Design of Test Parts to Characterize Micro Additive Manufacturing Processes. *Procedia CIRP*, (2015) 34, pp. 223–228

[77] Davoudinejad, A., Pedersen, D.B., Tosello, G., Dimensional Accuracy and Surface Finish in Additive Manufacturing, Evaluation of polymer micro parts produced by additive manufacturing processes by using vat photopolymerization method. *International Conference of the European Society for Precision Engineering and Nanotechnology* (2017)

PART 4

Multimaterial Micro Processing

12 Micro Powder Injection Molding

Volker Piotter

12.1 Introduction

Compaction of powder materials by sintering has been used by mankind for many centuries to manufacture parts and goods mainly out of non-metallic, inorganic substances. Additionally, in recent times, the procedure is applied more and more for metallic materials as well.

Among these sintering-based approaches, powder injection molding (PIM) has become a veritable manufacturing process for many sectors of industry producing components out of metal or ceramic materials [1–5]. In principle, the process technology is based on polymer injection molding but is combined additionally with sintering the complexly shaped powder component into a solid body.

The idea behind this technology is not entirely new but, instead, has a rich background of history: General Motors used a similar method for the commercial production of ceramic insulators for spark plugs as early as in 1937 [6]. Additional previous attempts were the so-called Rivers and Wiech methods.

Today, a large variety of products made by metal (MIM) and ceramic (CIM) injection molding has become widely established in industrial practice. For example, the global market volume for metal injection molded products reached ca. US$2000 M in 2014, with Asia covering the largest portion, followed by Europe and North America. Taking into account an estimated Compound Annual Growth Rate (CAGR) of more than 7%, the MIM market is expected to reach ca. US$3100 M in 2020 [7].

The reasons for these various applications, which are increasing rapidly, lie in the considerable economic performance in mass production, and in the ability of manufacturing even complex component geometries close to the final contours–i.e., with little or, as in many cases, no reworking. One major benefit is the wide range of materials which can be processed by PIM techniques. For instance, alloyed steels, ferrous materials, non-ferrous metals, hard metal alloys, or high-performance ceramic materials may be used.

As a further benefit to be mentioned, molding combined with sintering offers one of the highest levels of materials utilization compared to other manufacturing processes. Last but not least, slight modifications make it possible to fall back on the machine technology of polymer injection molding, which is also used in plastics processing and is therefore less expensive.

Industrial applications of PIM technology range from making cores for casting high-power turbines to components for industrial machines, micro electronics, medical technology, automotive devices, and many other applications. One example taken from information technology is zirconia sheaths for optical fibers (usually called ferrules), which must be produced on a rather high accuracy level at as narrow as possible tolerances. PIM technology ensures high economic performance as a result of short manufacturing times and a high degree of automation [8].

It has been a logical consequence to utilize the above-mentioned benefits of powder injection molding for micro fabrication as well. A good demonstration for this trend is given by the example shown in Figure 12.1. Before going into micro specific aspects, however, a description of the general process chain and its particularities will be given in the following section.

Figure 12.1 Micro metal injection molded part to be used as camera component; thickness of protruding ring structure: 0.1 mm, tolerance: ± 0.0075 mm, material: SUS360. Courtesy of Taisei Kogyo Co. Ltd., Osaka, Japan

12.2 Process Description

For the shaping of complex particle scaffolds made out of metals or ceramics, powder injection molding utilizes the well-established plastics injection molding technology. In this regard, it is an advantageous combination of polymer injection molding and powder or ceramic processing, supplementing the conventional techniques of processing these materials, such as press sintering, slip casting, or shaping by cutting.

The PIM process chain consists of a number of steps. To allow metal and ceramic powders, respectively, to be processed by thermoplastic techniques, these powders are compounded with an organic binder system consisting chiefly of thermoplastic polymers, waxes, and additives. The filler content of a typical powder-binder blend is approximately 50–60 vol % for ceramic feedstocks and roughly 55 vol % up to a maximum of 70 vol % for metal systems. This highly filled mixture is then processed into a so-called green product on an injection molding machine; up to a certain content, the injection molding process is comparable to that used for thermoplastic materials. For the PIM-specific differences, see Section 12.3.

After shaping by injection molding, the component has to be compacted into the final sintered part with lowest possible porosity. For this purpose, the binder is removed from the green product in such a way that a porous powder skeleton is obtained. A residual binder content, the so-called backbone polymer, and/or cohesion (oxide layers) provide sufficient inherent stability to keep the geometry of this skeleton more or less within its original dimensions. Subsequently, the existing powder scaffold is sintered to its final density; this procedure implies a considerable volume shrinkage of the molded part, roughly corresponding to the earlier binder volume fraction. The sintered density attainable is about 97% to about 99% of the maximum theoretical level.

It has to be clearly emphasized that each of the process phases described above influences the ultimate achievable part quality. Generally speaking, defects introduced into the sample at one stage of the process cannot be corrected in subsequent phases.

12.2.1 Feedstocks for Powder Injection Molding

12.2.1.1 Binder

As already explained, the main function of the binder is to enable shaping by injection molding and to temporarily preserve the shape of the green parts until the onset of sintering. Normally, the binders used for feedstocks for powder injection molding are composed of several organic components to ensure that the requirements imposed by the process in terms of viscosity, green stability, and binder removal characteristics are sufficiently met. For this purpose, various organic substances such as waxes (e.g., paraffin wax, synthetic hydrocarbon wax, oxide polyethylene wax), thermoplastic materials (e.g., polyethylene (PE), polypropylene (PP), polyoxymethylene copolymer (POM), polyethylene glycol (PEG), polyvinylbutyral (PVB)) as well as specially tailored additives stabilizing the dispersion are combined to form an appropriate binder system. In principle, any binder suitable for powder injection molding should exhibit the properties outlined below [9, 10]:

- Good wetting characteristics of the powder surfaces
- Low tendency to powder-binder segregation
- Good flowability
- High strength to ensure dimensional stability during demolding
- Thermal stability at injection temperature
- No chemical reaction with the inorganic filler
- Non-polluting characteristics, especially the debinding step
- Recyclability
- Medium-term durability of the feedstock

12.2.1.2 Powder Particle Properties

Of course it is not really surprising that, for producing detailed parts with high accurate structural details and good surface qualities, the powders which should be used have a quite fine particle size. Additionally, small dimensions of powder particles shall be applied also in the interest of sintering. To ensure a sufficient flowability and high packing density, the particle shapes should be spherical, rounded, or at least equiaxed. Irregular particles are a cause of a reduced degree of critical powder filling, which leads to higher shrinkage in sintering and worse dimensional tolerances.

Typical powders used for MIM feedstocks (e.g., 17-4PH, 316L) are atomized using gas (water may be applied in certain cases) and, accordingly, have spherical particles and mean powder particle sizes ranging from smaller than 10 μm to larger than 60 μm. Small powder fractions can be obtained by classifying.

Ceramic powders suitable for powder injection molding mostly have powder particle sizes of < 1 μm and clear agglomerations. In principle, such agglomerates may be crushed during compounding when the shear forces occurring in the mixing process exceed the binding forces of the agglomerates [11]. If sufficiently high shear forces are applied, soft agglomerates can be crushed, while hard agglomerates are clearly more difficult or even impossible to de-agglomerate and thus have to be eliminated by, e.g., sieving [12].

12.2.1.3 Feedstock Preparation and Rheological Properties

As PIM is aimed at the manufacturing of high value-added products, optimum feedstocks are required. For this purpose, it is essential that a maximum de-agglomeration of the powder particles is achieved and that the powder is equally and homogeneously dispersed in the organic material. Suitable machines for feedstock preparation, especially with fine-grained powders for micro applications, are kneaders or different kinds of extruders.

The main differences between the two machines are the continuous mode of the extruder compared to the batch mode of a kneader and the higher shear rates an extruder is able to generate. As the use of an extruder leads to several liters of feedstock per hour, it is normally used for the production of large amounts of feedstock. The powder and binder components are filled in at the dosing section of the machine, the mixing process is performed while the material is transported through the machine, and at the extruding section, the feedstock is released continuously. To obtain an accurate compound, the use of gravimetric dosing is necessary whereas to achieve mostly equal granulate sizes, rotating blades are used for breaking the extruded strand. For feedstock production, co-rotating twin-screw extruders or shear roll compactors are used.

12.2.2 Debinding

As mentioned, the duty of the organic binder is to provide sufficient flowability for shape-giving; therefore, it has to be extracted afterward. The particular challenge of this so-called debinding step is the nearly complete elimination (except for backbone-polymer) of the organic binder without influencing the geometry of the part and the chemical purity of the material [13]. Potential debinding strategies usually comprise debinding by solvent or thermal extraction of the organic substances. For example, German and Bose [8] identified eight possible debinding strategies.

In liquid-phase debinding, the green body is placed in a suitable solvent. It is usually aided by heat or pressure. Thermal debinding is performed by, e. g., degradation, evaporation, or liquid extraction (wicking) of the organic binder often supported by the capillary forces of a powder bed.

In practice, often two or more debinding methods are combined to achieve successive binder removal and thus gentle debinding and short process times. Thermal debinding, for example, might be supported by a preliminary solvent extraction step.

Depending on the binder system, the debinding times may range between a couple of hours and several days. Luckily, as this time rises sharply with increasing wall thickness of the components, debinding times required for micro-sized parts are among the shortest in PIM technology.

12.2.3 Sintering

Last but not least, it is necessary to densify the remaining powder scaffold; i.e., the debindered part (now called brown body) has to be compacted by sintering. The procedure is performed under a vacuum, or inert gas, or in a reducing atmosphere, such as hydrogen, in order to avoid oxidation of the powder. Solidification of the material can be achieved; up to 96 to almost 100% of the theoretical density, depending on the material and the powder morphology. Depending on the powder or powder composition, it may encompass both liquid and solid sintering steps. The thermodynamic driving force is the reduction in surface energy, which is why extended sintering processes are likely to cause distortion. Compaction during sintering causes linear shrinkage by approximately 15 to about 23%, mainly depending on the powder type used and the original powder loading. The material properties of the sintered part, such as mechanical or electromagnetic features, mainly correspond to the densities achieved. The mechanical properties of sintered parts are in the range of those of press-sintered products and are only insignificantly inferior to those of components produced in the conventional way, for example, by melting metallurgy.

The interdependencies between feedstock composition, process parameters, part dimensions, and dimensional tolerances have been crucial for the whole process performance since the early days of PIM [14–19]. A detailed study of the influence of injection molding parameters can be found in, e.g., [20]. The accuracy of the sintered components can be improved by increasing the powder loading in the feedstock. However, this approach is limited, as the maximum degree of powder filling causes the viscosity to rise so dramatically that mold filling is no longer attainable. Another strategy for increasing the powder loading in feedstocks is the production of molding compounds with bimodal or multimodal powder particle distributions enabled by the use of ultrafine or even nanoscale powders. If the difference in particle sizes of two powder fractions is large enough, this approach allows the packing density of powder mixtures to be clearly increased [21].

12.3 Powder Injection Molding of Micro Components (MicroPIM)

As explained before, the unique nature of MicroPIM is to combine two special injection molding variants, namely, powder process technology and replication in micro dimensions [22–31]. Therefore, it shows particular differences compared to plastics injection molding as well as to macroscopic powder processing. The main aspects are described in the following sections.

12.3.1 Dissimilarities between Powder and Plastics Injection Molding

- Mixing solid particles and liquid binder usually leads to crucial material combinations. Therefore, thc risk of powder-binder segregation during injection is always latent and often causes subtle effects and difficulties.
- The relatively low strength of the molded green parts demands smooth and gentle demolding velocities.
- The ejector tops must have large areas and there must be large numbers of them to minimize the forces and moments, respectively, again owing to the relatively weak green compacts.
- Because of the high content of metal and ceramic powder fillers, respectively, the rheological materials data of the feedstocks differ greatly from those of unfilled polymers. Usually, the consequence is an increase in viscosity.
- The thermal conductivities of powder feedstocks exceed those of matrix polymers by a factor of up to 12, depending on the powder loading. On the other hand, heat capacity is lower, so freezing takes place quite fast. As this represents a detrimental effect for mold filling, it is advantageous with respect to avoiding large temperature gradients between surface and bulk. All these differences have a particular impact on the process management of the injection molding step. For example, due to the accelerated freezing processes, the lengths of the mold distances and the minimum wall thicknesses are limited.
- During injection, there is enhanced formation of cooled peripheral layers and early sealing of the sprue. This effect has consequences for mold design: the gates of MIM/CIM molds should be larger than those used for thermoplastic molds.
- Mainly due to the large shrinkage during sintering, the dimensional precision is limited, i. e., the nominal dimension can usually be met only in a range of roughly ± 0.3% to ± 0.5% or even worse. Values of up to ± 0.1% might be obtained for selected component dimensions and after thorough process optimization [32].
- As for nearly all powder metallurgical processes, the final devices are not perfectly dense. For example, typical PIM parts reach 95 to approximately 99% of the theoretical density.
- The surface properties depend highly on the particles sizes of the powders used.

Powder injection molding can be performed by using modified units of the kind also employed in widely conventional injection molding in industry. This means that not only injection molding machines but also periphery equipment and handling facilities, which must only be modified for the special features of PIM fabrication, are available at a comparably low cost.

However, it is not surprising that wear phenomena are an important subject in PIM. Still, the literature in this field shows hardly any specialized publications on the subject; i.e., reference to wear problems is made in a few papers, such as [33] or, with respect to molding accuracies, in [34].

The plastification unit of the injection molding machine should be wear-resistant, i.e., at least have hardened surfaces withstanding the abrasive action of the molding compounds. Quite often, aggregates with hard chromium plating or made entirely of hard metal are used. An additional benefit of the latter method is that it leads to lower contamination of the molded feedstock.

12.3.2 Dissimilarities between Macro- and MicroPIM

Similarly to plastics processing, the replication of metal and ceramic material in micro dimensions reveals distinctive differences [35–38]. In general, it has to be emphasized that all rules for design and process conduct in macroscopic PIM have to be considered to a greater or at least to the same extent in micro dimensions:

- Elimination of air in the tools is usually performed via very thin venting gaps. This method, however, is often not suitable for MicroPIM because the dimensions of the microstructures themselves correspond to those of these gaps. As an appropriate solution, the core of the tool can be sealed and evacuated. Evacuation, in addition, avoids undesired entrapment of air bubbles, which would lead to considerable weld lines.
- The extremely small dimensions of the micro-sized parts and the high degrees of precision of the respective tools require special manufacturing methods (e.g., LIGA), at least for the mold inserts that may be considerably complex and expensive.
- The micro cavities in PIM tools often show abrupt changes in channel diameters, increasing the risk of jetting, powder-binder segregation, or dead-sector effects.
- A typical feature of many micro parts is the high aspect ratio, i.e., the ratio of height to width of the microstructures. As one of the results, the material is exposed to high shear forces, which might lead to decomposition or mechanical degradation of the respective organic binder systems.
- It is easily understood that powder injection molding of extremely thin cavities requires ultrafine particles, in certain cases even nano particles [39–41]. Powders of such size are usually more expensive than the rougher fractions commonly used. Of course, material consumption in MicroPIM is much lower, which certainly relativizes the cost argument. The most important challenge when using ultrafine powders, however, is the large increase in viscosity caused by the disproportionally high specific surface of such small particles.
- Depending on the tool layout and the injection molding machine used, the material quantities injected are mostly very low. Luckily, some companies (Arburg,

Wittmann Battenfeld, etc.) have already developed and launched special micro injection molding machines or special injection molding units to be mounted on conventional machines [42].

- Micro-sized parts usually exhibit high surface-to-volume ratios, so freezing of the injected feedstock takes place much faster [43]. This effect can be compensated by an increased injection speed only to a very limited extent. Therefore, the central section of the tool is often heated before injection to avoid premature material freezing:
  Temperature at start of injection: $T_{tool} \geq T_{gfeedstock}$.
  To avoid deformation of the component, tools must be cooled before demolding starts.
  Temperature at start of demolding: $T_{tool} \ll T_{gfeedstock}$.
  This so-called variotherm process causes cycle times (typical: ≤ 50 seconds up to a few minutes) longer than the usual ones.
- Handling procedures: Subsequent processing and/or assembly of the small components demands high precision and often requires special equipment [44]. For example, it has to be taken into account that, due to their light weight, singular micro parts sometimes stick to the gripper instead of falling down.

The particularities and restrictions described above may lead to the impression that MicroPIM is an impracticable option. However, technological challenges of the process resulted in distinctive research attempts [35, 37, 45, 46] and a remarkable performance could be achieved. As impressive examples, two micro ceramic gear wheels produced by using a lithographically made (x-ray LIGA) mold insert are shown in Figure 12.2.

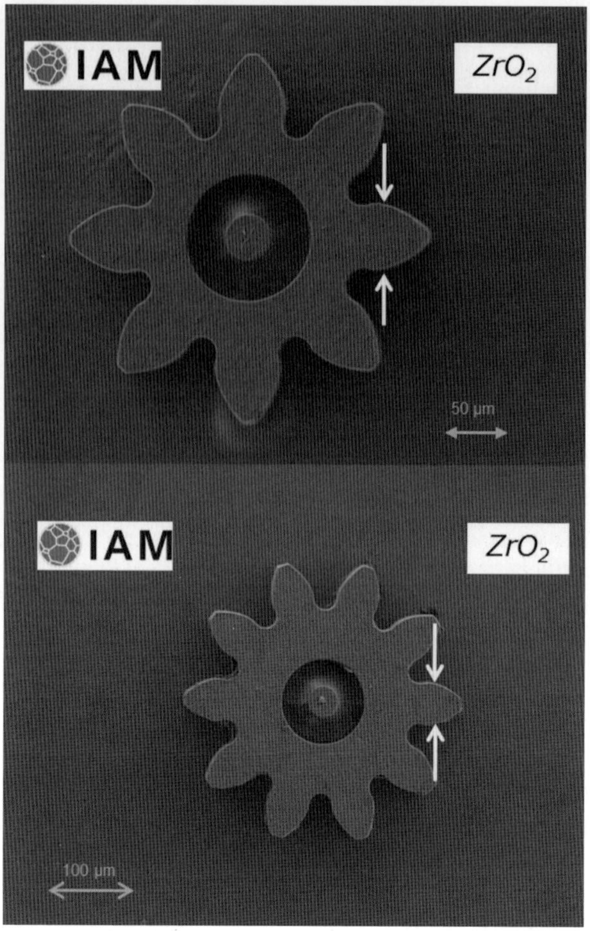

Figure 12.2 Ceramic gear wheels with different numbers of teeth made by micro powder injection molding using a LIGA mold insert; teeth width: ca. 50 µm. Courtesy of Karlsruhe Institute of Technology

The main outcomes of recent research and development are summarized as key data in Table 12.1. One essential parameter obvious from the data is the particle size of the powder used, which has a significant impact on both the achievable smallest dimensions as well as their surface quality [47–49].

Examples for MicroPIM parts already in practical use are given by Figure 12.3 and Figure 12.4.

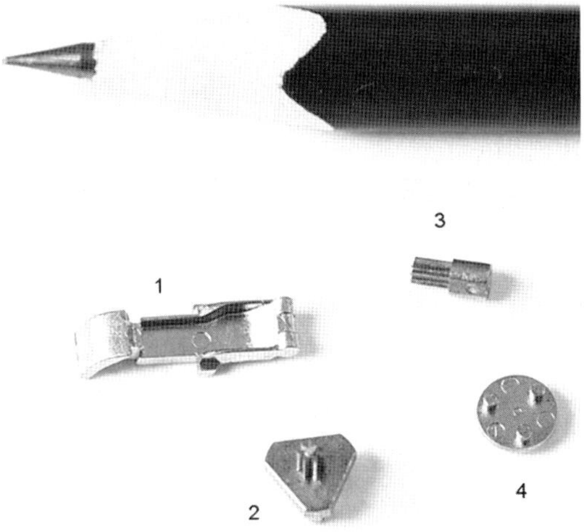

Figure 12.3 1: plug-in element for glass fiber connectors, material 7%NiFe; 2, 3: components for dental electro engines, material 17-4PH; 4: gear wheel carrier for micro electro engines, material AlSi4140. Courtesy of Parmaco Metal Injection Molding AG, Fischingen, Switzerland

Figure 12.4 Helical gear to be used for linear motor guides; diameter: 2.9 mm, module: 0.3 mm, tolerance: ± 0.01 mm, material: SUS360. Courtesy of Taisei Kogyo Co. Ltd., Osaka, Japan

Table 12.1 State of Micro Injection Molding Technology for Different Classes of Materials, Characterized by the Most Important Parameters

	Min. lat. dimensions [μm]	Min. details [μm]	AR*, salient structures	AR, immersed structures	Tolerances [%]	Surface roughness**, R_{max}/R_a [μm]
Metals	30	10	10	> 10	± 0.5	7/0.8
Ceramics	< 10	< 3	< 15	15	± 0.1 to ± 0.4	< 3/0.2
Polymers	<< 10	< 0.08	> 20	25	± 0.05	0.05/ < 0.05

* AR = aspect ratio, corresponds to ratio of flow lengths to wall thicknesses
** as a function of mold insert type used

12.4 2C Powder Injection Molding

Micro system technology products are used in a wide range of applications satisfying plenty of complex functions in a quite narrow space. It is not surprising that this multiplicity leads to quite various demands on the properties of the materials to be applied. In many cases, these demands can often not be met by one material choice alone. As a logical consequence, again in compliance with polymer injection molding, different process variants for processing two or more different materials have been developed [50]. Among these, great opportunities are expected for two-component micro injection molding (2C-MicroPIM), which represents an adequate method for series fabrication of components made of two or more materials.

Promising combinations of materials were described in [51], such as metals like Fe/316L, Fe/FeSi, FeSi/Fe, or FeNiMo/WCNi. Further publications about this topic followed soon [52–55]. Among others, two types of samples—namely, a layered composite as well as a core/shell composite—were studied and successfully produced for combinations of all-metal materials. Further promising material pairings for 2C-MicroPIM may, for example, consist of a combination of magnetic and non-magnetic metals. Other combinations are hard-tough or tight-porous materials.

There are again different possibilities to combine materials in one injection molding cycle. With respect to the merger of the feedstocks, one can distinguish between subsequent or simultaneous injection. In the case of the latter, (at least) two feedstocks filled with different powders have to be merged in a double-gate tool to obtain a composite green part.

Feedstocks must meet different, partly even contradicting, functions. They must allow for the complete filling of the micro-structured areas through low viscosity, deformation-free demolding due to high mechanical strength, pressureless debinding due to low swelling, distortion-free sintering due to homogeneous shrinkage, and highly adhesive compounds of at least two different ceramics or metals.

The most crucial challenge, however, is given by the adjustment of debinding and sintering parameters which have to be compatible for both powders to prevent excessive internal stresses. In a certain way, the risk of internal stresses promotes 2C-PIM to be applied in micro dimensions; however, even in such dimensions, compromises have to be made concerning compaction steps. The most important parameters are powder loading (which affects the sintering shrinkage), the type of powder (which affects the sintering activity), the sintering temperature, and the thermal expansion of the materials. If these requirements are appropriately met, even entirely different classes of materials—e.g., steels and ceramics—can be combined. Interesting examples have been produced and investigated by, e.g., Fraunhofer Institutes IKTS and IFAM [56, 57].

Similarly to polymer multicomponent technology, the 2C-PIM procedure can be configured to obtain both mobile and immobile bonds too [58]. To reach this quite challenging goal, not only the different thermal expansions but, furthermore, the different sintering properties of the partial feedstocks can be utilized. For immobile bonds, the powder contents and the sintering temperatures have to be nearly identical to achieve equal shrinkage rates and temperatures at which shrinkage starts. Examples of immobile bonds have been described in, e.g., [59].

It has to be mentioned that creating mobile bonds requires more engineering prudence because the inner section of a typical 2C-green body junction has to have a significantly higher shrinkage rate than the surrounding outer section. As a further important prerequisite, the sintering temperature of the inner section has to be lower so that while heating up, shrinkage of the inner section starts first and definite free-sintering of the partial volumes can be achieved. With respect to micro applications, gear wheel/shaft combinations consisting of different kinds of ceramics represent a further example of mobile bonds. Here, the difference in sintering temperatures has to be determined previously by dilatometric measurements to determine the right choice of powder type and size. Additionally, the injection pressure during mold filling has to be strictly controlled to avoid any kind of intermingling of the two partial volumes [60].

In addition to the above-mentioned approaches, 2C-MicroPIM can be performed in different sub-variants. As one option, instead of merging melt flows, pre-fabricated inserts in the form of foils or films could be covered by an injection molded body on the rear side [61]. Such an inmold-labeling (IML) approach using PIM feedstocks might use foils previously manufactured by, e.g., slip casting, foil casting, or rolling. Prior to inserting them into the injection molding tool, the foils can be printed, punched, embossed, or subjected to another preliminary treatment. In this way, it is possible to generate color patterns or lateral structures on the surface of the PIM body. Another advantage results from the fact that submicron powders or nano powders can be introduced into the foils without the drawbacks, i.e., a considerable increase in viscosity, usually occurring if PIM feedstocks have been filled with nano powders [62, 63].

Similarly to 2C-PIM of immobile bonds, the major challenge associated with PIM inmold-labeling consists in the development of an adequate sintering process. This sintering process has to allow for dense compacts of both partial volumes as well as for a solid junction between them.

12.5 Simulation of MicroPIM

12.5.1 MicroPIM Simulation by Use of Commercial Programs

As in all other commercial branches, vendors and customers of micro system products have a vital interest in the optimum performance of each particular device. For this purpose, parts have to be designed not only with respect to functionality; furthermore, material and manufacturing aspects and restrictions have to be taken into account as well.

At first glance, designing a MicroPIM part is subject to the same basic rules as designing a macroscopic powder injection molding part. The excessive heat loss due to the larger surface-to-volume ratio and the influence of relatively high shear rates can be considered severe differences. Additionally, the limitations given by the special technologies for mold insert fabrication and powder size distribution have to be considered. If complying to design rules is not satisfying, the utilization of computer programs simulating at least the injection molding process represents an appropriate strategy for further optimization. Computer assistance enables the detection and often the removal of weak spots at an early stage during parts development. In addition, it reduces the start-up phase of tools and the time until the introduction of the final products [64]. Concerning micro system technology in particular, it has to be taken into account that it is hardly possible to subsequently modify the micro-structured molding tools, so computer programs are advantageously applied to avoid design errors already at the design stage. Further on, process simulation can improve the components themselves technologically.

Several commercial software programs, such as:

- MOLDFLOW®, Moldflow Corp., Framingham, MA, USA
- Moldex3D®, CoreTech System Co. Ltd., Chupei City, NRC
- Sigmasoft®, Sigma Engineering GmbH, Aachen, Germany
- PIMsolver®, CetaTech, Sacheon, South Korea
- SIMUFLOW®, C-Solutions Inc., Boulder, CO, USA

are available for simulation of plastics injection molding. Generally, these programs are based on the finite element method and require preparatory volume modeling by means of 3D CAD/CAE systems.

The simulation programs require in advance the determination of the rheological, thermal, and mechanical data of the applied feedstocks. As demonstrated by many experiments so far, particular care must be given to the assessment of data of modified or newly developed micro injection molding feedstocks.

Profound utilization of software packages, as mentioned before, allows for the calculation of mold filling as a function of pressure, temperature, and material type.

The programs can detect in advance and avoid or at least reduce the areas where shear velocities are crucially high (of course, such areas are not unusual in micro cavities). As further advantage, many subroutines allow for the calculation of cooling and shrinkage processes as well as of special variants such as compression injection molding.

On the other hand, powder injection molding and especially MicroPIM reveal effects and particularities that cannot be depicted by commercial simulation programs so far. The reason is that they are based on single-phase material models, so phenomena like, e. g., powder-binder segregation [65] cannot be sufficiently simulated unless additional features are provided. Additionally, certain properties of the powder particles–e. g., higher inertia due to higher density–cannot be determined sufficiently and effects typical of PIM such as disproportionate formation of strands and folds, wall friction, and yield points often cannot be assessed correctly.

In particular for MicroPIM, the occurring shear rates are often above average and changes in cross sections of the flow channels are mostly quite abrupt, so micro system technology applications get more affected by the weaknesses described above than macroscopic PIM. Moreover, the higher surface/volume ratios of micro components must be considered to be aggravating all surface-dependent effects. Further on, it must be mentioned that the commercial software tools do not consider any of the typical special micro injection molding methods such as variotherm process control or tool evacuation. For this purpose, modified subroutines or even new simulation tools adapted to the needs of micro technology have to be developed in the future.

To summarize, having all the above-mentioned limitations in view, a comprehensive simulation of the MicroPIM process is quite problematic. Instead, profound knowledge and much experience are essentially required to be able to interpret simulated calculations and rheological processes in PIM melts. Although the process of micro injection molding is not described exhaustively, the specifically determined material data and experienced interpretation of the results obtained already allow for a helpful utilization of simulation tools within parts and process development phases.

12.5.2 MicroPIM Simulation Using Modified or Newly Developed Software Programs

Due to the difficulties and restrictions explained above, various approaches have already been started to improve the currently available simulation programs. Since the common quasi-laminar parabolic profiles proved to not precisely represent the actual PIM flow characteristics, it makes progress to select other flow profiles. Significantly better simulation results could be obtained by means of advanced material models using much more bulk profiles, e. g., Hershey-Buckley profiles.

From the commercial point of view, the most efficient way of reducing the time and effort of programming is to implement special features into existing software tools. Therefore, many software producers prefer investigating PIM-related effects indirectly to developing completely new material models. The occurrence of segregations, for example, can be deduced from the calculated shear rate profile.

A comprehensive example is described in [66]. Advanced filling simulation allows for the passable determination of the areas of high shear rates. These are well-known for being responsible for the occurrence of powder-binder segregations. As further improvement, Sigma Engineering incorporated additional equation terms describing the collision- and viscosity-induced particle flow, respectively. By this modification, a significantly better calculation of the occurring shear-induced powder-binder segregation could be achieved [67]. Furthermore, using a so-called autonomous optimization approach, it became possible to optimize part design and parameters by a low number of simulation runs to minimize shear gradients thus to significantly reduce particle segregation [68].

An even more complex and in a certain way really holistic approach, however, would be the implementation of multiphase material models. Several research facilities have developed approaches that may turn out to be attractive [69].

As an example for such comprehensive approaches, the R&D work carried out at IMTEK, University of Freiburg, shall be described here. A new material model, based on the smoothed particle hydrodynamics (SPH) method [70], has been generated. To enhance this model for sufficient MicroPIM prediction, two new features have been incorporated, i.e., an inherent yield stress and shear-induced powder segregation.

Simulation of inherent yield stress effects has been performed by means of a bi-viscosity approach [71]. For the simulation of shear-induced powder segregation during injection, important effects had to be considered. Due to different collision frequencies, the powder particles migrate toward areas of lower concentration and toward areas of lower shear rates. In addition, due to viscosity inhomogeneities, the powder particles migrate toward areas of lower viscosities.

After incorporating these approaches into the SPH model, a correct prediction of powder distribution in MicroPIM parts was achieved, as proven by CT measurements using synchrotron radiation [72].

This is not, of course, the only approach. For nearly the same objective–i.e., the creation of a PIM-specific simulation routine–new equation systems have been developed and comprehensively investigated at the University of Besançon. More detailed descriptions can be found in [73–75].

It shall not be forgotten that the above considerations only refer to the simulation of injection molding as the major shaping process. For the remaining process steps–especially for sintering–software routines are being developed or have already achieved a high state of the art too. Future concepts of entire process visualization must be aimed at providing integrated modelling and simulation schemes for all process steps from feedstock preparation over injection molding to sintering. Only in that way, comprehensive and reliable information on the performance and efficiency of the complete manufacturing process can be given to the users.

It is obvious that such holistic systems will require much development work and will lead to voluminous software packages; however, that shall be no problem in times of extensive computer utilization.

12.6 Summary and Outlook

In a nutshell, PIM offers the outstanding possibility of combining the excellent properties of metals and ceramic materials with the industrially established injection molding process.

The great attractiveness of PIM technology for commercial use can be summarized as follows:

- Large bandwidths of metals and ceramic materials can be processed, making it possible to obtain parts which can sustain various and high constraints
- The above includes processing of functional materials, such as new alloys, hard/soft magnets, piezo ceramics, etc.
- Outstanding economic efficiency in the case of large and medium-sized series production
- Complexly shaped parts can be manufactured very close to the final contours by only a small number of process steps
- A large variety of sub-variants for manufacturing of, e.g., multimaterial devices
- The basic machinery equipment is available commercially

The final conclusion for the utilization of PIM in Micro System Technology (MST) regards PIM as a reliable manufacturing process. It can be stated that it represents a key technology for producing micro components sustaining high loads and stresses, of geometries of different degrees of complexity, from metal alloys or ceramics, in medium to large quantities. It is easy to understand that micrometer- or, better, sub-micrometer-sized powders of preferably spherical shape are required for precision technology components which require very little or no reworking.

At present, several major trends can be discerned in the further development of powder injection molding. Of course, the reduction of failure rates in powder injection molding requiring improvements in all process steps is of major importance. Especially, the tendency toward powder-binder segregation, which cannot be suppressed completely and causes distortion and/or defects in the component, must be recognized and avoided as far as possible. The reliable detection of failures in the green body (cracks, cavities, etc.), preferably online during the process steps, represents another challenge for the future of PIM technology.

Again, powder-binder segregation is probably the main reason why the final accuracies, currently ranging between ± 0.1% and ± 0.5% of the nominal sizes, are not sufficient for all kinds of high value-added products. Approaches to improvement are an enhanced utilization of simulation routines to homogenize the powder distribution in the green bodies, or to increase the powder loadings to reduce the relative shrinkage rates.

As already mentioned, one of the big benefits of PIM is the wide range of metallic and ceramic materials; and even for MicroPIM, various feedstocks are commercially available. Current R&D approaches, for example, comprise the development of water-soluble, environmentally friendly binder components.

An important driving force for progress in PIM technology is the increase of the range of materials, allowing potential applications to be opened up, which is reflected, e.g., in the use of metals beyond ferrous materials, such as tungsten, titanium, and alloys of these. Further keywords under this aspect would be functional materials [76], refractory metal alloys, high-entropy alloys, and compositionally complex alloys.

At this point, it shall be mentioned that due to the low material consumption, MicroPIM offers a certain attractiveness for material development and testing, not least for the investigation of new feedstock concepts [77–80].

Last but not least, the various approaches toward manufacturing multicomponent parts, such as two-component powder injection molding, in-mold labeling with PIM feedstocks, or sinter bonding, have to be recognized. It shall not be forgotten that the aim of multicomponent PIM is not only to reduce assembly expenditure of highly complex products, but also to manufacture new functional units. Therefore, the various methods of multicomponent powder injection molding offer clear economic and technical potentials for sophisticated applications today and in the future.

References

[1] German, R. M., *Powder metallurgy and particulate materials processing: The processes, materials, products, properties, and applications*, Metal Powder Industries Fed., Princeton, N. J. (2005)

[2] Moritz, T., Lenk, R., Ceramic injection moulding: a review of developments in production technology, materials and applications, *Powder Injection Moulding International* Vol. 3, No. 3, Inovar Communications Ltd. (2009), pp. 23–34

[3] German, R. M., *Metal Injection Molding*, Metal Powder Industries Fed., Princeton, N. J., ISBN 978-0-9819496-6-6 (2011)

[4] http://www.epma.com/doc_details/249-metal-injection-moulding-a-manufacturing-process-for-precision-engineering-components-3rd-edition (2013), EPMA webpage, accessed on 11/07/2016

[5] Williams, N., Metal Injection Moulding: Building on solid foundations in the medical sector, *PIM International* Vol. 11, No. 1 (2017), pp. 43–57

[6] Schwarzwalder, K., Injection Molding of Ceramic Materials, *American Ceramic Society Bulletin* 28-11 (1994), pp. 459–461

[7] Murray, K., MIM Market Overview, Euro MIM Group Open Meeting, PM2016 World Congress, Hamburg, 9–13/10/2016, published by EPMA under: http://www.epma.com/document-archive-world-pm2016/576-world-pm2016-sectoral-group-meetings-2016-overview-of-the-european-mim-market/file, accessed on 24/03/2017

[8] German, R. M., Bose, A., *Injection molding of metals and ceramics*, Metal Powder Industries Federation, Princeton, USA, ISBN 1-878-954-61-X (1997), p. 80ff

[9] Mutsuddy, C., Ford, R. G., *Ceramic injection molding*, Chapman & Hall, ISBN 0 412 53810 5 (1995), pp. 65–137

[10] Hanemann, T., Weber, O., Polymethylmethacrylate/polyethyleneglycol-based partially water soluble binder system for micro ceramic injection moulding, *Microsyst. Technol.* 20 (2014), pp. 51–58

[11] Song, J. H., Evans, J. R.G, The assessment of dispersion of fine ceramic powders for injection moulding and related processes, *Journal of the European Ceramic Society* 12 (1993), pp. 467–478

[12] Boehm, H., Blackburn, S., Agglomerate breakdown in fine alumina powder by multiple extrusion, *Journal of Materials Science* 29 (1994), pp. 5779–5786

[13] German, R. M., *Powder injection molding*, Metal Powder Industries Federation, Princeton, N. J. (1990), pp. 5ff

[14] Billiet, R., The Challenge of Tolerance in P/M Injection Molding, in Howard I. Sanderow, ed. Annual Powder Metallurgy Conference proceedings, San Francisco, California, Metal Powder Industries Federation, Princeton, N. J. (1985), pp. 723–742

[15] Zauner, R., Binet, C., Heaney, D., Piemme, J., Variability of feedstock viscosity and its correlation with dimensional variability of green powder injection moulded components, *Powder Metallurgy* 47(2) (2004), pp. 151–156

[16] Greene, C. D., Heaney, D. F., The PVT effect on the final sintered dimensions of powder injection molded components, *Materials & Design* 28 (1) (2007), pp. 95–100

[17] Nishiyabu, K., Andrews, I., Tanaka, S., Accuracy evaluation of ultra-compact gears manufactured by the microMIM process, *Powder Injection Moulding International* Vol. 2, No. 4, Inovar Communications Ltd. (2008), pp. 60–63

[18] Knoepfle, C., Maetzig, M., Walcher, H., Influence of process parameter on the quality of MIM parts, Proc PM2016 World Congress, Hamburg, 9–13/10/2016, published on USB by EPMA, ISBN: 978-1-899072-48-4

[19] Osada, T., Control the Deformation of MIM Parts by the Powder Size Distribution. Proc PM2016 World Congress, Hamburg, 9–13/10/2016, published on USB by EPMA, ISBN: 978-1-899072-48-4

[20] Beck, M., Piotter, V., Ritzhaupt-Kleissl, H.-J., Hausselt, J., Statistical analysis on the quality of precision parts in ceramic injection moulding, Proc. of 10th Euspen Conference 2008, Zurich, Vol. 2, ISBN 978-0-9553082-5-3 (2008), pp. 179–183

[21] German, R. M., *Particle packing characteristics*, Metal Powder Industries Federation, Princeton, New Jersey, USA, ISBN 0-918404-83-5 (1989), p. 110

[22] Rota, A., Imgrund, P., Petzoldt, F., Micro MIM – a Production Process for Micro Components with Enhanced Material Properties, Proceedings of Euro PM 2004, European Powder Metallurgy Association, ISBN 1899072 15 2 (2004), pp. 467–472

[23] Petzoldt, F., Micro powder injection Moulding – challenges and opportunities, *Powder Injection Moulding International* Vol. 2, No. 1, Inovar Communications Ltd. (2008), pp. 37–42

[24] Piotter, V., Bauer, W., Hanemann, T., Heckele, M., Mueller, C., Replication technologies for HARM devices: status and perspectives, *Journal of Microsystem Technologies* Vol. 14, Springer (2008), pp. 1599–1605

[25] German, R. M., Materials for Microminiature Powder Injection Molded Medical and Dental Devices, *International Journal of Powder Metallurgy* Vol. 46, 2 (2010), pp. 15–18

[26] Attia, A. A., Alcock, J. R., A review of micro-powder injection moulding as a microfabrication technique, *J. Micromech. Microeng.* 21 043001, Vol. 4 (2011)

[27] Zhang, J., Sahli, M., Gelin, J.-C., Khan-Malek, C., Experimental analysis of the evolution of the physical properties of pyramidal-shaped metallic replicas made using the MIM process, *Int. J. Adv. Manuf. Technol.* 68 (2013), pp. 1063–1074

[28] Liu, L. et al., Filling of Microarray Fabricated by Micro Powder Injection Molding, *Materials Science Forum* 898 (2017), pp. 1171–1176

[29] Cho, H., Lee, J., Park, S. J., Fundamental Experiments for Fabrication of Copper Micro Heat Pipe by Powder Injection Molding, Proc PM2016 World Congress, Hamburg, 9–13/10/2016, published on USB by EPMA, ISBN: 978-1-899072-48-4

[30] Han, J. S. et al., Development of Manufacturing Technology for the Fabrication of High-Aspect-Ratio Piezoelectric Structures using Micro-Powder Injection Molding, Proc PM2016 World Congress, Hamburg 9–13/10/2016, published on USB by EPMA, ISBN: 978-1-899072-48-4

[31] Piotter, V., Klein, A., Mueller, T., Plewa, K., Manufacturing of integrative membrane carriers by novel powder injection moulding, *Microsyst Technol.* 22 (2016), pp. 2417–2423

[32] Islam, A., Giannekas, N., Marhöfer, M., Tosello, G., Hansen, H. N., A Comparative Study of Metal and Ceramic Injection Moulding for Precision Applications, Proc. of 4M/ICOMM2015 Conference, Research Publishing, Singapore, ISBN: 978-981-09-4609-8 (2015), pp. 567–570

[33] Loehe, D., Hausselt, J. et. al., *Advanced Micro and Nanosystems Vol.3 Microengineering of Metals and Ceramics*, ed.: H. Baltes, O. Brand, G.K. Fedder, C. Hierold, J. Korvink, O. Tabata, Wiley VCH-Verlag, Weinheim, ISBN: 3-527-31208-0 (2005), Chapters 10–12

[34] Schneider, J., Zum Gahr, K.-H., Herz, J., *Tribological Characteriazation of Mold Inserts and Materials for Microcomponents*, In: Advanced Micro and Nanosystems Vol. 4. Microengineering of Metals and Ceramics, ed.: H. Baltes, J. Korvink et al., Wiley VCH, Weinheim, ISBN 3-527-31493-8 (2005), pp. 579–603

[35] Loh, N. H., Tor, S. B., Tay, B. Y., Murakoshi, Y., Maeda, R., Fabrication of micro gear by micro powder injection molding, *Microsystem Technologies*, Springer Verlag, 14 (2007), pp. 43–50

[36] Fu, G., Tor, S., Loh, N. H., Tay, B., Hardt, D. E., A micro powder injection molding apparatus for high aspect ratio metal micro-structure production, *Micromechanics and Microengineering*, IOP Publishing, 17 (2007), pp. 1803–1809

[37] Nishiyabu, K., Tanaka, S., Small is better if testing MIM nano theories, *Metal Powder Report* No. 3, Elsevier (2008), pp. 28–32

[38] Tirta, A., Prasetyo, Y., Baek, E.-R., Choi, C.-J., Study of Micropart Fabrication via 17-4 PH Stainless Nanopowder Injection Molding, *Journal of Nanoscience and Nanotechnology* Vol. 11 (2011), pp. 249–255

[39] Baek, E.-R., Supriadi, S., Choi, C.-J., Lee, B.-T., Lee, J.-W., Effect of Particle Size in Feedstock Properties in Micro Powder Injection Molding, Proceedings of 2006 Powder Metallurgy World Congress, Korean Powder Metallurgy Institute, ISBN 89-5708-121-6 94580, Vol. 1 (2006), pp. 41–42

[40] Schmidt, H., Rota, A. C., Imgrund, P., Leers, M., Micro metal injection moulding for thermal management applications using ultrafine powders, *Powder Injection Moulding International* Vol. 3, No. 2 (2009), pp. 54–58

[41] Rajabi, J., Muhamad, N., Sulong, A. B., Effect of nano-sized powders on powder injection moulding: a review, *Microsyst. Technol.* 18 (2012), pp. 1941–1961

[42] Maetzig, M., Walcher, H., Haupt, U., New Injection Moulding Equipment for PIM Microparts, Proc.

Euro PM2011, Vol. 2, European Powder Metallurgy Association, Shrewsbury, UK, ISBN 978-1-899072-21-7 (2011), pp. 195–200

[43] Kuhn, S., Burr, A., Kübler, M., Deckert, M., Bleesen, C., Study on the replication quality of micro-structures in the injection moulding process with dynamical tool tempering systems, *Microsyst. Technol.* 16 (2010), pp. 1787–1801

[44] Freundt, M., Brecher, C., Wenzel, C., Hybrid universal handling systems for micro component assembly, *J. Microsystem Tech.* 14 (2008), pp. 1855–1860

[45] Heaney, D., Mass Production of Micro Components Utilizing Lithographic Tooling and Injection Molding Technologies, First International Conference on Micro-Manufacturing, ICOMM 2006, No.: 61 (2006)

[46] Okubo, K., Tanaka, S., Ito, H., Nishiyabu, K., Manufacturing of 316L Stamper for Imprinting by Micro Sacrificial Plastic Mould Inserted, Proc. Euro PM2011, Vol. 2, European Powder Metallurgy Association, Shrewsbury, UK, ISBN 978-1-899072-21-7 (2011), pp. 183–188

[47] Liu, L., Loh, N.H., Tay, B.Y., Tor, S.B., Murakoshi, Y., Maeda, R., Micro powder injection molding: Sintering kinetics of microstructured components, *Scripta Materialica* 55, Elsevier (2006), pp. 1103–1106

[48] Piotter, V., Hanemann, T., Heldele, R., Müller, M., Müller, T., Plewa, K., Ruh, A., Metal and ceramic parts fabricated by microminiature powder injection molding, *Int. Journal of Powder Metallurgy* Vol. 46, No. 2 (2010), pp. 21–28

[49] Piotter, V., A review of the current status of MicroPIM, Part 1: Materials, processing, microspecific considerations and applications, *Powder Injection Moulding* Vol. 5, No. 3, Inovar Communications Ltd. (2011), pp. 27–36; Part 2: Screw injection units, simulation and process variants, *Powder Injection Moulding* Vol. 5, No. 4, Inovar Communications Ltd. (2011), pp. 25–30

[50] Johnson, J.L., Tan, L.K., Suri, P., German, R.M., Design Guidelines for Processing Bi-Material Components via Powder-Injection Molding, *Journal of Min. Met. Mat. Sc.* 55 (10) (2003), pp. 30–34

[51] Pischang, K., Birth, U., Gutjahr, M., Becker, H., Steger, R., Investigation on PM-composite Materials for Metal Injection Molding, Advances in Powder Metallurgy & Particulate Materials 1994, Vol. 4, Proc. of PM^{2}TEC Conf. (1994), pp. 273–284

[52] Alcock, J.R., Logan, P.M., Stephenson, D.J., Metal co-injection moulding, *Journal of materials Science Letters* 15, Chapman & Hall (1996), pp. 2033–2035

[53] Moritz, T., Two-component CIM parts for the automotive and railway sectors, *Powder Injection Moulding International* Vol. 2 No 4, Inovar Communications Ltd. (2008), pp. 38–39

[54] Petzoldt, F., Multifunctional parts by two-component Powder Injection Moulding (2C-PIM), *Powder Injection Moulding International* Vol. 4, No. 1 (2010), pp. 21–27

[55] Li, Y., He, H., Liu, P., Zhang, J., Design and Manufacture of Gears with a Skin-Core Structure by Metal Co-Injection Moulding, *Powder Injection Moulding International* Vol. 4, No. 1 (2010), pp. 50–54

[56] Dutra, G., Mulser, M., Calixto, R., Petzoldt, F., Investigation of material combinations processed via Two-Component Metal Injection Moulding (2C-MIM), *Materials Science Forum*, Trans Tech Publications, Switzerland, Vols. 727-728 (2012), pp. 248–253

[57] Moritz, T., Mannschatz, A., Müller-Köhn, A., Kucera, A., Current opportunities and challenges for CIM technology, *Ceramic forum international: CFI* 89, No. 10, ISSN: 0173-9913 (2012), E21–E25

[58] Maetzig, M., Walcher, H., Assembly moulding of MIM materials, Proc. Euro PM 2006 – Powder Metallurgy Congress & Exhibition, 23–25/10/2006, Published by EPMA, ISBN 978-1-899072-33-0, Vol. 2 (2006), pp. 43–48

[59] Ruh, A., Dieckmann, A.-M., Heldele, R., Piotter, V., Ruprecht, R., Munzinger, C., Fleischer, J., Haußelt, J., Production of two-material micro assemblies by two-component powder injection molding and sinter-joining, *Journal of Microsystems Technology* 14, Springer (2008), pp. 1805–1811

[60] Ruh, A., Hanemann, T., Heldele, R., Piotter, V., Ritzhaupt-Kleissl, H.-J., Hausselt, J., Development of Two-Component Micropowder Injection Moulding (2C MicroPIM): Characteristics of Applicable Materials, *International Journal of Applied Ceramic Technology* 8(1) (2011), pp. 194–202

[61] Mannschatz, A., Härtel, A., Müller-Köhn, A., Moritz, T., Michaelis, A., Wilde, M., Manufacturing of Two-colored Co-sintered Zirconia Components by Inmold-labelling and 2C-Injection Molding, cfi/Ber, DKG 91 (2014) No. 8, E1–E5

[62] Baumann, A., Brieseck, M., Hoehn, S., Moritz, T., Lenk, R., Development in multi-component powder injection moulding of steel-ceramic compounds using green tapes for inmould label process, *Powder Injection Moulding International*, Vol 2 No. 1; Inovar Communications Ltd. (2008), pp. 55–58

[63] Vorster, E., Piotter, V., Plewa, K., Kucera, A., Micro Inmould Labelling Using PIM-Feedstocks, Proceedings of Powder Metallurgy 2010 World Congress, EPMA, ISBN: 978 1 899072 13 2, Vol. 4 (2010), pp. 505–510

[64] Ahn, S., Chung, S.-T., Park, S.J., German, R.M., Application of a simulation tool in powder injection Moulding (PIM), *Powder Injection Moulding International* Vol. 3, No. 4, Inovar Communications Ltd. (2009), pp. 60–65

[65] Jenni, M., Schimmer, L., Zauner, R., Stampfl, J., Morris, J., Quantitative study of powder binder separation of feedstocks, *Powder Injection Moulding International* Vol. 2, No. 4 (2008), pp. 50–55

[66] Thornagel, M., MIM-Simulation: Benefits of Advanced Rheology Models; Proc. Euro PM2011, Vol. 2, European Powder Metallurgy Association, Shrewsbury, UK, ISBN 978-1-899072-21-7 (2011), pp. 201–206

[67] Thornagel, M., Schwittay, V., Hartmann, G., Powder-Binder Segregation: PIM-Simulation at Breakthrough, Proceedings of Euro PM 2014 conference, published by EPMA Shrewsbury, UK, ISBN 978-1-899072-45-3. Recession can also be found in PIM International Vol. 8 No. 4 (2014), pp. 67–69

[68] Gebauer, T., Optimizing Particle Segregation in the Metal Injection Molding Process, Proc PM2016 World Congress, Hamburg 9–13/10/2016, published on USB by EPMA, ISBN: 978-1-899072-48-4

[69] Kauzlaric, D., Lienemann, J., Pastewka, L., Greiner, A., Korvink, J.G., Integrated process simulation of primary shaping: multi scale approaches, *Microsystem Technologies* 14 (2008), pp. 1789–1796

[70] Monaghan, J.J., Smoothed particle hydrodynamics, *Reports on Progress in Physics* 68 (2005), pp. 1703–1759

[71] O'Donovan, E.J., Tanner, R.I., Numerical study of the bingham squeeze film problem, *Non-Newtonian Fluid-Mechanics* 15 (1984), pp. 75–83

[72] Kauzlaric, D., Pastewka, L., Meyer, H., Heldele, R., Schulz, M., Weber, O., Piotter, V., Hausselt, J., Greiner, A., Korvink, J.G., Smoothed particle hydrodynamics simulation of shear-induced powder migration in injection moulding, *Phil. Trans. R. Soc. A* 369 (2011), pp. 2320–2328

[73] Barriere, T., Quinard, C., Kong, X., Gelin, J.C., Michel, G., Micro Injection Moulding of 316L Stainless Steel Feedstock and Numerical Simulations, Proceedings of EuroPM 2009 Conference, Copenhagen, European Powder Metallurgy Association, ISBN 978 1 899072 07 1 (2009), pp. 325–330

[74] Kong, X., Barriere, T., Gelin, J.C., Micro-PIM Process with 316L Stainless Steel Feedstock and Numerical Simulations. Proceedings of Powder Metallurgy 2010 World Congress, EPMA, ISBN: 978 1 899072 13 2, Vol. 4 (2010), pp. 669–676

[75] Claudel, D., Sahli, M., Gelin, J.C., Barriere, T., An Inconel based feedstock and its Identification of rheological constitutive model for powder injection moulding. Proc. of 4M/ICOMM2015 Conference, Research Publishing, Singapore, ISBN: 978-981-09-4609-8 (2015), pp. 322–325

[76] Han, J.S. et al., Fabrication of high-aspect ratio micro piezoelectric array by powder injection molding, *Ceramics International* 42 (2016), pp. 9475–9481

[77] Choi, J.P. et al., Design of trimodal Fe micro-nanopowder feedstock for micro powder injection molding, *Powder Technology* 317 (2017), pp. 1–5

[78] Choi, J.P. et al., Analysis of the rheological behavior of Fe trimodal micro-nano powder feedstock in micro powder injection molding, *Powder Technology* 319 (2017), pp. 253–260

[79] Oh, J.W. et al., Effect of nanopowder ratio in bimodal powder mixture on powder injection molding, *Powder Technology* 302 (2016), pp. 168–176

[80] Oh, J.W. et al., Influence of nano powder on rheological behavior of bimodal feedstock in powder injection molding, *Powder Technology* 311 (2017), pp. 18–24

13 Multimaterial Micro Injection Molding

Aminul Islam

13.1 Introduction

Standard micro injection molding is an established and widely used process, with special variants of micro molding processes, such as multicomponent micro injection molding, micro insert molding, micro powder injection molding etc., becoming more and more popular. These especial processes are seen as production methods that provide advantages beyond the capabilities of standard micro molding processes. Demand for multicomponent micro parts is growing rapidly, especially in the field of automotives, electronics, medical instruments, optical industries, consumer electronics, and so on. Micro/nano manufacturing products, and their associated production equipment, are expected to represent a market of over 420 billion euros by 2035 [1]. Multimaterial and multifunctional integrated products are becoming, in most cases, key economic and technological factors for the majority of advanced applications. The success of future products and processes will depend highly on manufacturing systems that can reliably and economically combine different materials to produce multicomponent and multifunctional micro products. By combining different materials, it is possible to manufacture components featuring a wide spectrum of properties. The continuous trend toward miniaturization and multifunctionality embedded in products calls for more and more functionally versatile multimaterial components for future micro systems. As handling and assembly are difficult and expensive procedures in micro technology [2, 3], methods to reduce mounting efforts have high economic significance. By merging shaping and assembly processes into a single step by multimaterial (also known as multicomponent) molding, significant economic progress can be made in micro manufacturing.

Considering the potential of multicomponent micro injection molding, this chapter will focus on the different aspects of multimaterial micro injection molding. Special emphasis will be placed on two-component micro injection molding as this is the basic process of multicomponent micro injection molding. The challenging task for two-component molding is to find a material pair that fulfills the diverse requirements for engineering applications and at the same time has a reasonably good bond and sharp interface between the two polymer materials. Many fascinating applications of two-component or multicomponent polymer parts are restricted due to the weak interfacial adhesion and interface quality of the polymers. A thorough understanding of the factors that influence the bond strength of polymers is necessary for multicomponent polymer processing. This chapter focuses on the parameters that influence the quality of polymer–polymer bonding and the interface and analyses the relations between the bonding and the interface quality with a special focus on micro scale applications. The results and discussion presented in this chapter will provide a guideline for multicomponent micro molding suitable for a wide range of industrial applications.

13.2 Multimaterial Molding and Multimaterial Micro Molding

13.2.1 Introduction and Applications

Multimaterial and multicomponent molding processes are also called multi-C and multi-K molding, respectively, where C and K stand for "Component" and "Kunststoffe" (Kunststoffe means plastics in German). The multicomponent molding process is defined in many different ways. According to Gouker et al., multimaterial or multicomponent molding is the molding process that creates a part with two or more materials via multi-stage molding [4]. Multicomponent molding involves two or more injection units on the same machine filling cavities in a sequence—when the first set of cavities is filled with the first material, parts are moved or rotated to accept the next shot. "Multi" literally means more than one. The presence of an already molded component during the second or subsequent molding stages makes multimaterial injection molding different from the conventional molding process [5]. Multicomponent molding can be generalized as a molding process that deals with more than one molding material. Injection molding of two or more different materials, or injection molding of the same material in two different colors, or molding the same materials in foam–solid or virgin–regrind combinations can be defined as multicomponent molding processes [4–6].

Some examples of multicomponent parts used in everyday life are presented in Figure 13.1. Multicomponent molding is often used to increase product functionalities, aesthetic appeal, stability, product differentiation, and marketability of traditional designs [7, 8]. In a multicomponent part with a hard–soft plastic combination, the hard component provides the strength and stiffness whereas the soft component acts as a sealant or insulation or provides good haptics. In most cases, the main purpose of combining plastics in an injection molding process is to save costs by reducing the number of processing, handling, and assembly steps or joining processes [9, 10]. Furthermore, the use of bonding agents and primers required for joining the two components can be eliminated completely. One common example of a multicomponent part is a toothbrush handle where two or more materials are combined in hard and soft combinations. The hard plastic provides the strength needed for the toothbrush handle and the soft material provides the necessary grip. Moreover, the aesthetics of the toothbrush is increased with the combination of materials in different colors.

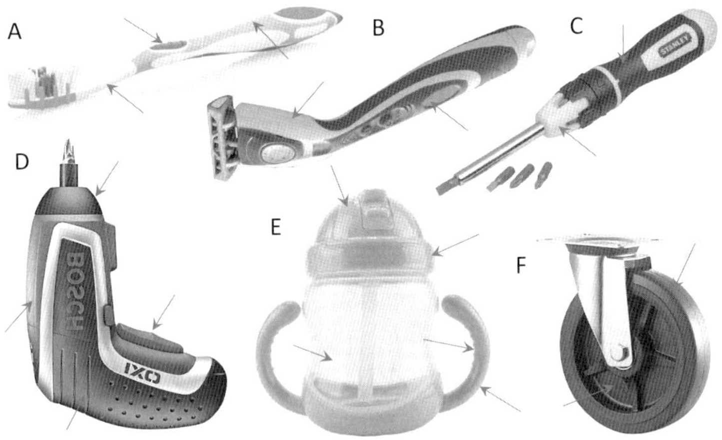

Figure 13.1 Some common examples of multicomponent plastic parts (the arrows indicate different plastic materials combined in the parts). The sources of pictures A, B, C, D, E, and F are references [11], [11], [12], [13], [14], and [15], respectively

Multicomponent micro injection molding is the molding of multicomponent parts on a micro dimensional scale. If the overall part size is in the mm range or the part weight is in the mg range, the process can be called multimaterial micro molding. If the overall part size is not in the mm range but some features of the part are in the µm range or the surface roughness is in the sub-µm range, the molding of the multicomponent parts can still be characterized as multicomponent micro injection molding. Some examples of multicomponent micro parts are presented in Figure 13.2.

3K connector for micro applications 3K micro gear wheel 2K parts for hearing aid applications

Figure 13.2 Examples of multicomponent micro parts. The sources of pictures B and C are references [16] and [17], respectively

The primary advantages of the multicomponent process are the ability to combine different materials' properties in single products and save the cost of assembly operations. Several studies have indicated that the assembly costs make up 40–50% of the manufacturing costs of the conventional products [18]. On the other hand, the assembly cost of micro products can reach 80% of their overall cost due to the predominance of manual assembly operations and the difficulties associated with the handling of the micro parts [19]. Multicomponent processes offer a wide range of opportunities to use materials cost-effectively and combine materials for optimum functionality. While doing these, multicomponent processes can eliminate the cost of the secondary assembly operations both for micro and macro applications. Figure 13.3 schematically illustrates the advantages of multicomponent methods compared with the traditional method.

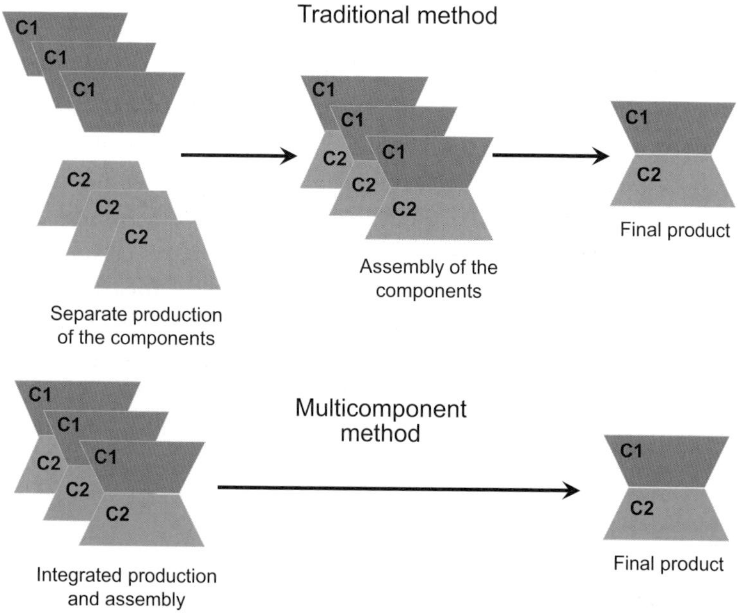

Figure 13.3 Schematic representation showing a comparison between a traditional production process and a multicomponent production process

Reflecting on the history of the multicomponent process, it can be seen that in 1962, G. Carozzo patented the first multicomponent process for injection molding of automotive taillights with multiple colors [15]. Co-injection molding, with high-pressure and sequential injection, was developed by ICI (England) in the 1970s to produce a foam-core "sandwich" structure [20]. In 1980, Jules M. Hock and Donald S. de Vries developed a novel over-molding technique for a piston with a rigid core and integrated flexible seal [21]. The widespread industrial uses of the multicomponent process started around the mid-'80s.

13.2.2 Application Areas

The applications of multicomponent molding are widespread. In the European market, about 20% of the market share of injection molded parts is taken up by multicomponent molding, which has a growth rate of about 15%. The growth rate of the conventional molding market is about 7%. The significant difference with the multicomponent market growth indicates that more and more plastic products are being redesigned to harness the advantages of multicomponent molding. The important application areas of multicomponent molding are:

- Automotive
- Machine tools
- Medical equipment
- Cosmetics and toys
- Household and consumer electronics

Over the past few decades, advances in plastic technology have grown exponentially for the invention of new materials and efficient processes. The global engineering plastics market was worth about $57.2 billion in 2014 and is expected to reach $91.78 billion by 2020 [22] and this growth in the plastics technology will be in the realm of the three 'M's–mega, micro, and multi [23]. This means that plastic-based large scale structures, micro parts, and multimaterial parts will drive future development. So, when they are multi and micro, this holds the key for future industries. Multicomponent micro parts will be widely used in the near future for applications such as medical implants, drug delivery and diagnostic devices, micro systems, and micro mechanics, etc. [24].

13.2.3 Advantages and Disadvantages

Multicomponent molding is an exciting technology that offers many advantages. All the known advantages of conventional multicomponent injection molding are valid for multicomponent micro injection molding, for example:

- Integrations of different material properties and functions in a single part
- Versatile options for assembly of the components, reduction of the number of components
- Reduction of process steps, tolerance chain, and assembly effort
- Realization of moveable connections, decoration, aesthetics, and marking
- 3D design freedom, breakdown of a complex 3D shape
- Shorter cycle and machine set-up time (compared to two "mono material" machines)
- Enhanced quality assurance, reproducibility, and less factory space

Besides the above facts, there are some special advantages of multicomponent micro injection molding such as:

- Handling and assembly of the micro parts present big challenges in many manufacturing scenarios due to the small size of the components. These problems can be reduced in micro manufacturing with the use of multicomponent molding where intermediate handling of the parts is eliminated and different components of the parts are assembled inside the mold during the molding process.
- The high surface area to volume ratio of micro parts changes the wettability, friction, adhesion, surface-to-surface interaction, thermal characteristics etc. These can be advantageous for multicomponent molding. For example, a polymer material pair with poor natural adhesion may be ill-suited to macro products, but it may still work very well in multicomponent micro molding because of the size effect and the large surface [25].
- The multicomponent molding technique can be a promising solution to overcome ejection problems associated with high aspect ratio microstructures. The micro parts that are difficult to eject from the mold due to the smaller dimensions can be molded on a carrier plate made with a plastic material that will have a weak adhesion with the micro parts. The ejector placement and the ejection of the part can be done easily with the aid of a carrier plate and, later, the micro parts will be easily separated due to the weak adhesion with the carrier plate [25].

Multicomponent molding also has some shortcomings both in conventional and micro scale molding. The primary disadvantage of multicomponent molding is the cost. Both machines and mold are complex and expensive. For example, machines involving sliding, or 360°, rotary platens may be about twice the cost of a conventional injection molding machine. For multicomponent micro injection molding, the complexity of the mold making is increased compared to single component micro injection molding, hence the cost. So, it is essential that the production volume of a multicomponent micro part is high enough to support the investment. A standard multicomponent molding machine and a 3K mold for automotive lighting are presented as examples in Figure 13.4. Multicomponent molding requires specialist knowledge about part design, mold making, materials, machines, and the process, which can be considered as another disadvantage of the process.

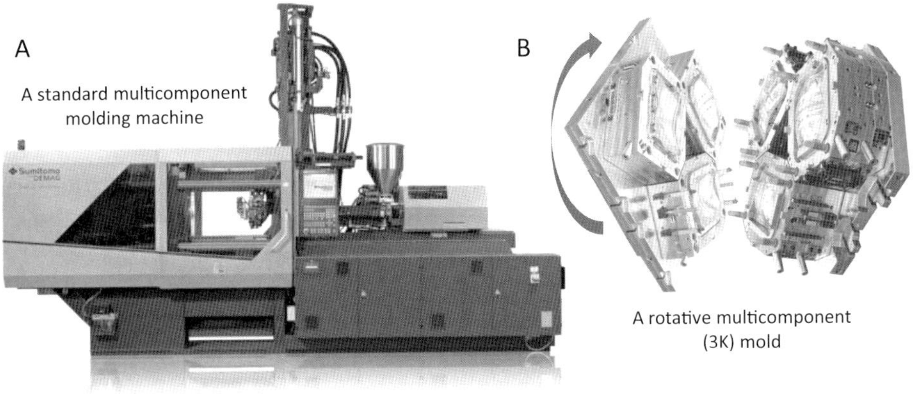

Figure 13.4 A standard multicomponent injection molding machine from Sumitomo [26], and a rotative multicomponent mold for automotive lighting [27]

It has previously been mentioned that the high surface-to-volume ratio of micro parts can be advantageous for multicomponent molding, but it can also be a disadvantage in some cases. It can be difficult to obtain a sharp and well-defined interface in the case of multimaterial micro parts. Micro parts or features that are supposed to be integrated with another material during multicomponent molding can be easily deformed, distorted, or melted by the heat coming from the second material due to their high surface-to-volume ratio. The larger surface facilitates heat transfer into the micro parts. The thermal inertia of the micro parts is much smaller compared to the macro parts [28]. So, micro parts can melt more quickly than larger parts, which will have relatively low surface-to-volume ratios.

13.2.4 Variants of Multimaterial Molding

Based on the manufacturing process, multimaterial molding can be divided into main sub-groups like co-injection molding, sequential molding, and over-molding. The first two types of injection molding are done in the same machine and mold. The over-molding is done by a cavity transfer process in a single mold machine system or in two different machines and molds. There are many other variants of multicomponent molding based on the mold mechanism and how different materials are injected into the mold. Some important variants of multicomponent molding are [29]:

- Bi-injection: simultaneous or sequential injection of two components into one cavity.
- Core-back technique: injection of two components into one cavity one after the other; the space for the second component is accessed by shifting a valve.
- Rotary technique: transfer to the new injection point is by axial or vertical rotation with the rotary function provided in the machine or in the mold.
- Sandwich technique: different plastics are molded together to produce a skin-core structure.
- Transfer technique: the preform is transferred manually or by means of a robot into the second cavity or to the second machine.

13.3 Two-Component Micro Molding

The basic process of multicomponent molding is two-component molding where two different plastic materials, in particular their associated properties, are combined in the same product. Two-component molding is also known as 2C molding or 2K molding. Two-component (2K) injection molding enables manufacturers to produce demanding multifunctional parts meeting different—and often contradictory—requirements. Parts with integrated functions (e.g., sliding bearings), multilayer parts (e.g., sandwich parts), in-mold assembled parts like hinge or moveable joints (by selecting materials that do not adhere to one another) etc., can be produced by 2K molding. 2K molding is widely used for hard-soft material combinations to provide good grip and tactile feelings. With a hard-soft material combination, mechanically sealed or air-tight parts are also realized. It is also one of the most industrially suitable processes for manufacturing so-called molded interconnect devices (MIDs) [30]. Two of the most important technical considerations both for macro and micro scale 2K molding are [31]:

1. A high polymer-polymer bond strength (in most applications, good bonding is critical to the mechanical performance of the final product and also to prevent plastic parts from breaking during ejection after molding).
2. A sharp and well-defined interface between the two plastic materials (for aesthetic reasons and also for technical reasons like the selective metalization of polymers to make electrical structures, a sharp and well-defined interface between materials is required).

The challenging task for two-component molding is to find a material pair that fulfills the diverse requirements for engineering applications and at the same time has a reasonably good bond between the two polymers in the pair and an acceptable interface. In the following sections, a thorough investigation to understand the factors affecting polymer–polymer adhesion and sort out important parameters that could be used as tuning factors for bond strength is presented. Moreover, the dilemma of a choice between polymer–polymer bonding and interface quality are discussed.

13.3.1 Polymer–Polymer Bonding

Many fascinating applications of two-component or multicomponent polymer parts are restricted due to the weak interfacial adhesion of the polymers. A thorough understanding of the factors that influence the bond strength of polymers is necessary for multicomponent polymer processing. Adhesion between two polymer materials is a complex phenomenon. The adhesion between two reactive polymers is governed by a chemical reaction (when the necessary conditions for the reaction are present) between the molecules of two polymers at the interface and the strength achieved by this process is high [31]. In a case with the same grade of polymers, the adhesion is governed by intermolecular diffusion at the interface. If the material pair is held together under favorable pressure and temperature, the molecules start to diffuse at the interface from either side and after a certain time the interface disappears, and the adhesion strength obtained by this process is the same as the strength of the material [33]. But the mechanism of polymer adhesion becomes complex when two different polymers interact at the interface, and the degree of adhesion is governed by the entanglement between the molecules of two polymers at the interface [31]. What is important for multicomponent polymer processing to achieve a good adhesion is the formation of the interfacial zone/mixing zone where two materials can mix together. The structure and strength of the interface formed are crucial for the bond strength of the multicomponent plastic parts. A wide mixing zone usually results in a strong interface whereas a weak interface is formed when the mixing zone is absent or narrow. Formation of the interfacial zone/mixing zone in multicomponent polymer processing is schematically presented in Figure 13.5.

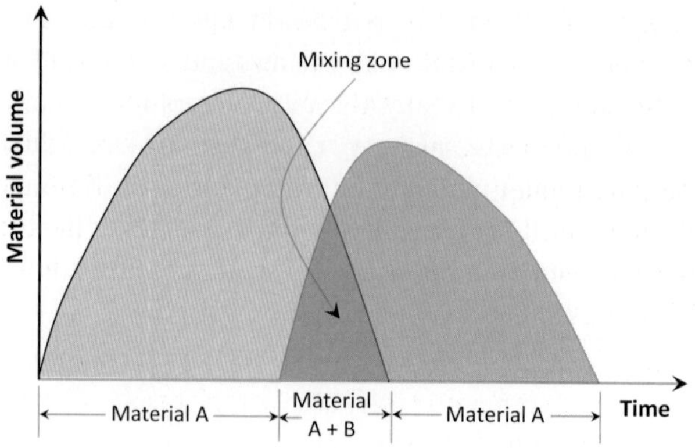

Figure 13.5 Schematic representation of the formation of a mixing zone (interface) between two polymers during multicomponent molding (edited from [34])

For multicomponent polymer processing, the adhesion mechanism and the quality of adhesion can be different based on the molding process. For example, when a co-injection molding technique is used, where two materials are injected at the same time into the mold from two different sides, the materials are more likely to have good bonding than in the over-molding process, where a second material is injected when the first material is already in the solid state. The worst bonding scenario is associated with the cavity transfer process of multicomponent molding. During 2K molding by a cavity transfer process, one part is molded first and then it is transferred to another cavity where a second polymer material is injected to over-mold the first part. In this section and in the following experimental cases, 2K molding with a cavity transfer process will be the focus. There are several reasons to choose a cavity transfer process as a test case. If two plastic materials achieve good adhesion in a cavity transfer process, it is highly likely that the material pairs will have good adhesion in the other variants of multicomponent processes. Cavity transfer molding (when one material is in the solid state while the other material is injected) is valid for many other variants of multicomponent molding, such as sequential molding, rotational molding, coreback molding etc.

In the case of the cavity transfer process, the adhesion at the interface of the two polymers may be explained by one or more of the following facts:

- When a polymer part is over-molded with a second polymer, the polymer melt flows inside the micro grooves of the substrate (first shot part) surface and makes a mechanical lock. This is very much governed by the surface roughness of the first shot part and the flow behavior of the second shot polymer melt. The geometry of the 2K molded part can contribute to the mechanical locking of the two polymers. If the part contains drafts or undercuts, then it is almost impossible to separate the two polymers without breaking the part.

- The diffusion theory of adhesion is particularly applicable for identical polymer materials and is not exactly valid for 2K molding. Here, the time is not sufficient for complete diffusion; and in many cases of 2K molding, parts consist of two different grades of polymers and it makes diffusion more unlikely to happen. However, if the part remains inside the mold for a long time after the second polymer is injected, and the two polymers have un-crosslinked molecules with chemical similarity, then a small degree of molecular diffusion is possible depending on the temperature range and time.
- In 2K molding by a cavity transfer process, the most probable case of adhesion can be molecular entanglement at the interface rather than the complete diffusion of molecules at the interface. The adhesion can be very much affected by the degree of molecular entanglement of different molecules at the interface. During 2K molding, the two polymers are brought together under heat and pressure. So, the thermally excited molecules of the second shot melt can entangle with the first shot polymer chain if they are in the melt state because of the interface temperature attained. A higher degree of entanglement will make the interface stronger in the solid state.
- In the case of 2K molding, one can roughly imagine the first shot polymer as a solute and the second shot melt as a solvent. If the second shot melt is a suitable solvent for the first shot polymer, then the solid substrate will melt at the interfacial zone and the two polymers will mix to form a combined solid state of materials.

13.3.1.1 Hypotheses on Adhesion of 2K Molded Polymers

Two-component injection molding, in a broad sense, is a welding (thermally induced bonding) of two polymers. Higher temperatures and pressures should affect the joining of the two polymers in a positive way. In that respect, the hypothesis is that the higher process settings (mold temperature, melt temperature) in the molding machine will be favorable for a higher bond strength between the first shot and second shot polymers. The thermal and rheological properties of the polymer melt can influence adhesion between the two polymers.

An obvious influence of the chemical properties of polymers on the bond strength will always be there. If the two polymers have similar chemical properties and solubility parameters, they are expected to form better bonds with each other. Environmental factors like humidity, thermal history, and a harsh environment can affect the bond strength during the manufacturing of two-component parts and also in the service conditions.

The surface energy of the 2K polymer part is an important factor for the adhesion of two polymers. A higher surface energy of the substrate improves the flow and contact of the second shot polymer melt and maximizes the adhesive performance (see Figure 13.6, picture A). On the other hand, a higher surface tension of the second shot polymer melt prevents the complete wetting of the substrate and makes it difficult to flow inside the micro grooves of the substrate surface and the result is a weak adhesion between two subsequently molded polymers (see Figure 13.6, picture B).

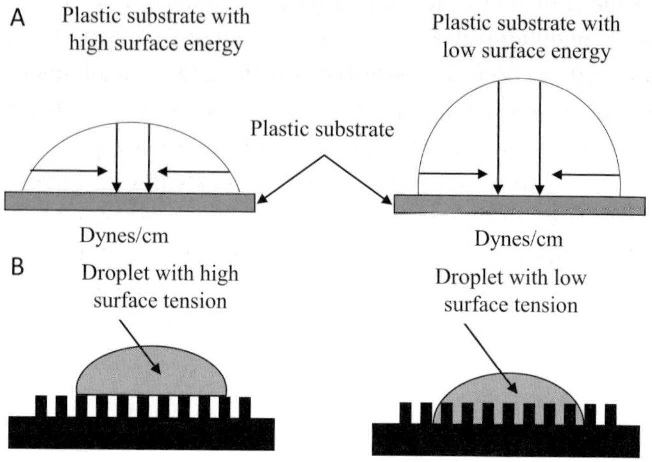

Figure 13.6 Influence of substrate surface energy on polymer–polymer adhesion (a high surface energy of the substrate increases the spreading of the second polymer melt and increases the adhesion)–image A [31]. Influence of surface tensions of the second shot polymer melt on the polymer–polymer adhesion (a high surface tension of the second shot melt prevents the melt flow inside the micro cavities of the polymer substrate and inhibits the adhesion)–image B [31]

The roughness of the inlay part (first shot part) can increase the mechanical locking of the second polymer and increase the interaction area of the interface. The roughness can facilitate heat transfer and can increase the localized melted zone so that the polymers can melt together to form a good bond. Different part geometries can influence the fracture behavior and crack propagation mechanism, and that will influence the energy absorption before the two-component part break at the interface. The design of the plastic part can also influence the bond strength. The drafts and undercuts, shrinkage of the second polymer and weld line (two melt fronts meet face to face) or meld line (two melt streams meet side by side) formations can influence the polymer–polymer adhesion. Fillers and additives may influence the bond strength; especially, glass fibers can increase the surface roughness and change the material stiffness, affecting the bond strength.

A weak boundary layer, such as contamination, air traps, moisture etc., can influence the bond strength of two materials. Removing the weak boundary layers from the inlay part can maximize the adhesion. This is an important consideration for injection molding by a cavity transfer process. During the transfer of the first shot part to the second cavity, it can be contaminated with moisture, dust particles, and other impurities (for example, sweat in the case of manual handling) and these can affect the bonding with the second polymer to a considerable extent. Moreover, special environmental conditions, like a corrosive environment or high temperature exposure, can affect the bond strength of the post-molded 2K parts.

The adhesion mechanism of 2K molded parts is complex and is a combination of many effects or phenomena. For a strong adhesion between the two polymers, some molecular interactions are required, which could be a chemical reaction like covalent bonding or molecular diffusion, or an entanglement of polymer chains at the interface. When it comes to the point of adhesion between two polymers in a micro scale product, some special considerations should be taken into account. Micro scale parts have a larger surface area to volume ratio than macro scale parts. So, the wettability, friction, and adhesion between the second shot polymer melt and the first shot substrate are more critical at the micro scale. In the case of micro parts, the interface temperature at the second shot can easily melt the first shot part to form a good mix of the two materials or it may even completely destroy the part geometry.

13.3.1.2 Experimental Investigations and Results

An extensive investigation aimed at finding suitable polymer pairs for 2K molding with a special focus on micro applications was performed by Islam and colleagues and the details for the investigation are presented in [31, 35–41]. The objective of the investigation was to find polymer pairs that had good adhesion and could form a sharp and well-defined interface between the pairs. For the experiments, an ISO recommended tensile test specimen was chosen as reference geometry. The final two-component test specimen design is shown in picture A (Figure 13.7); the parts were molded by a cavity transfer process in a modified Engel-135 machine (picture B of Figure 13.7). During the first step of injection molding, the first shot part was made and in the second step, this was combined with a different plastic material. Picture C (Figure 13.7) shows a finished 2K molded test specimen made by PEI1000 and PEEK material combination. The tensile tester used to determine the bond strength of two polymers was LLOYD 500, produced by Lloyd instruments (picture D of Figure 13.7), and the maximum capacity of the machine was 2500 N. All the tensile tests were performed at 5 mm min^{-1} test speed.

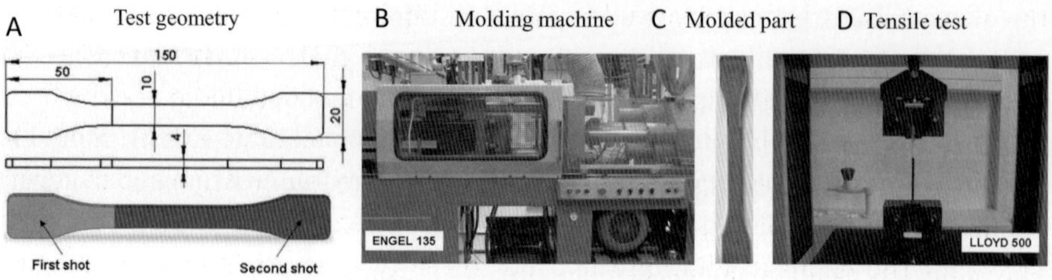

Figure 13.7 Two-component test geometry (A) [31], molding machine used for the experiment (B) [31], molded test part (C) [31], and the tensile test set-up (D) [31]

Table 13.1 contains a list of materials used for the experiments. The materials were selected based on their suitability for use in micro- and high-performance engineering applications. Some commonly used engineering plastics were also chosen for the experiments.

Table 13.1 List of Materials Used for Polymer–Polymer Bond Strength Investigation

No	Name	Abbreviation	Trade name	Grade	Manufacturer
1	Polyetherimide	PEI	Ultem	1000	GE
2	Polyetherimide	PEI	Ultem	2312EPR	GE
3	Polyetheretherketone	PEEK	Victrex	150GL30	Victrex
4	Polyoxymethylene	POM	Hostaform	C27021	Ticona
5	Liquid crystal polymer	LCP	Vectra	E820i	Ticona
6	Liquid crystal polymer	LCP(Pd)	Vectra	E820i Pd	Ticona
7	Liquid crystal polymer	LCP(LDS)	Vectra	E820i LDS	Ticona
8	Polybutylene terephthalate	PBT	Pocan	DP7102	Lanxess
9	Polybutylene terephthalate	PBT	Vestodur	GF30FR LDS	Degussa
10	Polyphenyleneether blends	(PPE+PA+GF)	Noryl	GTX810	GE
11	Polyphenyleneether blends	(PPE+PA)	Noryl	GTX964	GE
12	Polyphenyleneether blends	(PPE+HIPS)	Noryl	GFN1520V	GE
13	Polystyrene	PS	Polystyrol	143E	BASF
14	Polystyrene	PS	Polystyrol	158K	BASF
15	Polystyrene	PS	Polystyrol	158KGf30	BASF
16	Acrylonitrilebutadien-estyrene	ABS	Terluran	997VE	BASF
17	Polycarbonate	PC	Lexan	500R	GE

The above-mentioned materials were paired in different combinations to make 2K test parts, which were characterized by tensile testing. One important thing that was noticed during the tensile test was that the polymer combinations either with LCP or POM usually imparted poor adhesion with other polymer materials. The results are presented in Figure 13.8. POM and LCP are two widely used materials in micro injection molding due to their excellent mechanical properties and suitability for the micro molding process. But these two materials may not be good options for multicomponent micro injection molding when good bonding is required. If LCP or POM are required in multicomponent molding, a mechanical locking system (for example, drafts, undercuts etc.) in the design is advised. Figure 13.9 presents the tensile test results from 2K parts made with two different polymer materials (excluding LCP and POM in the material combinations). The highest strength was obtained in the case of PEI1000–PEEK molding (37 MPa); PEI1000 also adheres strongly with PC. ABS has reasonably good adhesion with many other polymers.

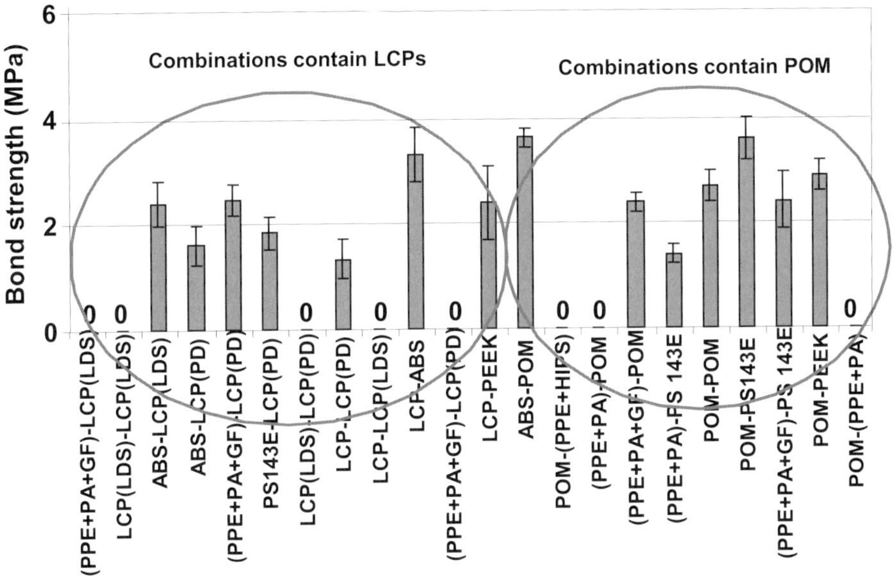

Figure 13.8 Bond strength of 2K molded tensile bars—parts molded with LCP and POM in combination with other materials (the error bars indicate the standard deviation calculated from five data points) [31]

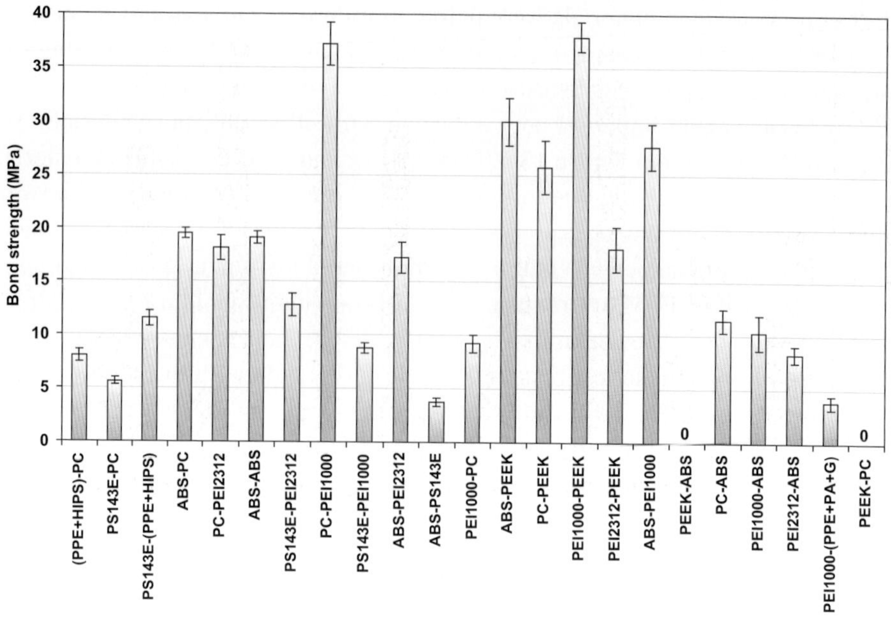

Figure 13.9 Bond strength of 2K molded tensile bars with various material combinations (the error bars indicate the standard deviation calculated from five data points) [31]

An interesting fact to notice in Figure 13.9 is the adhesion of two different polymers at different shot sequences in the 2K molding process. For example, when ABS is molded at the first shot and is combined with PC in the second shot, the bond strength is about 20 MPa. In the opposite shot sequence, meaning when PC is molded in the first shot and combined with ABS in the second shot, the bond strength obtained is about 11 MPa. So, the same material combination but different shot sequence makes a huge difference in the bonding of the two materials and many examples of these are presented in Figure 13.10. To precisely understand why this happens, it is important to understand the interface temperature of the two materials when the second shot material is injected.

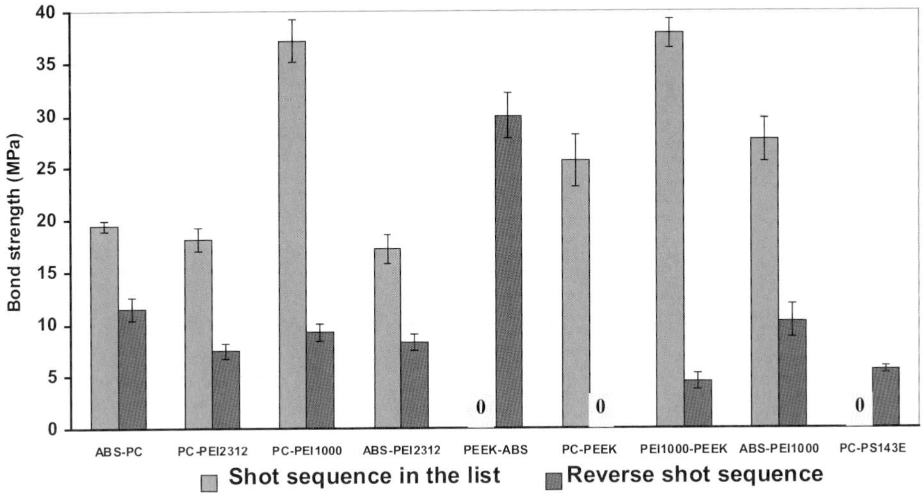

Figure 13.10 Bond strength as a function of shot sequence (material sequences along the *x*-axis indicate the initial shot sequences and the opposite sequences indicate the reverse shot sequences–the error bars indicate the standard deviation calculated from five data points) [31]

The interface temperature of 2K molding can be defined as the temperature of the surface where two materials meet. Usually, the temperature of the second shot melt is much higher than the inlay part's temperature (the part molded in the first shot). In the best-case scenario, the inlay part's temperature can be equal to the mold temperature in the cavity transfer process unless the first shot part is preheated to a higher temperature than the mold temperature. When the melt touches the surface of the inlay part, the melt temperature goes down and the inlay part's temperature goes up. At some point, these two temperatures reach a steady-state condition and this steady-state temperature can be defined as the interface temperature of the 2K molding. If the interface temperature is high enough to melt the surface of the inlay part, then the two materials can melt together and have an entanglement among the polymer chains, which should lead to a higher adhesion force (for some cases, even a sufficient interface temperature cannot create good bonds due to the special chemical or surface properties of both or either of the materials). The interface temperature is affected by the melt and mold temperature, and by the thermal and heat transfer properties of the polymers. In the 2K molding situation, the interface temperature can be estimated by the following formula [42]:

$$T_i = \frac{b_1 T_1 + b_2 T_2}{b_1 + b_2} \qquad (13.1)$$

$$b = \sqrt{k \rho C_p} \tag{13.2}$$

where

T_i = interface temperature

b_1 = thermal effusivity of the inlay part

b_2 = thermal effusivity of the melt

T_1 = mold temperature

T_2 = second shot melt temperature

k = thermal conductivity of the material

ρ = material density

C_p = specific heat capacity of the material

For 2K molding, the interface temperature phenomenon is schematically presented in Figure 13.11. During the process, when the melt touches the inlay part, the melt temperature (T_2) goes down, the mold temperature T_1 is raised, and then these two temperatures reach an equilibrium point, which is the interface temperature (T_i) of the two materials. If T_i can melt the surface of the inlay part, then the two materials can have better bonding. The thermal effusivity (b) of the material determines the interface temperature between the plastics through a weighted average of the two temperatures. The unit of the thermal effusivity is $Ws^{½}\ K^{-1}\ m^{-2}$. It is an important thermophysical property, which characterizes the thermal impedance of matter and indicates the ability of the matter to exchange thermal energy with the surroundings [42].

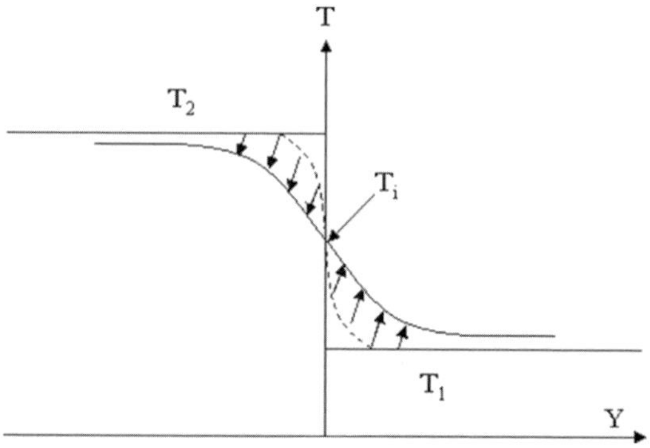

Figure 13.11 Interface temperature of the inlay part and the second shot melt (T_1 is the inlay part temperature, T_2 is the second shot melt temperature, and T_i is the interface temperature between the first shot and second shot polymer–a higher T_i ensures better mixing of the two polymers) [42]

To validate the claims made about interface temperature, PEEK-PEI1000 was used as a reference. The combination had poor bond strength when PEEK was molded first and PEI1000 was injected in the second shot (4 MPa). The same material pair was much stronger simply by reversing the shot sequence (37 MPa). To explain this, the interface temperature of the inlay part and second shot melt was calculated. The material data were edited from the CES EduPack [43], Moldflow database [44], and from the supplier data sheets for the second shot melt and for the first shot solid inlay. A semi-crystalline polymer has a crystalline melting point, which is sufficient to ensure the mixing of the polymer in 2K molding. But an amorphous polymer does not have a melting point. It only has the glass transition temperature, which indicates the softening of the polymer but is not sufficient to ensure mixing with other polymers. So, the amorphous melting point was taken from the average of the glass transition temperature and the recommended lower processing temperature of the amorphous polymer. According to the author's experience, the calculated amorphous temperature was sufficient to ensure mixing of the polymer with another polymer melt. The formula used for the calculation of the amorphous melting point was ½ (glass transition temperature + recommended lower process temperature).

The relationship between interface temperature and bond strength for the PEEK and PEI1000 material combination is schematically presented in Figure 13.12. In the first case of the PEEK-PEI1000 combination, the calculated interface temperature was 231 °C and the melting point of the inlay PEEK was 343 °C. On the other hand, with the PEI1000-PEEK combination, the interface temperature was 297 °C and the melting point of the inlay part was 282 °C, which explains why the bond strength was much lower in the first case than in the second case. The same explanation is valid for all material combinations that are presented in Figure 13.10, and showed a significant difference in the bond strength based on shot sequences.

First shot	Second shot	Bond strength (MPa)
PEEK	PEI1000	4

First shot	Second shot	Bond strength (MPa)
PEI1000	PEEK	37

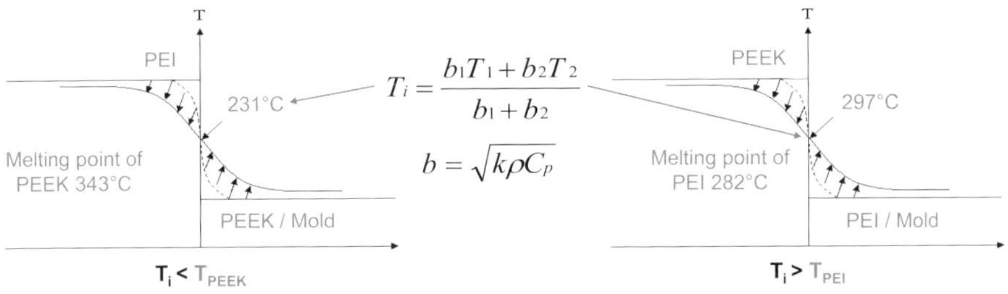

Figure 13.12 The interface temperature of the PEEK-PEI1000 combination for both shot sequences and their influence on the bonding of the two materials

13.3.1.3 Other Factors Affecting Bonding between Polymers

Besides the interface temperature, a large number of factors affects the bonding between two polymers. Some of these factors are highlighted in picture A of Figure 13.13. Among the injection molding process parameters, the process temperatures are the most important parameters to modulate the bond strength. Higher mold and melt temperatures increase the interface temperature of the inlay part and the second shot melt. Higher interface temperatures facilitate the melting of the inlay part at the interface so that the two polymers can melt together (as discussed in the previous section). Higher interface temperatures also increase molecular the inter-diffusion and entanglement rate at the bond interface. Injection pressure and holding pressure have some influence on the bond strength. Higher injection pressures and holding pressures increase the mechanical locking of the second shot polymer melt at the interface. Injection speed affects the interfacial adhesion as a higher injection speed increases the melt temperature due to shear heating and decreases the melt viscosity. The experimental results presented in [31] show that holding and cooling times affect the bond strength when they are set to insufficient values and any further increment of these two parameters away from the required values does not have any positive effect on the bond strength. The comparative influences of the various molding parameters are edited from [31] and presented in picture B of Figure 13.13.

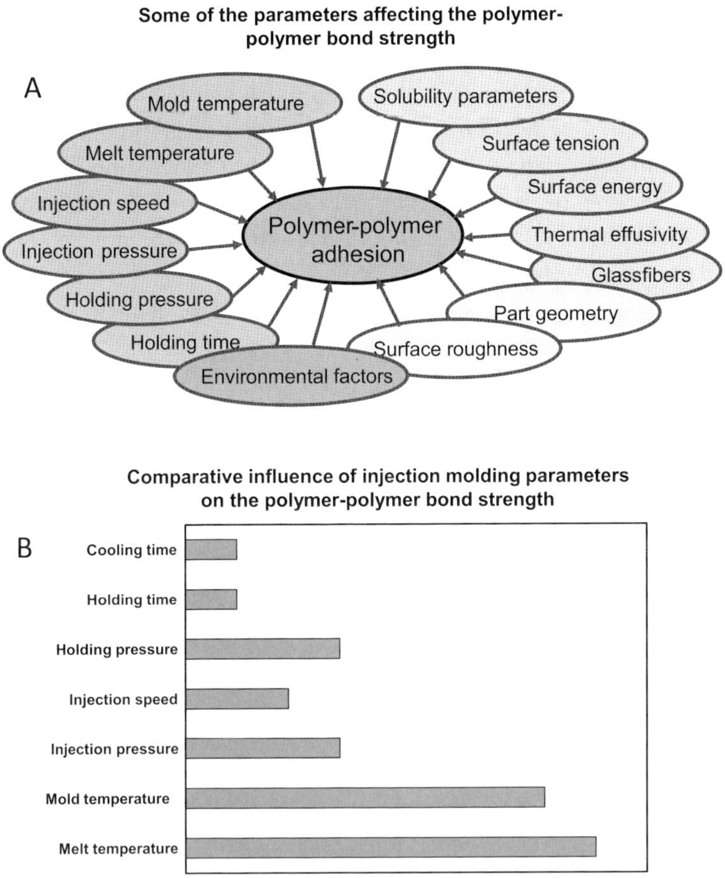

Figure 13.13 Factors affecting polymer–polymer adhesion (A), relative effects of the injection molding process parameters on the adhesion of two polymers (B) [35]

13.3.1.4 Effects of Surface Roughness on the Bonding of Two Polymers

The surface roughness of the inlay part or the part made in the first shot can play a vital role for polymer–polymer adhesion as it plays a vital role in polymer to metal adhesion for an insert molding scenario [45]. In the investigation presented in [35], to find the influence of the interface roughness on the bond strength of two component parts, three different roughnesses were intentionally generated at the interface surface of the ABS inlay parts by grinding. Three different rough surfaces had S_a values (average roughness values) of 0.6 µm, 1.7 µm, and 3.4 µm, respectively. Samples with pre-defined roughnesses were over-molded in the second shot by PC. The tensile test results showed a big influence of the interfacial roughness on the bond strength (picture A of Figure 13.14). From the plot, it is clear that in a certain range, the interfacial roughness can cause a dramatic change in the adhesion. The rough surface increases the mechanical interlocking of the melt on the inlay surface. It also increases the interfacial contact area of the two polymers. An increased surface area affects the bonding in two possible ways: it increases the heat transfer rate from the melt to the solid part and provides a bigger area for mechanical interactions. Besides mechanical locking, the rough surface facilitates localized melting of the inlay part. The peaks of the rough surface act as extended heat transfer elements and can easily be melted with the second shot melt. Figure 13.14 (picture B) schematically shows how the roughness peaks concentrate the heat flux and create a localized melting zone.

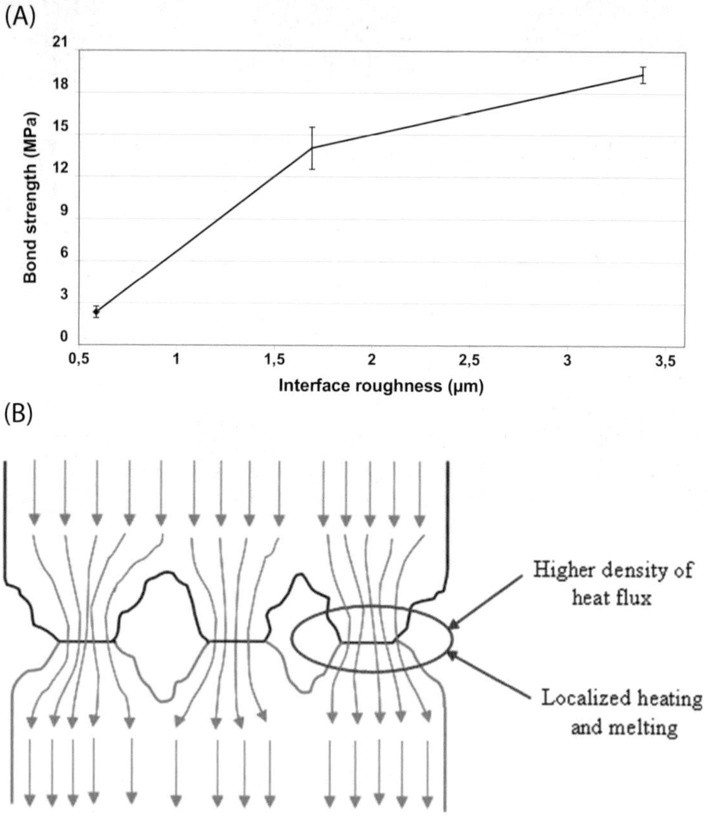

Figure 13.14 Bond strength vs. interface roughness (A) [35], effects of roughness on localized melting (B) [35]

Figure 13.15 illustrates the effects of roughness on the bond strength of two-component injection molded parts. It presents comparative pictures of the broken surfaces of ABS and PC after the breaking of injection molded 2K (ABS–PC) parts by tensile testing. The broken interfaces show a higher degree of weaker material (ABS) traces on the stronger material (PC) side in the case of the higher roughness. This high degree of penetration of one material inside the other increases the mechanical locking, maximizes the mixing, and consequently increases the bond strength.

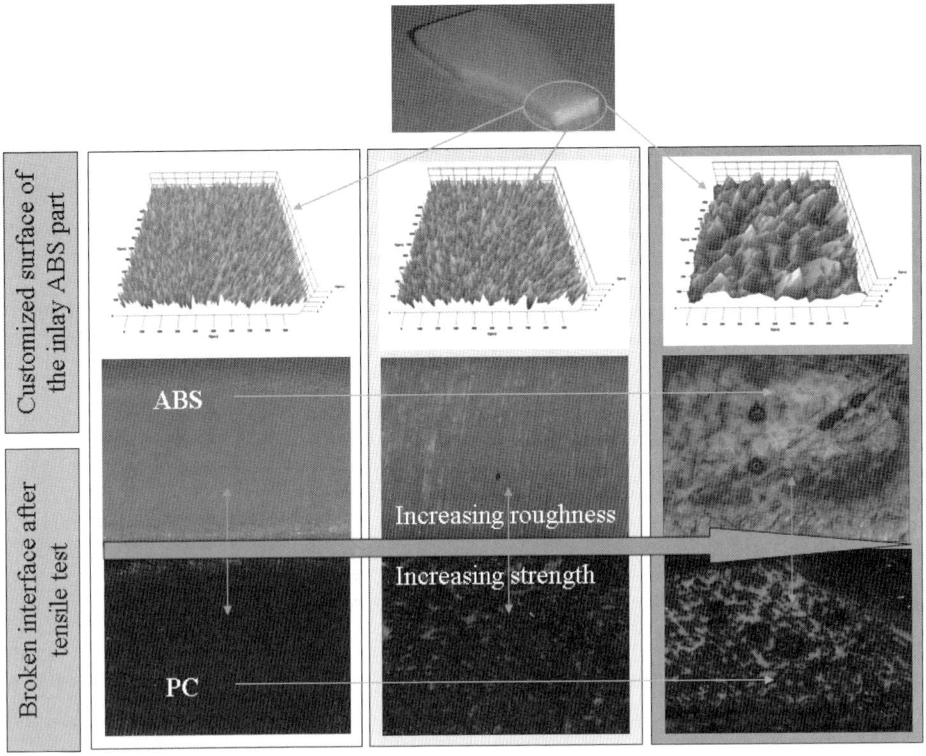

Figure 13.15 Effects of interface roughness on polymer–polymer adhesion—with an increased roughness, an increased amount of traces of one material is visible on the other material at the broken surface [35]

13.3.1.5 Effects of Environmental Factors

The influence of environmental factors on the mechanical and chemical performances of plastics is a well-studied area. With the growing application of two-component injection molding, the environmental effect on polymer–polymer bond strength is of huge interest, especially for applications where the parts are exposed to humidity, a cyclic temperature, or a corrosive environment. One particular example is molded interconnect devices (MIDs). Both for macro and micro scale MIDs, the environmental factors demand special attention because the post-molded parts undergo special environmental conditions during the chemical metalization process (electroless metal plating of plastic parts). For 2K parts used in sealing or fluid flow applications, parts exposed to variable temperatures and environmental conditions etc., environmental reliability is a concern. In references [31, 35], extensive investigations on the effects of some special environments (listed in Table 13.2) are presented. These tests were specially formulated keeping the MID applications in mind, but the results are relevant for many other applications. The moisture test simulates general service conditions where the parts are exposed to high moisture. To simulate the long-term effect of moisture exposure and to shorten the test time, the parts were exposed to boiling water for a short time. The corrosion stress simulates the metalization environment and the temperature test simulates basically the service conditions. The results from this investigation provide comprehensive information and recommendations concerning the loss of polymer–polymer bond strength during the metalization process. Three combinations of materials were selected for the environmental tests: ABS–PC, PC–PEI1000, and PEI1000–PEEK.

Table 13.2 Methods for Different Environmental Tests

Test name	Recipe
Moisture test	Boiling of 2K samples in water (100 °C) for 10 min
Corrosion test	2K samples in 25% (v/v) H_2SO_4 solutions for 30 min
Temperature test	Samples are at -18 °C for 72 h and in +80 °C for 30 min

Figure 13.16 (picture A) shows the effects of different test conditions on different material pairs with respect to the references (when the parts are not treated with special environmental conditions). Figure 13.16 (picture B) indicates the comparative effects of different test environments on different material pairs. All the combinations are adversely affected by boiling water and the percentage reduction in bond strength correlates with the weight percentage of water absorption by the two polymers, which means that the higher the water absorption by the two polymers, the higher the reduction in bond strength (shown in picture A of Figure 13.17). The most interesting result was observed in a temperature test. In this test, ABS–PC and PC–PEI1000 were weakened but the PEI1000–PEEK combination gained bond strength. To understand this abnormality, more tests were done with PEI1000–PEEK.

Figure 13.17 (picture B) shows the temperature test results from the PEI1000–PEEK combination. It was found that the greater the amount of time spent under the temperature treatment, the better the adhesion. The material pair was strongest at 160 °C for 1 h. At 240 °C for 1 h, the bond strength decreased drastically. Possibly, the reason for the increased bond strength of PEI1000–PEEK under temperature treatment could be co-crystallization at the interface between PEI1000 and PEEK. PEEK is a semi-crystalline material and it has a glass transition temperature of 143 °C. A temperature annealing slightly above its T_g helps PEEK to form more crystals and form a stronger bond with PEI1000. On the other hand, at 240 °C, PEI1000 starts softening and loosens its bond with PEEK.

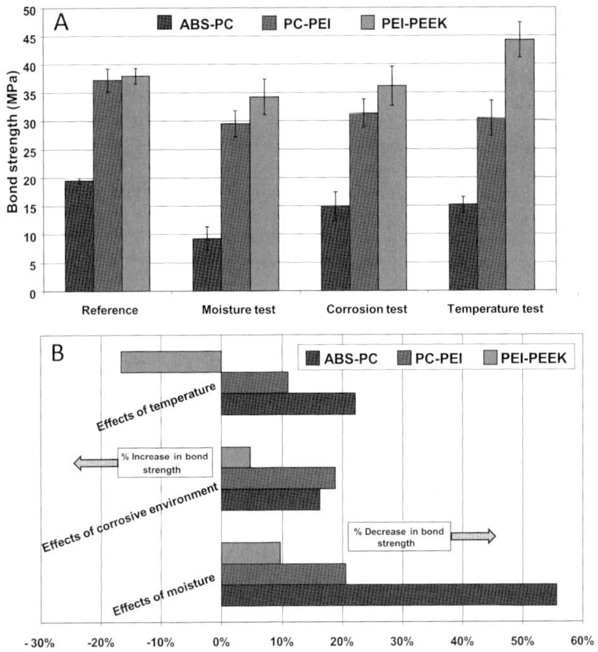

Figure 13.16 Effects of environmental factors on bond strength (plot A), effects expressed as a percentage of the original bond strength (plot B) [31, 35]

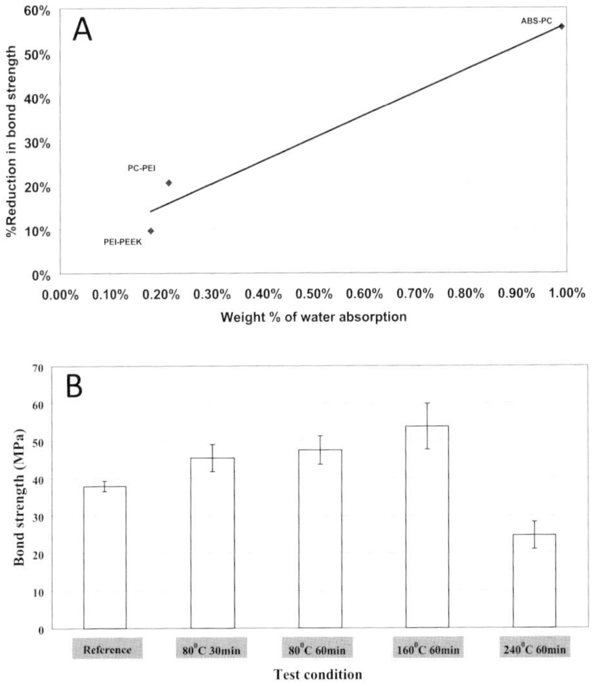

Figure 13.17 (A) Effects of water absorption on bond strength, (B) effects of temperature annealing on PEI1000-PEEK samples [35]

13.3.2 Polymer–Polymer Interface

An interface between two polymer materials is formed when they are brought into contact, usually with the application of heat and pressure. Two polymer–polymer interfaces are presented in Figure 13.18. For multicomponent molding, the interface of the materials is a critical quality parameter. The two main types of interfaces available for multicomponent polymer processing are microscopic and macroscopic interfaces. A polymer–polymer interface can also occur as a combination of the interfaces mentioned.

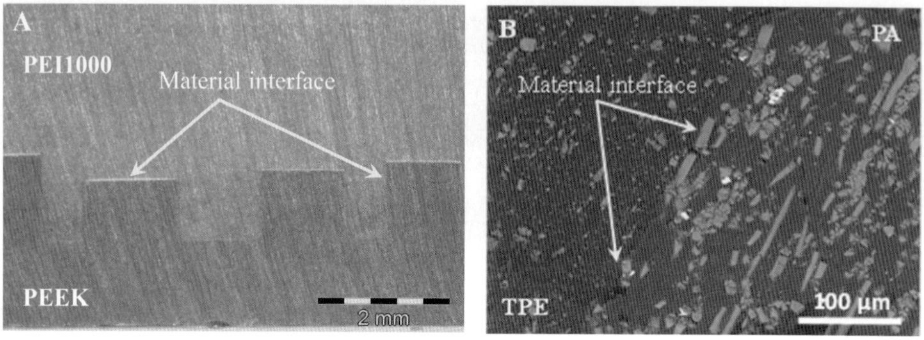

Figure 13.18 Pictures showing polymer–polymer interfaces: interface between two hard plastics (PEI1000–PEEK)–picture A, interface between a hard and soft plastic (PA–TPE)–picture B

According to [36–38], a microscopic interface is the chemical interface formed as a result of bonding along the mating surface between two materials. In particular, the driving force for a microscopic interface is cross-polymerization of the differing molecules of the two polymers. The extent and strength of the chemical interface is a complex function of many variables, including: molecular properties (e.g., size and weight), material processing temperatures, and viscosities. Macroscopic, or "interlocking", interfaces are formed by mechanically locking the two materials together via some form of geometric interface. Such types of interfaces can be used to join two chemically incompatible materials (e.g., plastic and metal). Additionally, interlocking interfaces can be used to selectively control the relative movement of the separate materials for various degrees of freedom. In some cases, multicomponent parts are designed with a combination of the above-mentioned interfaces. For example, it is possible to form an interface with molecular bonding along a geometric interlocking surface. According to Bruck et al. [39], a combination of microscopic and macroscopic interfaces performed much better in tensile loading than flat bonded interfaces, which in turn, performed much better than non-bonded interlocking interfaces. This suggests that when component strength is an important consideration, combination interfaces should be used whenever possible and financially viable.

13.3.2.1 Challenges for 2K Injection Molding: Bonding and Interface Dilemma

Most engineering applications of two-component injection molding call for high bond strength between the two polymers; on the other hand, a sharp and well-defined interface between the two polymers is required for applications like selective metalization of polymers, parts for micro applications, and also for the aesthetic purpose of the final products. The investigation presented in [46] indicates a huge dilemma between obtaining reasonably good bond strength and at the same time keeping the interface quality suitable for applications. The process conditions required for a sharp and well-defined interface are exactly the opposite of those that are congenial for higher bond strength. So, for the production of two-component injection molded parts, there is a compromise to be made between the interface quality and the bond strength of two polymers.

In all variants of 2K injection molding, to achieve a reasonable adhesion between the two polymers is a difficult task. The situation turns even more critical when a sharp and well-defined interface between the two polymers is equally as important as the adhesion between the two; for example, in molded interconnect devices by two-component injection molding. For this specific application, two-component plastic parts need a strong interfacial adhesion and at the same time, the two polymers need a sharp and well-defined interface to ensure complete selective deposition of metal during the metalization process [30, 47]. A sharp and distinguishable interface of two materials is the ultimate way to prevent a short circuit of the MID during the application. Like MIDs, many other products call for a sharp interface and high bond strength of two polymers to achieve the required quality of the final products. Another important reason to have a sharp interface is the aesthetics of the final parts. Without a sharp and well-defined interface, the 2K parts will look bad and they will not be well-accepted by users. Figure 13.19 shows examples of parts molded with well-defined interfaces (pictures A and B) and test parts molded with well-defined and not well-defined interfaces (picture C). Achieving a sharp and well-defined interface is much more difficult for micro products compared to macro dimensional parts due to the size effect at the interface zone and other difficulties associated with the micro mouding process.

Figure 13.19 Plastic parts showing sharp and well-defined interfaces (pictures A and B, picture source [48]), 2K molded test parts showing a well-defined and not well-defined interface (picture C)

A comparative analysis between the interface study and polymer-polymer bond strength investigations (presented in references [46, 47]) reveals a dilemma between the two parameters for 2K molded plastic parts. The relation between the polymer-polymer bonding and interface quality is illustrated with the example of a PEEK-PC material combination. There is no bonding between the materials for the PEEK-PC combination (when PEEK is molded in the first shot and PC in the second shot–the two materials stay together only because of the mechanical locking in the part), whereas in the reverse sequence–PC-PEEK (PC is molded first and PEEK in the second shot)–the bond strength is 26 MPa. Figure 13.20 shows microscope pictures of the polished section of PEEK-PC (PEEK molded in the first shot and PC molded in the second shot). The right section contains the most critical feature of the 2K part (200 μm wide channel with aspect ratio 8). Figure 13.21 shows microscopic pictures of a part made with a PC-PEEK material combination (PC molded in the first shot). The interface study shows that this material pair has a better interface quality when PEEK is molded first. It means that the shot sequence with poor bond strength yields a sharp and well-distinguishable interface of the two polymers. When a polymer pair has a weak adhesion, the interface contains a gap (clear separation line with cracks, which can be seen in picture B of Figure 13.20) of several micrometers between the two polymers. This micro gap between the two polymers makes the border sharp and distinguishable after injection molding. But when two polymers adhere strongly, the gap between the two materials is reduced or disappears (as in picture B of Figure 13.21).

PEEK-PC (PEEK First shot, PC-Second shot): Bond strength 0 MPa

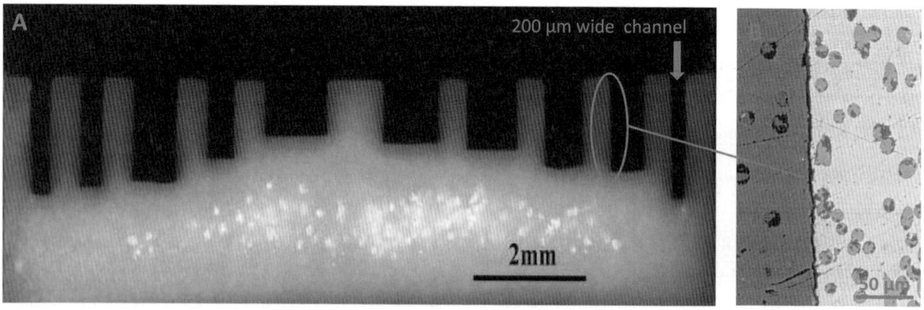

Figure 13.20 Microscope pictures of the polished section of a 2K part made with PEEK-PC (PEEK molded in the first shot and PC molded in the second shot). There is a sharp and well-defined material interface but no bonding between the materials

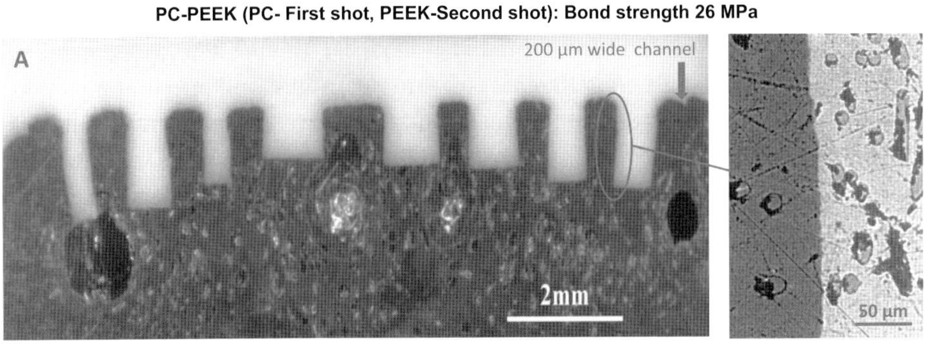

Figure 13.21 Microscope pictures of the polished section of a 2K part made with PC–PEEK (PC molded in the first shot and PEEK molded in the second shot). There is a sharp and well-defined material interface but no bonding between the materials

For stronger bonds between two polymers during 2K molding, the interface temperature at the second shot should be high enough so that the surface of the inlay part (first shot) is melted and the two polymers can melt together to form a better bond [42]. On the other hand, for geometrical correctness and interface quality of the 2K part, the inlay part should withstand the heat and pressure of the second shot melt so that it can retain its shape and keep the geometry undistorted. In the experimental case, the heat deflection temperature (HDT) of PC is lower than that of PEEK (HDT for PC and PEEK are 146 °C and 315 °C, respectively). When PEEK is used as the first shot material, it can withstand the injection molding conditions of the second shot PC so that the interface is well-defined where the un-deformed geometries of the features and sharp edges between the interfaces are observed, whereas the bond strength is poor due to insufficient melting and mixing of the two polymers. In the reverse shot sequence (PC first shot and PEEK second shot), the two polymers can melt together and achieve a stronger bond, but the geometrical features of the inlay PC are destroyed due to its poor thermal and mechanical properties during the second shot injection of PEEK. Previous investigations have also shown that higher process settings in the injection molding machines facilitate the bond strength between two polymers. Current investigations on the interface between two polymers suggest that higher settings of injection molding parameters may destroy the inlay geometry and affect the sharpness of the interface negatively.

For compatible material pairs, if a sufficient interface temperature is provided during the second shot of the molding process, a wide interfacial zone (mixing zone) is formed, due to which the two materials are difficult to distinguish, but this mixing zone makes the bonding strong (pictures A and B of Figure 13.22). The absence of this mixing zone contributes to the sharp and well-distinguishable interface and makes the bond between the two mating materials poor (pictures C and

D of Figure 13.22). The shape and dimension of this mixing zone are highly dependent on the flow direction of the second shot material, the position of the molding gate, and injection parameters like pressure, temperature, and speed of injection. For multicomponent micro parts, this mixing zone between the two materials can relatively be so large compared to the overall dimension of the parts that no separate entities or features of the parts will be left, making the parts unacceptable for any applications.

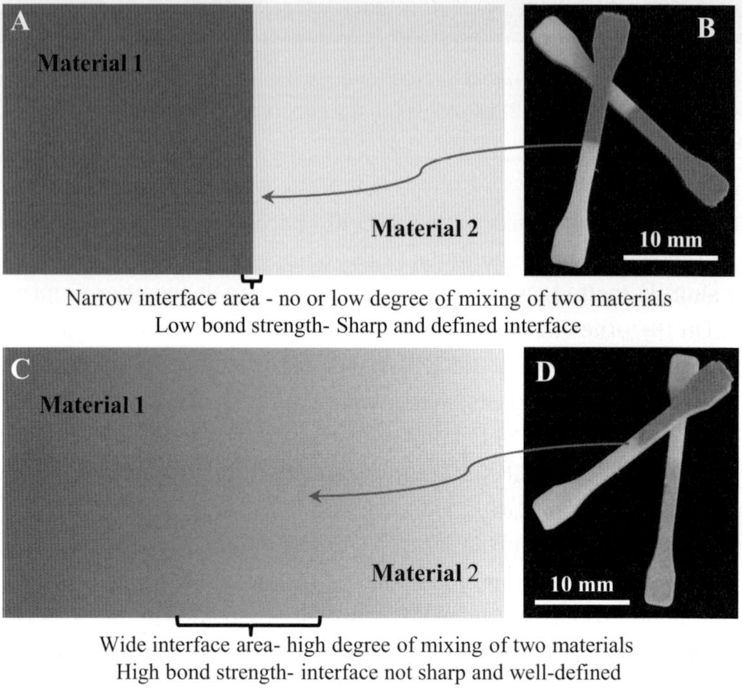

Figure 13.22 Low mixing sharp and well-defined material interface (pictures A and B), high mixing wide but unsharp material interface (pictures C and D)

Finally, it can be concluded that the requirements for producing a high quality interface between two polymers are opposite to the requirements for higher interfacial adhesion between the two polymers. Choosing between polymer–polymer interface quality and polymer–polymer bond strength is a dilemma. So, injection molding process conditions, material shot sequence, and part geometry etc., should be customized depending on whether bond strength or interface quality is the priority. The quality of the polymer–polymer interface depends heavily on the injection molding process conditions. The melt temperature, mold temperature, and the injection speed can effectively modulate the interface quality between two polymers. Insufficient mold and melt temperature lead to inaccurate filling of the second shot melt; on the other hand, higher mold and melt temperature settings deform the first shot inlay part and destroy the features and geometries. Too high an injection speed deforms the inlay part, especially when the temperature settings are high.

13.3.2.2 Special Considerations for the Polymer–Polymer Interface of Micro Parts

The ability of features with smaller dimensions to retain their shape during the second shot is inferior due to their high surface area to volume ratio and lower mechanical strength. In general, micro parts or micro features have a high surface area to volume ratio, which means they have a bigger surface to facilitate heat conduction, and the parts or features can easily be melted, which might not be the case in macro dimensions. Sharp corners in the first shot part can be destroyed during injection of the second polymer and the geometry gets rounded and deformed (picture B of Figure 13.23) due to heating by the second shot polymer melt if the material selection, shot sequence, and molding parameters are not selected correctly. First shot polymers with high mechanical strength and high heat deflection temperatures impart better geometrical correctness during multicomponent molding (picture A of Figure 13.23).

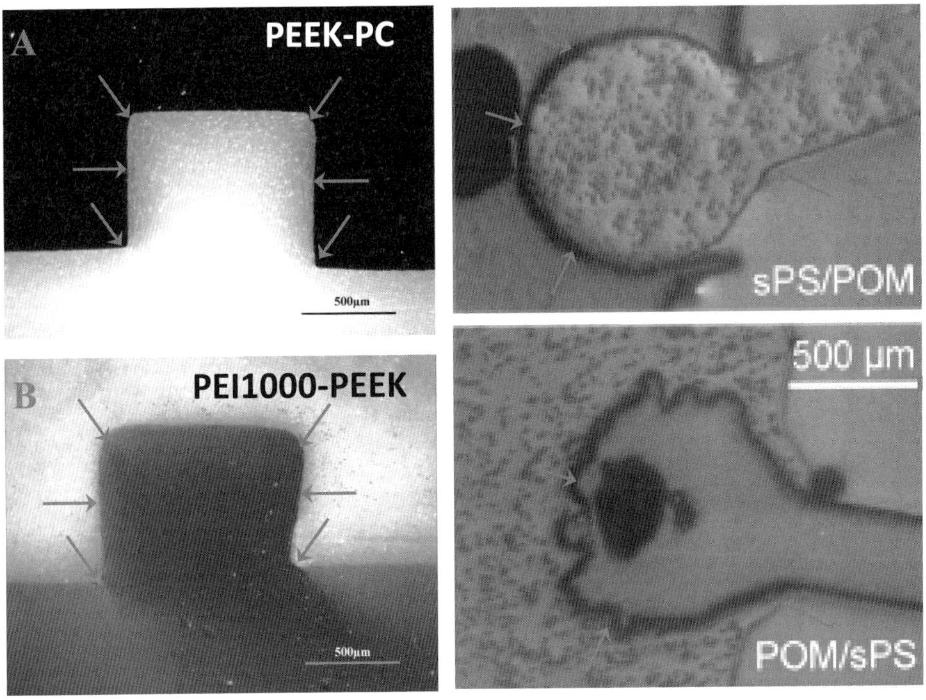

Figure 13.23 Comparative pictures of distortion and deformation of 2K parts on the micro scale (pictures A and B), comparative pictures of well-defined and not well-defined interfaces in micro parts (pictures C and D; pictures' source: [49])

For improved replication of micro parts, higher mold and melt temperature settings, as well as a higher injection speed, are always recommended. But, for multicomponent micro injection molding, care must be taken when choosing injection parameters for the second shot material. Higher process settings, like a high mold and melt temperature, high injection speed, and pressure simply destroy the inlay part geometry as the micro parts are mechanically weak and possess poor shape retention ability. Micro parts have a higher surface area to volume ratio and low thermal inertia. These make micro components more susceptible to heat and pressure. A first shot material with a higher heat deflection capacity and second shot with a lower process temperature can be wise choices from a replication quality and border sharpness point of view. The material shot sequence is another simple but important consideration. Pictures C and D of Figure 13.23 show the difference in part quality that can be made simply by changing the material shot sequence in the case of multimaterial micro molding. When the same material pair has a different bond strength based on different shot sequences, the shot sequence at which the two materials have lower bond strength would be the preferable shot sequence for the materials for 2K or multi-K molding when a sharp and well-defined interface is required. For multimaterial micro molding, selection of the material shot sequence and injection molding process conditions should be customized so that a good interface can be ensured. At the micro scale, bonding between the polymers becomes less critical than the interface quality, as some acceptable degree of bonding between the polymers can be relatively easily obtained in micro parts due to the special geometrical and thermal characteristics of the parts. A lot of attention is needed during the design of the mold for multicomponent micro molding, especially on the placement of the gate so that it does not influence the deformation of the first shot part during the injection of the second material.

13.4 Adhesion Modification for Multicomponent Molding

As mentioned before, for most applications of 2K molding, good adhesion between the components is desired. Achieving good bonding for some polymer combinations can be technically challenging. Some suggestions of how to enhance the adhesion between two polymer materials are listed below:

- Choosing material combinations based on their chemical compatibility will enhance the bonding. A quick check of the chemical compatibility can be made with the help of the solubility parameters of the materials. The investigations presented in [31] show that when the solubility parameters are close for two materials, that indicates that the two materials have chemical similarity and there is a good chance that the two materials will mix together and bond together during the 2K molding if the right temperature conditions are provided.

- Interface temperature is another important aspect to check for the selection of material combinations for 2K molding. The interface temperature between the first shot and the second shot should be high enough so that it can melt the surface of the first parts; in that case, the two materials can melt together and the chemical compatibility of the materials can play a role. If the two materials are not melted together, no matter what, the bonding will be poor. Sometimes, the material shot sequence can make a big difference in the interface temperature and can result in good bonding between the materials.
- High settings of process parameters in the second shot molding (like high temperature, pressure, speed etc.) will have a positive influence on the bonding of the two materials.
- The surface roughness of the inlay part or the first shot part can play a vital role in enhancing the bonding. A high interface roughness may result in high bonding between two polymer materials.
- Sometimes, the shrinkage of the material can be used in favor of the bonding between the two materials. For example, for the part presented in picture A of Figure 13.24, if the material for M1 is chosen so that it has higher shrinkage compared to the material for M2, M1 will exert a mechanical force on M2 due to the shrinkage difference and the two materials will be mechanically locked.

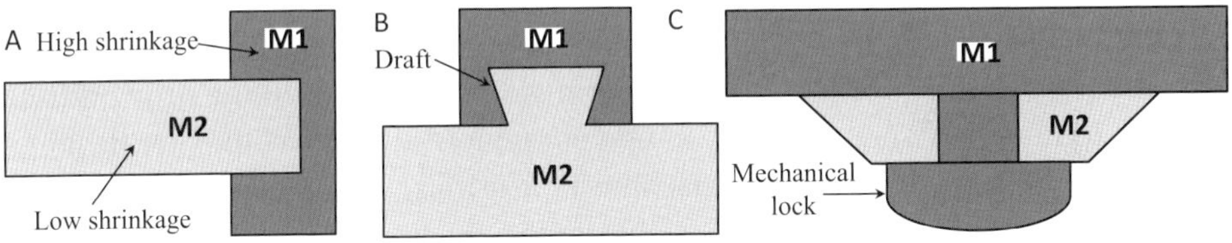

Figure 13.24 Concept for the use of material shrinkage in favor of the bond strength (picture A), examples of draft and mechanical locking to enhance the bonding between two materials (pictures B and C, respectively)

- Using draft angles or mechanical interlocks (presented in pictures B and C of Figure 13.24) can be ways to enhance connection for some material combinations where there is no natural or chemical adhesion, or the interface temperature is not high enough to melt the two materials together. In this case, the first component is designed so that when the second component is injected, it flows around the undercuts or drafts in the first component and the two materials are mechanically locked. Mechanical interlocks can be used to improve attachment to the substrate in areas of high stress or abrasion. It is important to remember that the design of a draft or mechanical lock for micro scale parts is challenging because of the small size of the component.

- Use of venting both in the first and second shot can help in bonding. The venting can release the trapped air between the two materials, which otherwise can affect the bonding negatively.
- Clean and humidity controlled molding environments are always recommended for 2K or multicomponent molding because humidity, dirt, oil, and other contaminants can destroy the bonds between the two materials.
- Proper surface modification may lead to a significant improvement in the adhesion of the polymer materials [51]. Different types of surface treatments, like corona treatment, plasma treatment, heat treatment, coatings etc. [52], can increase the bonding between the polymer materials. Many polymeric materials inherently have a low surface energy that results in poor surface adhesion. The correct surface preparation of these materials can increase the surface energy and improve surface adhesion properties. However, surface treatments in-between shots of multicomponent molding can be technically challenging for micro dimensional parts and can impose a high complexity on the process.

13.5 Other Quality Issues for Multicomponent Micro Molding

Besides bond strength and interface quality between two materials, the other important quality issues for multicomponent molding are underfilling or overfilling and the flashing of either material. To prevent underfilling, overfilling, or flashing, the shot-to-shot accuracy of the machine is a critical factor. A machine with a closed-loop injection speed is the right choice for multicomponent molding. A smart and robust mold design for multimaterial molding can make a substantial difference in the quality of multicomponent molded parts. Furthermore, it can be helpful for the adhesion of multicomponent molding to supplement the thermal and chemical bonding between two materials with a mechanical joint achieved by using undercuts or by similar design.

The multicavity mold should be well-balanced. Hot runner manifolds must be balanced, too, for multicomponent molding. Mold temperature is an important factor for multicomponent injection molding. Accurate control of the mold temperature is essential when the mold is run for the second shot.

The physical mechanism of warpage for multicomponent molding is more complicated than that of a single component part. The filling/packing history introduces volumetric shrinkage in molded parts. For multicomponent molding, the difference in the shrinkage properties of different materials may introduce a huge amount of stress in the material interface. In addition, when considering the combination of material selections for the first shot (including the insert) and second shot, and the geometrical effects of designs, the parameter space (i.e., the process window) will become huge and complex to optimize. The cooling stage can introduce a thermal unbalance phenomenon to the final product. The product geometries, process conditions, and molded materials affect the quality of the final products significantly. For the multicomponent parts, the warpage of the final injected products not only results from the injection history, but also from the part geometry and materials [50].

For multicomponent molding, the delay time between injections is very influential as well as the injection pressure, the melt temperature of each polymer, and the mold temperature [53]. Thicker parts have more thermal mass and lose heat more slowly, which is not the case with the micro parts. That is why, for the thin-walled parts or micro scale parts, it is necessary to have short cycle times between shots to ensure adhesion between materials before cooling.

In addition to the bonding and interface difficulties, multicomponent processes are liable to standard injection molding defects such as flash, bubbles (voids), brittleness, flow lines, discoloration, nonfill, excessive shrinkage, sink marks, warpage etc. Challenges particular to multicomponent processes are material incompatibility, gate placement, demoldability, and ejection system problems, etc. [54]. Micro parts have characteristically high surface-to-volume ratios and, correspondingly, a faster cooling rate than larger components [55]. To fill features at the micro scale, high temperatures and high injection speeds are required, leading to correspondingly high stresses and shear rates during molding [56]. This variable thermomechanical environment influences the nucleation and growth rate of crystalline entities and, therefore, the mechanical properties of the resultant multicomponent micro parts. These phenomena of micro molding become even more complex for multimaterial micro molding where more than one material interacts and parts have to go through high thermal gradients and differential material shrinkage.

13.6 Conclusion

Mass production of small, high-precision, high-value-added multicomponent plastic parts will play a vital role in future micro manufacturing industries. Multimaterial micro plastic parts comprise parts made with two or more plastic materials with micro dimensions, weighing a few milligrams, or larger parts with micro/nanoscale surface features. Among current plastic processing technologies, multicomponent injection molding is one of the most important processes for its extensive use in industrial applications where different polymer materials are combined on purpose. In spite of high machine and mold costs, multicomponent injection molding is a viable industrial process because of its high production volume, reduced cycle time, eliminated secondary operations, reduced amount of floor space required by eliminated machines, and part staging and assembly areas. The final output of multicomponent molding is a streamlined manufacturing process that increases the functionalities and aesthetic value of the final products—more importantly, these can also be achieved today in micro scale manufacturing.

Micro manufacturing is a challenging process simply due to the small dimensions of the parts, and, with multimaterial micro manufacturing, the complexities are multiplied. Two important quality issues for multicomponent parts are bonding and interface quality. For micro parts, achieving a sharp and well-defined interface can be really difficult. For multimaterial micro part production, special attention is needed as the parts are mechanically weak due to the smaller dimensions, they possess a larger surface compared to the volume of the part, they sometimes have a very high aspect ratio etc. The micro molding process is usually a high speed and high pressure process. Material selection, shot sequence, and process parameter adjustment should be customized so that the first shot part can survive the heat and pressure coming from the injection of the second shot and retain the original size, shape etc. At the same time, achieving a reasonable adhesion between the materials must be ensured.

Several investigations show that there are several factors that can influence the bonding and the interface quality of multicomponent parts. Among the injection molding process conditions, melt and mold temperatures are vital process parameters that can influence bond strength. Injection speed, pressure, and holding pressure also have some effects on the bond strength of multicomponent polymer parts. The interface temperature during the second shot molding is vital for the adhesion of the two polymers. If the two polymers cannot melt together, the adhesion will be poor. Besides this, the surface roughness of the first shot part and environmental factors, such as moisture, have a profound influence on the bonding of two materials. Interface roughness increases mechanical locking and a localized melted zone, so the materials can have a higher rate of entanglement and higher strength. Environmental factors, like moisture, a corrosive environment, and thermal history affect the bond strength of the post-molded parts. Thermal annealing may increase the bond strength between polymers by enhancing polymerization or crystallization or even by removing the thermal stresses from the molded multicomponent parts. Recognition of the role of the main factors controlling interfacial adhesion and the quality of the polymer–polymer interface is extremely important for producing successful multicomponent products. There is a reverse connection between the bond strength and the interface quality of the two polymer materials. For a strong interface, the material pair loses its interfacial sharpness and aesthetic quality. The conditions that increase the bonding between the two polymer materials are very likely to deteriorate the interface sharpness. The material shot sequence of multicomponent molding can influence both bonding and interface quality.

Multicomponent micro injection molding has reached a promising technological level where the process can fulfill the technical requirements of highly demanding applications. Research and development efforts need to focus especially on controlling the thermo-physical behavior of molding materials, their warpage behavior, and controlling thermal stress in the parts. New material combinations (hard-hard, hard-soft etc.) should be developed specifically for multicomponent micro molding to achieve reasonable bonding and sharp interfaces during molding. Process simulations of multicomponent micro molding are still complex and are difficult to implement in industrial applications, therefore, additional efforts are needed in the near future to bridge the gap between scientific research, software validation, and actual product and process development. Micro insert molding combined with multicomponent technology can provide an interesting new approach for micro manufacturing of metal/multiplastic hybrid parts. Another interesting field of the multicomponent process is multicomponent micro powder injection molding (μPIM) [57] for the realization of magnetic/non-magnetic or conductive/non-conductive, porous/non-porous, hydrophilic/hydrophobic material combinations at the micro scale. Focus for future research should be placed on the areas mentioned so that the full potential of multicomponent micro molding processes can be exploited for a vast range of industrial applications.

References

[1] Nanotechnology–a UK Industry View (2013), UK Technology Strategy Board (TSB), KTN Report

[2] Eriksson, T.G., Hansen, H.N., Gegeckaite, A., Arentoft, M., Automated assembly of micro mechanical parts in a Microfactory setup. *Proceedings of the 5th international workshop on Microfactories: IWMF* (2006), Laboratoire d'Automatique de Besançon - Institut de Productique, Besançon, France

[3] Antonelli, D., Fantoni, G., Porta, M., Santochi, M., A Methodology for the Selection of Micro-Assembly Techniques. *Am. J. Eng. Appl. Sci.* (2010) 3(4), pp. 718–722

[4] Gouker, R., Gupta, S., Bruck, H., Holzschuh, T., Manufacturing of multi-material compliant mechanisms using multi-material moulding. *Int. J. Adv. Manuf. Technol.* (2006) 30, pp. 1049–1075

[5] Banerjee, A., Li, X., Fowler, G., Gupta, K., Incorporating manufacturability considerations during design of injection molded multi-material objects. *Research in Engineering Design* (2007) 17(4), pp. 207–231

[6] Bryce, D.M., *Plastic Injection Molding: Manufacturing Process Fundamentals*, first edition (1996), Society of Manufacturing Engineers, Dearborn, Michigan, US

[7] Goodship, V., Love, J.C, *Multi-Material Injection Moulding* (2002), Smithers Rapra Publishing, Shropshire, United Kingdom

[8] Asfari, L., Multi-shot injection molding process diagram, CC BY-SA 4.0

[9] GLS Corporation, GLS Overmolding Guide, Rev. 082704 (2004)

[10] Banerjee, A., Li, X., Fowler, G., Gupta, S., Incorporating manufacturability considerations during design of injection molded multi-material objects. *Res. Eng. Des.* (2007) 17(4), pp. 207–231

[11] Tepro, Technology in Progress, http://www.tepro.com.tr/index.php/en/teknolojiler/multinject-2

[12] B&Q Corporate, http://www.diy.com/departments/tools-equipment/hand-tools/screwdrivers-keys/DIY636168.cat

[13] Clas Ohlson, Sweden, http://www.clasohlson.com/uk/Bosch-IXO-IV-3.6-V-Cordless-Screwdriver/Pr183198000 Pinterest, The world's catalog of ideas https://www.pinterest.com

[14] Pinterest, The world's catalog of ideas, https://www.pinterest.com

[15] Global sources, http://www.globalsources.com

[16] PEZY group, http://www.pezy.nl/2k-and-3k-injection-moulding-platic-micro-product

[17] Oechsler, Multicomponent Materials, Process techniques, Tool making, Germany, http://www.oechsler.com

[18] Ananthasuresh, G. K., Kota, S., Designing compliant mechanisms. *ASME Mech. Eng.*, (1995) 117(11), pp. 93–96

[19] Antonelli, D., Fantoni, G., Porta, M., Santochi, M., A Methodology for the Selection of Micro-Assembly Techniques. *Am. J. Eng. Appl. Sci.* (2010) 3(4), pp. 718–722

[20] Carozzo, G., Injection-moulding composite articles. US Patent no: 3051994A (1962)

[21] Hock, J. M., Vries, D. S., One-piece plastic piston. US Patent no: 4214507 (1980)

[22] *Engineering Plastics Market–Segmented by Product, Application, and Geography–Trends and Forecasts (2016–2022)* (2017), Mordor Intelligence, https://www.mordorintelligence.com/industry-reports/global-engineering-plastics-market-industry

[23] Knights, M., Injection Molding LSR: Three 'M's of Innovation: Mega, Micro, and Multi, Plastics Technology (2007), http://www.ptonline.com/articles/injection-molding-lsr-three-%27m%27s-of-innovation-mega-micro-and-multi

[24] Brousseau, E. B., Dimov, S., Pham, D. T., Some recent advances in multi-material micro- and nano-manufacturing. *Int. J. Adv. Manuf. Technol.* (2010) 47(1–4), pp. 161–180

[25] Islam, A., Hansen, H. N., Two component injection moulding: Present and future perspectives. *Society of Plastic Engineers: Plastic Research (SPE)* (2009) 1, DOI: 10.1002/spepro.000049

[26] Sumitomo (SHI) Demag Plastics Machinery, https://www.sumitomo-shi-demag.eu/products/systec-multi-multi-component-injection-moulding-machines.html

[27] ERMO, Rotative Moulds, http://www.ermo-tech.com/en/lighting/technology/rotative-molds

[28] Kennerknecht, T., *Fatigue of Micro Molded Materials–Aluminum Bronze and Yttria Stabilized Zirconia* (2014), KIT Scientific Publishing, Karlsruhe, Germany

[29] Sumitomo (SHI) Demag Plastics Machinery GmbH, Multi-component technology, https://www.sumitomo-shi-demag.eu/processes/multi-component-technology.html, 2017

[30] Islam, A., Hansen, H. N., Tang, P. T., Jørgensen, M. B., Ørts, S. F., Two-component microinjection moulding for MID fabrication. *Plast, Rubber Compos.* (2010) 39(7), pp. 300–307

[31] Islam, A., *Two component micro injection moulding for Moulded Interconnect Devices* (2008), Department of Mechanical Engineering, Technical University of Denmark, PhD Thesis

[32] Imgrund, P., Rota, A., Wiegmann, M., *Investigation of material interfaces obtained by twocomponent micro-MIM and co sintering* (2006), Fraunhofer Institute for Manufacturing and Advanced Material, EPMA's EuroPM

[33] Zoetefief, F. W., *Multi component injection moulding* (1995), Eindhoven University of Technology, Netherlands, PhD Thesis

[34] Kim, J. K., Thomas, S., Saha, P. (Ed.), *Multicomponent Polymeric Materials*, Springer Series in Materials Science (2016) 223, DOI 10.1007/978-94-017-7324-9_4

[35] Islam, A., Hansen, H. N., Bondo, M., Experimental investigation of the factors influencing the polymer-polymer bond strength during two-component injection moulding. *Int. J. Adv. Manuf. Technol.* (2010) DOI: 10.1007/s00170-009-2507-8 2010

[36] Gouker, R., Gupta, S., Bruck, H., Holzschuh, T., Manufacturing of multi-material compliant mechanisms using multi-material moulding. *Int. J. Adv. Manuf. Technol.* (2006) 30, pp. 1049–1075

[37] Beaman, J., Bourell, D., Jackson, B., Jepson, L., McAdams, D., Perez, J., Wood, K., Multi-material selective laser sintering: empirical studies and hardware development. *NSF Design and Manufacturing Grantees Conference* (2000), Vancouver, BC, Canada, pp. 3–6

[38] Fowler, G. T., *Cost and performance evaluation models for comparing multi-shot and traditional injection moulding* (2004), University of Maryland, US, MSc Thesis

[39] Bruck, H. A., Fowler, G., Gupta, S. K., Valentine, T. M., Using geometric complexity to enhance the interfacial strength of heterogeneous structures fabricated in a multi-stage, multi-piece moulding process. *Experimental Mechanics* (2004) 44, pp. 261–271

[40] Islam, A., Hansen, H. N., Marhöfer, D. M., Angel, J. A., Dormann, B., Bondo, M., Two-component micro injection moulding for hearing aid applications. *Int. J. Adv. Manuf. Technol.* (2012) 62(5–8), pp. 605–615

[41] Islam, A., Hansen, H. N., Effects of glass fibers on micro moulded plastic parts. *Commercial micro manufacturing magazine*, Oct–Nov issue (2011), MST Global Ltd., Huntington, Chester, United Kingdom

[42] Agassant, J. F., Avenas, P., Sergent, J. P., Carreau, P. J., *Polymer Processing, Principles and Modelling* (1991), Carl Hanser Verlag, Munich, Germany

[43] Material database, CES EduPack 2006 (Software package)

[44] Moldflow simulation software package, MPI 5.1, Moldflow Corporation

[45] Tosello, G., *Precision moulding of polymer micro components: Optimization, Simulation, Tooling, Quality Control and Multi-Material Application* (2008), Department of Mechanical Engineering, Technical University of Denmark, PhD Thesis

[46] Islam, A., Hansen, H. N., Tang, P. T., Two component injection moulding: an interface quality and bond strength dilemma. *Proceedings from the 24th Annual Meeting of the Polymer Processing Society* (2008) Salerno, Italy

[47] Islam, A., Hansen, H. N., Tang, P. T., Jie, S., Process chains for the manufacturing of moulded interconnect devices. *Int. J. Adv. Manuf. Technol.* (2009) 42(9–10), pp. 831–841

[48] Shenzhen Corelmould Precision Co., Ltd., http://corelmould.com

[49] Michaeli, W., Rogalla, A., Ziegman, C., Micro Assembly Injection Moulding of Hybrid Microsystems. *Microsyst. Technol.* (2003) 9, pp. 427–430

[50] Huang, C. T., Yeh, C. H., Tseng, S. C., Geometrical effect and material selection in multi-component molding (MCM) development, ANTEC (2006)

[51] Pukánszky, B., Interfaces and interphases in multicomponent materials: past, present, future. *Eur. Polym. J.* (2005) 41, pp. 645–662

[52] Rory, A. W., *Plastic Surface Modification, Surface Treatment and Adhesion* (2015), Carl Hanser Verlag, Munich, Germany

[53] Caamano, J., Hard rules for soft-touch overmolding (2002) http://www.machinedesign.com/archive/hard-rules-soft-touch-overmolding

[54] Banerjee, A., Li, X., Fowler, G., Gupta, S., Incorporating manufacturability considerations during design of injection molded multi-material objects. *Res. Eng. Des.* (2007) 17(4), pp. 207–231

[55] Zhang, N., Choi, S. Y., Gilchrist, M. D., Flow induced crystallization of poly(ether-block-amide) from the microinjection molding process and its effect on mechanical properties. *Macromol. Mater. Eng.* (2014) 299(11), pp. 1362–1383

[56] Zhang, N., Gilchrist, M. D., The effects of miniaturization and processing on microinjection moldings. *Plastic Research Online* (2015), Society of Plastics Engineers (SPE), DOI: 10.2417/spepro.005865

[57] Piotter, V., Finnah, G., Prokop, J., Ruprecht, R., Hausselt, J., Multi-component microinjection moulding-trends and developments. *Int. J. Adv. Manuf. Technol.* (2010) 47, p. 63

Index

1. Micro Injection Molding Machines Technology

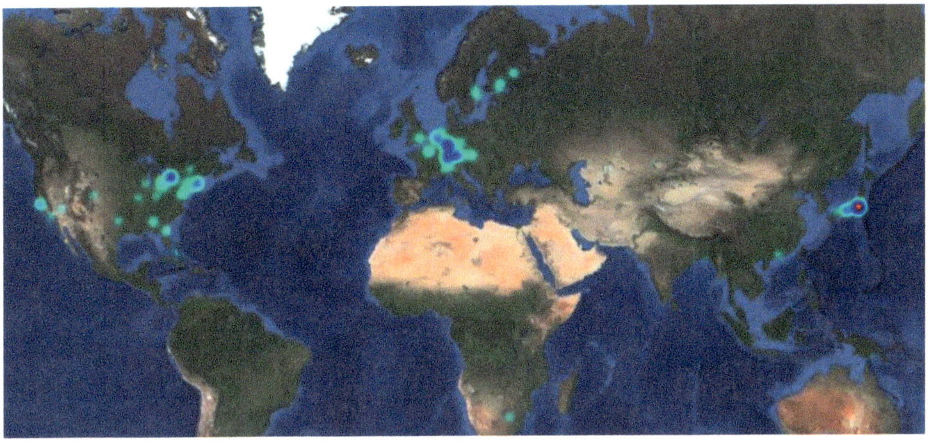

Figure 1.3 Geographical distribution of private assignees in the reference patent set

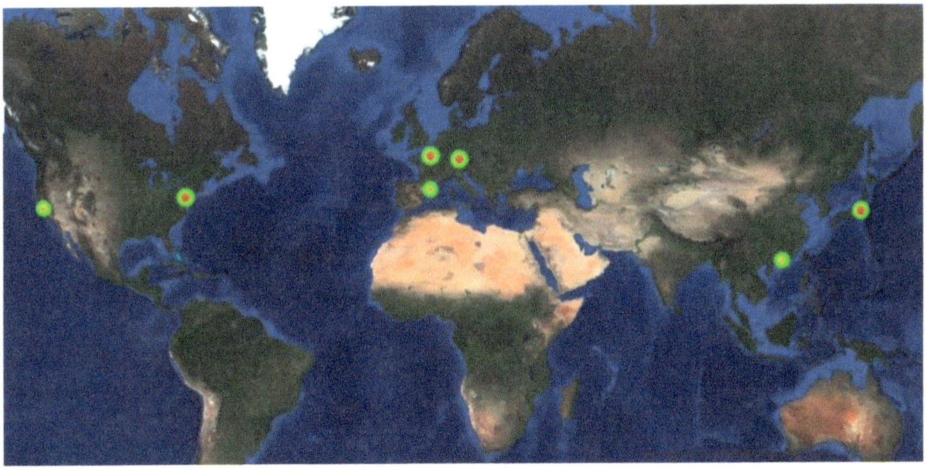

Figure 1.4 Geographical distribution of public assignees in the reference patent set

2. Micro Molding Process Monitoring and Control

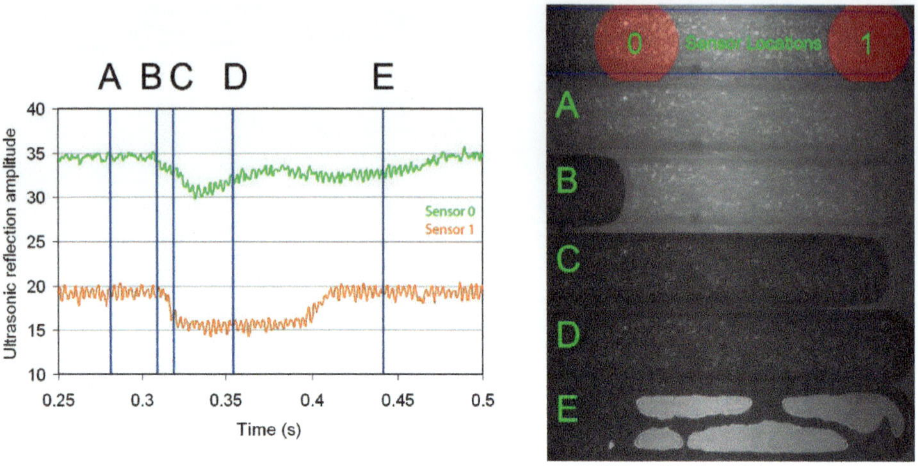

Figure 2.16 Synchronized ultrasound and high-speed visualization of cavity filling of a flat plaque component of width 2.0 mm and depth 250 μm

4. Surface Replication in Micro Injection Molding

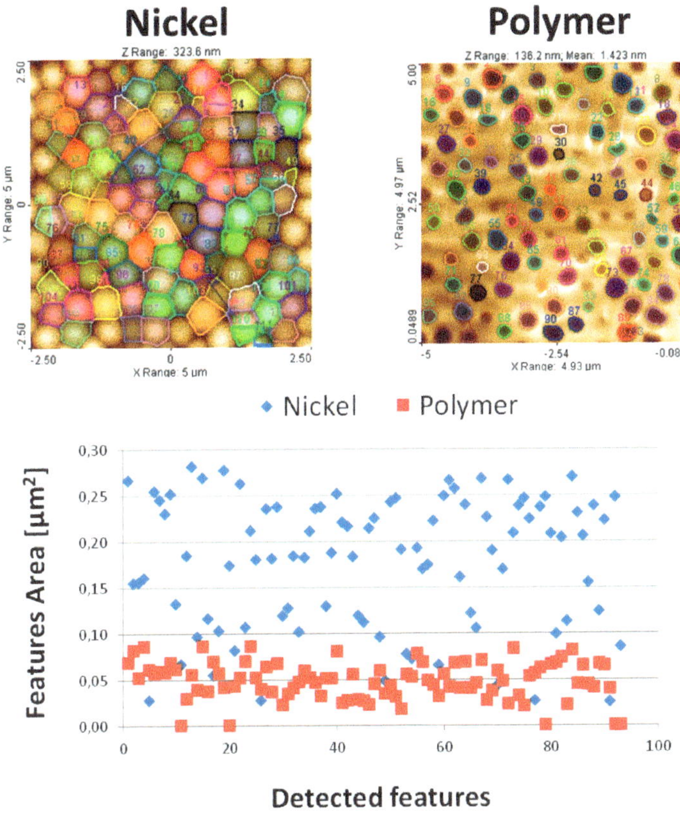

Figure 4.11 Detected nickel and polymer replicated features' dispersion using AFM. Evaluation area: 5 µm × 5 µm

6. Ultra-Precision Machining Technologies for Micro Injection Mold Making

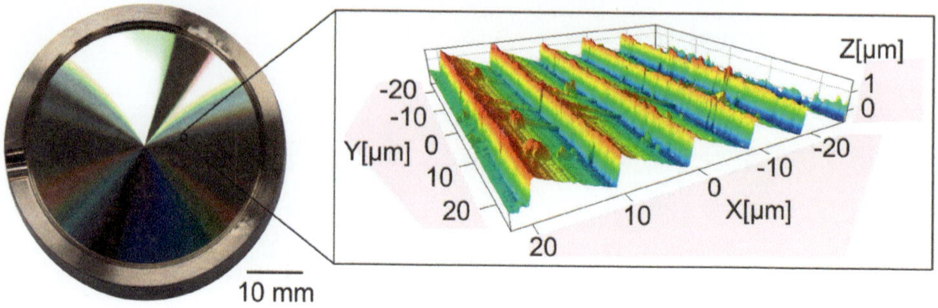

Figure 6.19 Diamond-turned mold insert and surface topography measured with a white light interferometer

7. Surface Treatment of Mold Tools in Micro Injection Molding

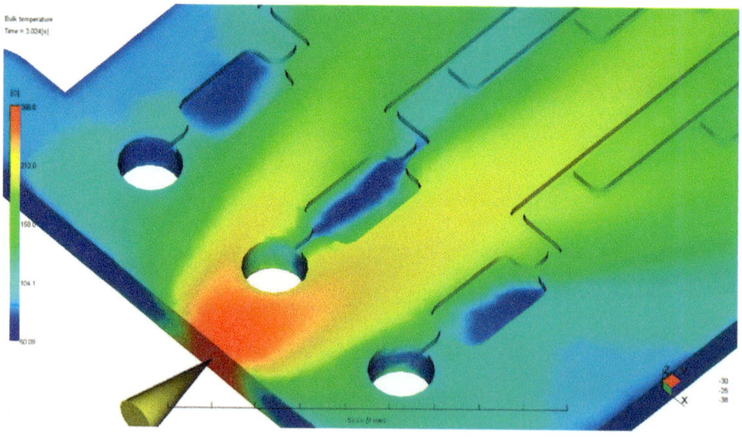

Figure 7.6 Bulk temperature distribution at the end of the fill for ABS

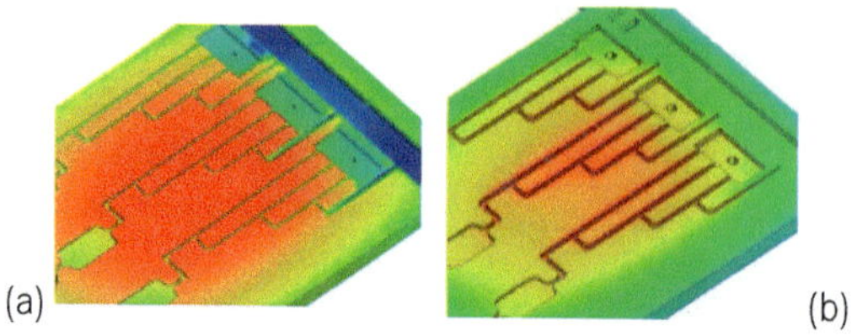

Figure 7.8 Simulation of part (a) and mold (b) surface temperature distribution after filling

9. Modeling and Simulation of Micro Injection Molding

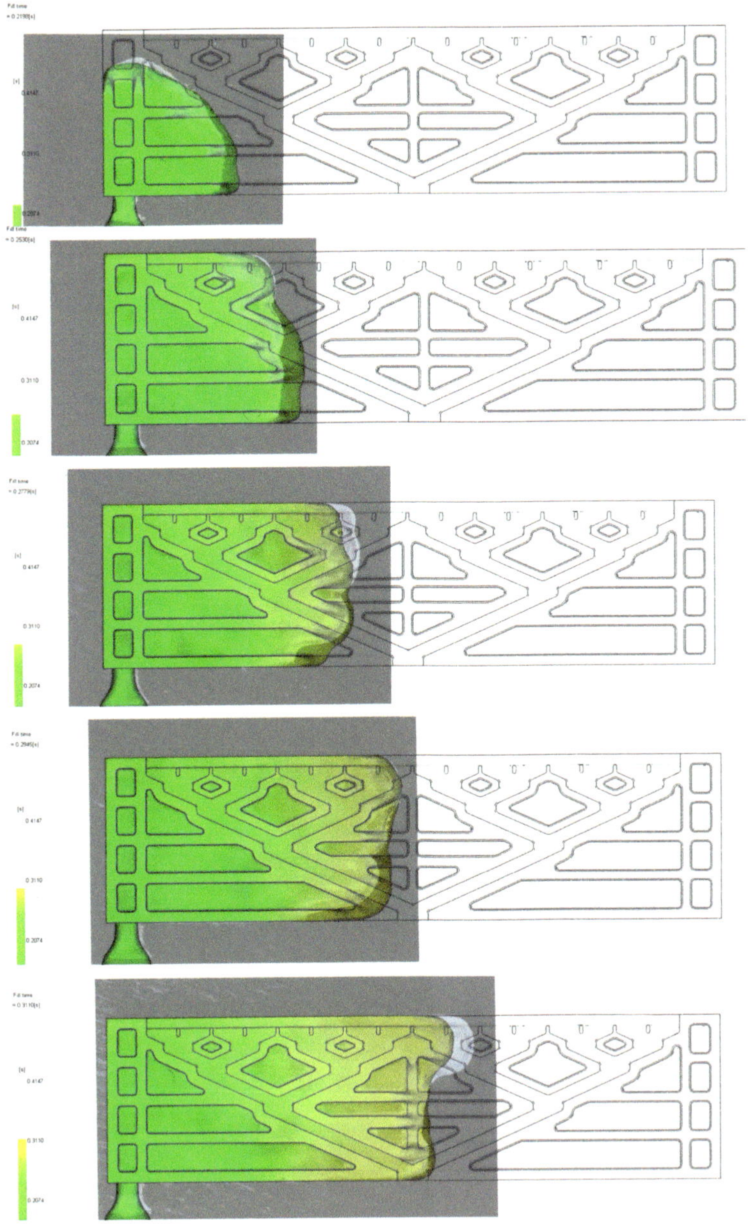

Figure 9.5 Comparison of molded short shots (pictures taken with a digital camera) and the simulated flow front pattern (pictures taken in Autodesk Moldflow©) on the global part level of a microfluidic part [60]

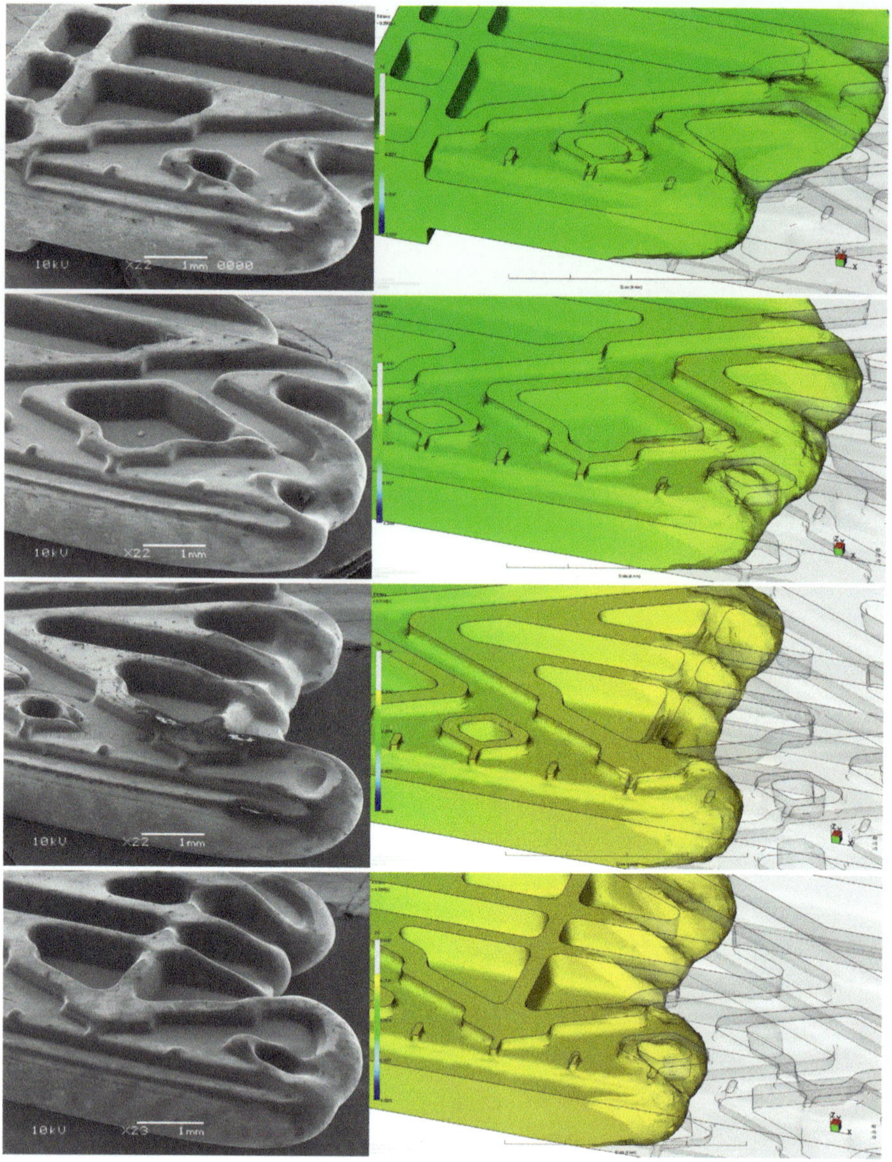

Figure 9.6 Comparison of molded short shots (pictures taken with a scanning electron microscope) and the simulated flow front pattern (pictures taken in Autodesk Moldflow©) on the local feature level of a microfluidic part [60]

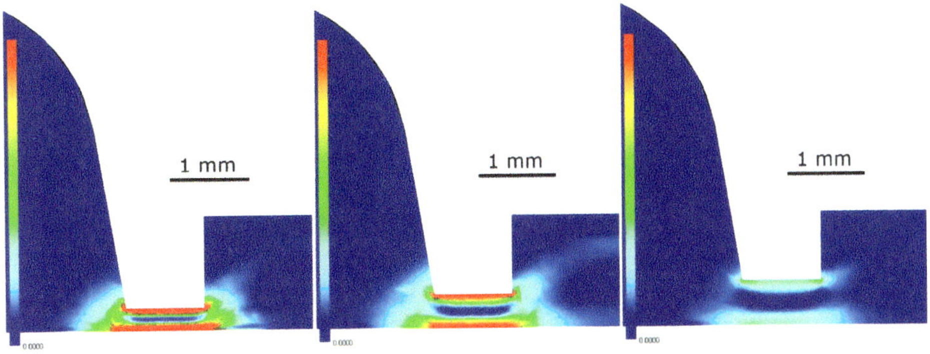

Figure 9.10 Predicted shear rate at the film gates of the microfluidic chip for different thicknesses (from left to right: 280 μm, 420 μm, 560 μm). Only the largest gate showed acceptable shear rates. Color scales from blue to red: shear rate from 0 s^{-1} to 40000 s^{-1} (allowed limit for used polymer) [60]

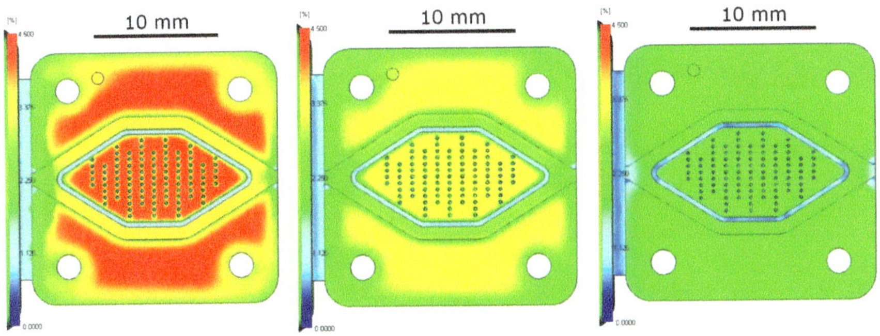

Figure 9.11 Predicted average volumetric shrinkage of the microfluidic chip for different film gate thicknesses (from left to right: 280 μm, 420 μm, 560 μm). The packing performance increases with gate thickness. Color scales from blue to red: shrinkage from 0% to 4.5% [60]

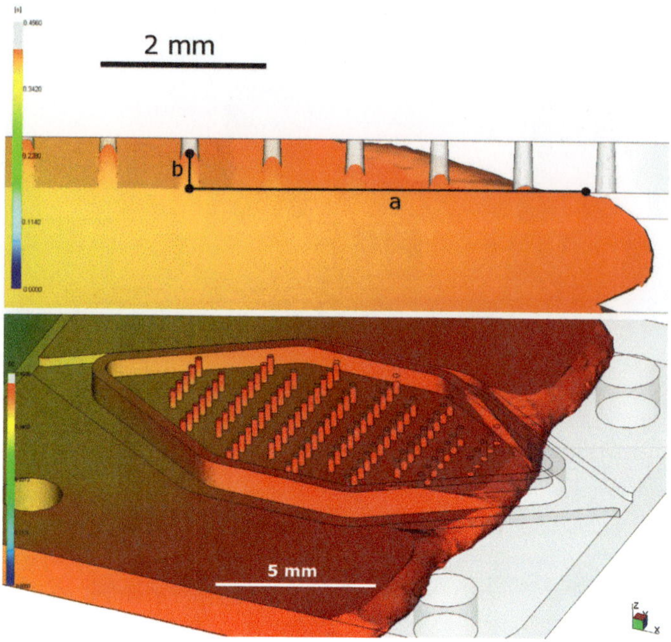

Figure 9.12 Cross-sectional (top [60]) and 3D view (bottom) of the flow inside the cavity of a micro structured part, displaying a distinct hesitation effect at the micro pillars: filling in the bulk direction is faster than in the direction of the micro pillars; in other words, the flow length, *a*, in the bulk direction is longer than the flow length, *b*, in the direction of the micro pillars